Holomorphic Hilbert Modular Forms

The Wadsworth & Brooks/Cole Mathematics Series

M. Adams, V. Guillemin, *Measure Theory and Probability*
W. Beckner, A. Calderón, R. Fefferman, P. Jones, *Conference on Harmonic Analysis in Honor of Antoni Zygmund*
G. Chartrand, L. Lesniak, *Graphs & Digraphs, Second Edition*
J. Cochran, *Applied Mathematics: Principles, Techniques, and Applications*
W. Derrick, *Complex Analysis and Applications, Second Edition*
J. Dieudonné, *History of Algebraic Geometry*
R. Dudley, *Real Analysis and Probability*
R. Durrett, *Brownian Motion and Martingales in Analysis*
R. Epstein, W. Carnielli, *Computability: Computable Functions, Logic, and the Foundations of Mathematics*
S. Fisher, *Complex Variables*
A. Garsia, *Topics in Almost Everywhere Convergence*
P. Garrett, *Holomorphic Hilbert Modular Forms*
J. Kevorkian, *Partial Differential Equations: Analytical Solution Techniques*
R. McKenzie, G. McNulty, W. Taylor, *Algebras, Lattices, Varieties, Volume I*
E. Mendelson, *Introduction to Mathematical Logic*
R. Salem, *Algebraic Numbers and Fourier Analysis* and L. Carleson, *Selected Problems on Exceptional Sets*
R. Stanley, *Enumerative Combinatorics, Volume I*
J. Strikwerda, *Finite Difference Schemes and Partial Differential Equations*
K. Stromberg, *An Introduction to Classical Real Analysis*

Holomorphic Hilbert Modular Forms

Paul B. Garrett
University of Minnesota

Wadsworth & Brooks/Cole Advanced Books & Software
Pacific Grove, California

Brooks/Cole Publishing Company
A Division of Wadsworth, Inc.

Printed in the United States of America
10 9 8 7 6 5 4 3 2 1

Library of Congress Cataloging in Publication Data
Garrett, Paul B.
Holomorphic Hilbert modular forms.

Bibliography: p.
Includes index.
1. Hilbert modular surfaces. 2. Holomorphic mappings.
I. Title.
QA573.G37 1989 516.3′5 88-33753
ISBN 0-534-10344-8

Sponsoring Editor: *John Kimmel*
Marketing Representative: *Cathy Twiss*
Editorial Assistant: *Jennifer Greenwood*
Production Editor: *Linda Loba*
Manuscript Editor: *Anne Draus*
Permissions Editor: *Carline Haga*
Interior and Cover Design: *Vernon T. Boes*
Typesetting: *Asco Trade Typesetting Ltd.*
Cover Printing: *Phoenix Color Corporation*
Printing and Binding: *Arcata Graphics-Fairfield*

Introduction

This book is intended as a systematic, self-contained, and straightforward introduction to a substantial part of the theory of holomorphic Hilbert modular forms, associated L-functions, and especially their arithmetic. As such, it is an introduction to the theory of automorphic forms in general, especially the arithmetic of holomorphic automorphic forms. Beginning from the most standard algebraic number theory and function theory, one can develop matters far enough to recover some of Shimura's results on special values of L-functions attached to Hilbert modular forms. Also within reach are the theorem on special values of L-functions of totally real number fields (after Klingen), and examples of results on special values of certain grossencharacter L-functions (due to Damerell over $\mathbb{Q}$ and to Shimura in general).

One of many reasons for the study of Hilbert modular forms is for applications to Dirichlet series. One emphasis of this book is the construction of Dirichlet series

$$D(s) = \sum_{n \geq 1} c_n/n^s$$

which have analytic continuations and functional equations. Many such are obtained as (Mellin) integral transforms of holomorphic Hilbert modular forms: the analytic continuations and functional equations follow almost immediately from the properties of such modular forms. Further, the existence of an Euler product factorization

$$D(s) = \prod_{p \text{ prime}} 1/P_p(p^{-s})$$

(where P_p is a polynomial) of a Dirichlet series arising from a modular form f is essentially equivalent to properties of f: f should be an eigenfunction for certain operators (the Hecke operators) on automorphic forms.

The first three chapters bear upon such matters. A considerably subtler issue is the determination of the nature of the values of such Dirichlet series $D(s)$ at special integers s. Some samples of results in this direction are given in sections 6.2–6.5. The discussion of the arithmetic of the special values of these Dirichlet series relies profoundly upon arithmetic properties of modular forms themselves.

It is possible to give a reasonably full account of these special-values results by virtue of new direct proofs (chapter 7) of a theorem (stated in 6.1) of which all these special values results are corollaries. One half of this theorem is part of a result of Shimura concerning the arithmetic properties of Fourier coefficients of Hilbert modular forms. The other half is an apparently new comparison of inner products of cuspforms, that allows an omission of some delicate matters concerning Hecke operators.

In the first five chapters I have attempted to collect in one place fundamental ideas and methods that arose during the period 1904–1961. The diffuseness of the literature and the desirability of illustrating the efficacy of "modern methods" I have taken as justification of this reprise. Chapters 6 and 7 bear upon more recent (and more arithmetic) developments. All but the method of 7.9–7.10 and the choice of presentation must be attributed to many other authors—notably, Blumenthal, Hecke, Siegel, Maass, Rankin, Klingen, Shimura, and Selberg.

There is no presumption of familiarity with the theory of elliptic functions and elliptic modular forms, (although some experience might prove helpful). Indeed, the beautiful but very special ideas and calculations arising from the theory of elliptic functions may not be the best paradigm for general expectations: there is no convenient generalization of the Weierstrass P-function. There are even more mundane coincidences that are implicitly exploited in the classical theory of elliptic modular forms; for example, that the rational integers are a Euclidean (hence principal) domain. But, as it turns out there is no genuine need to depend on such fortuitous coincidences, and because the general situation is important in its own right, I have tried to give a portrayal of Hilbert modular forms that is essentially blind to special features of the ground field. It is possible to do so coherently and palatably by considering modular forms on adele groups.

Currently, it may seem that the theory of elliptic modular forms (that is, Hilbert modular forms with ground-field $\mathbb{Q}$) is most immediately relevant to other parts of arithmetic. However, Klingen's determination of the arithmetic nature of the special values of L-functions of totally real number fields (see 6.2) is a pointed example of the necessity and utility of more general considerations. Also, the treatment here (6.3) of Damerell's and Shimura's results on grossencharacter L-functions makes essential use of the arithmetic properties of Fourier coefficients of Hilbert modular forms (due to Shimura). (See 6.1.) Further, the arguments of 6.3–6.5 rely upon comparisons of inner products of automorphic forms; 6.1 gives a new result in this direction.

The new proof (Chapter 7), which yields a part of Shimura's theorem on the arithmetic properties of the Fourier coefficients of Hilbert modular forms and gives the comparison of inner products, suggests another broadening of perspective and methods: the discernible arithmetic of certain (Siegel's) Eisenstein series on a larger group (a symplectic group of rank 2) is the starting point of the proof. Thus, Siegel's already striking idea (in [Si2]) for understanding the arithmetic nature of the Fourier coefficients of more general Eisenstein series becomes more significant. Shimura's original proof (in [Sh3]) of the theorem on Fourier coefficients depends on a special case of his profound results (in [Sh1]) on canonical models of arithmetic quotients, that in turn depend on quite serious results from algebraic geometry.

While the method of 7.9 and 7.10 seems incapable of recovering Shimura's theorems on generation of class fields, it is profitable to see that the arithmetic nature of the Fourier coefficients can be understood directly. Previous methods of comparison of inner products of cuspforms relied essentially on the delicate notion of "newform" (as in [M], [A], [C]); the idea here is less technical and is relevant in more general situations as well.

Another point hopefully illustrated here is the central role of Eisenstein series—in Rankin's integral representation of L-functions (4.10), in Shimura's applications (6.3–6.5), and in the present proof of the theorem of 6.1 (Chapter 7).

The first chapter studies Hilbert modular forms from a "classical" viewpoint—that is, as it would have been done prior to 1960. Siegel's book [Si1] also gives an introduction to this material. Most of this material merely imitates the theory of elliptic modular forms. An exception is the method of proof of the finite-dimensionality of spaces of cuspforms in 1.7; this method is due to Siegel and Maass. This method, applicable as well to elliptic modular forms, supplants the use in that theory of either residues or the Riemann-Roch theorem, neither of which is as helpful in general. Precise attribution of all historic sources is beyond my ability; the results here are due to many authors, going back as far as Blumenthal's pursuit of Hilbert's suggestion in 1904. Some of the more important sources are [B1], [B2], [H1], [H2], [H3], [KL1], [M], [P], and [Se]. With regard to some matters (Hecke operators and L-functions) the treatment in this chapter is only introductory, as these topics are best treated later in a different way. For the classical versions of these topics—that is, where the base field is $\mathbb{Q}$, one might consult [Sh2] or [Gu].

The second chapter introduces a general notion of automorphic form on $GL(2)$ over the adeles. The influential books [GGP] and [JL] make the usefulness of this viewpoint clear. This viewpoint makes any ground field tractable by uniform methods. For example, a classical treatment of Hecke operators for congruence subgroups over rings of integers not of narrow class number one is excruciatingly technical and not very helpful; an adelic treatment supplants this approach by a discussion even simpler than consideration of Hecke operators for $SL(2, \mathbb{Z})$. Further, in this setting the Euler product expansion of Dirichlet series obtained as Mellin transforms of modular forms arises as inexorably as in Tate's thesis's treatment of L-functions of number fields (as in [La] or [CF]). In 2.7 some references to recent developments regarding analytic continuation of higher L-functions are given. Appendices A.1, A.2, and A.4 give some relevant background on integration theory on homogeneous spaces, harmonic analysis on the adeles, and invariant differential operators on $SL(2, \mathbb{R})$.

Both for completeness and for clarity, the third chapter compares the classical and adelic versions of these things. An essential ingredient is a Strong Approximation Theorem for $SL(2)$; this theorem is proven in Appendix A.3, which specializes and simplifies a proof found in [Kn]. Although it is not hard to give a quite elementary proof over $\mathbb{Q}$ (for example, in Chapter 3 of [Sh2]), it is not easy to do so over arbitrary number fields; the original references for the ideas involved in even more general strong-approximation theorems are the papers [E1] and [E2]. I do not go far in discussing Hecke operators at bad primes; for such a discussion, see [A], [M], and [C].

Chapter 4 is concerned with Eisenstein series. Consideration of analytic and arithmetic properties of Eisenstein series goes back almost 200 years and is still central to number theory and automorphic forms. Most of this chapter is a study of the formal properties of Eisenstein series, which become even more transparent in an adelic formulation. The method of proof of analytic continuation and functional equation given in 4.2 reformulates an idea I first saw in Godement's exceptional article [Go]; perhaps the idea should be attributed to Maass and Selberg in some form. This avoids direct consideration of confluent hypergeometric functions and their p-adic analogues, although the latter have virtues of their own. Section 4.10 is a mildly adelic version of a part of Rankin's papers, [R1] and [R2]. (See also [P].) Although Rankin's first paper on these matters appeared in the 1930's, the potential of the general idea is not yet exhausted. Sections 6.3–6.5 pursue the arithmetic side of the idea (decisively illustrated in Shimura's papers [Sh4], [Sh5], and [Sh6]).

In Chapter 5 the treatment is again somewhat classical, for a non-trivial reason: theta series that are of half-integral weight cannot be considered as automorphic forms on an adelic $GL(2)$; indeed, the study of half-integral-weight automorphic forms is a subject in its own right and is not pursued here. The main point of this chapter is the proof that theta series are modular forms; the proof uses an adelic version of the Poisson summation formula. Theta series have been studied for a long time; some references are [B2], [KL2], and [Sc]. Theta series of integral weight reappear in 6.3 in connection with grossencharacter L-functions.

Chapter 6 discusses some developments since 1961. Section 6.1 states part of a fundamental theorem of Shimura's (from [Sh3]); the theorem concerns the arithmetic properties of Fourier coefficients of Hilbert modular forms. I cannot believe that this result was not known until 1975, considering that versions of this result over $\mathbb{Q}$ were known before 1900 via the theory of elliptic functions. (Certainly the other parts of the results of [Sh3], those regarding generation of class fields, could not have been known earlier.) Also, the theorem stated in 6.1 contains an equally important fact concerning comparison of inner products of cuspforms; this fact is indispensable for applications. Section 6.2 recapitulates Klingen's papers [K1] and [K2]; that is, it combines the result of 6.1 over $\mathbb{Q}$ with 4.8's calculation of Fourier coefficients of Eisenstein series to determine the arithmetic nature of special values of certain L-functions of totally real number fields. Sections 6.3–6.5 give a substantial class of examples of Shimura's results on special values of L-functions attached to Hilbert modular cuspforms ([Sh4], [Sh5], and [Sh6]). The results of 6.5 for ground-field $\mathbb{Q}$ were obtained in a different way in [D]. All these special-value results fit into the general conjectural pattern enunciated in [De].

The last chapter is devoted to a new proof of a part of Shimura's theorem (6.1) on Fourier coefficients and to a proof of the comparison results concerning inner products. Sections 7.1–7.8 follow Siegel's idea (in [Si2]) to prove the rationality of Fourier coefficients of some specific Eisenstein series on a symplectic group of rank 2. Such calculations can be done in much greater generality, as in [Sh7], for example, Section 7.9, following the idea of [G1], determines the restriction of this Eisenstein series to an imbedded copy of $SL(2) \times SL(2)$, embodied in the "Main Formula." Section 7.10 combines these two items to obtain the theorem of 6.1.

As complementary and supplementary reading, one might consult the books [Sh2], [Si1], [GGP], [JL], [Ge], and [Gu]. I have omitted many topics and applications, most of which would have the subject matter of this volume as prerequisite: algebraic geometry of Hilbert modular varieties, compactifications of arithmetic quotients, moduli of abelian varieties, generation of class fields, base change, non-holomorphic modular forms, half-integral-weight forms, converse theorems, representation theory, trace formulas, Kronecker limit formulas, congruence properties of modular forms, application to Hasse-Weil zeta functions, Siegel-Weil formulas, theta correspondences, triple-product L-functions, ...

I find it interesting to trace the arguments used here to obtain the rationality results. The arithmetic of Dirichlet L-functions over $\mathbb{Q}$ is quite old; in appendix A.5 the results are proven by residues and by the classical Poisson summation formula. From this result, following Siegel in 7.1–7.8 and using the method of 7.9–7.10, Shimura's theorem of 6.1 over $\mathbb{Q}$ is obtained, as is the comparison of inner products. Then from this result, following Klingen in 6.2, the special-values result for L-functions over totally real number fields is obtained. From this result in turn, 7.1–7.10 yield the theorem of 6.1 over arbitrary totally real number fields. Finally from this result, the results of 6.3–6.5 follow by Shimura's arithmetic application of Rankin's method. I take the viewpoint that all these arguments have the effect of reducing rationality questions to questions about Dirichlet L-functions over $\mathbb{Q}$: no new numbers are approached from first principles; rather, every step is a reduction. Diagramatically:

Dirichlet L-functions over $\mathbb{Q}$ (A.5) ⇘ Siegel's method (7.1–7.8) ⇓

rationality of Fourier coefficients of Siegel's Eisenstein series over $\mathbb{Q}$ ⇓

the Main Formula (7.9), methods of 7.10 ⇘

rationality properties for $GL(2, \mathbb{Q})$ (6.1 over $\mathbb{Q}$) ⇓

Eisenstein series Fourier expansion (4.8) ⇘

Klingen's method (6.2) ⇘

special values of Dirichlet L-functions over totally real number fields (6.2) ⇓

Siegel's method (7.1–7.8) ⇘

rationality of Fourier coefficients of Siegel's Eisenstein series over totally real fields ⇓

the Main Formula (7.9), methods of 7.10 ⇘

rationality properties for $GL(2)$ over totally real fields (6.1)

Rankin-Shimura method (4.10, 6.3–6.5) ⇓ ⇓

⇘ ⇓

special values of grossencharacter L-functions over totally imaginary quadratic extensions of totally real number fields,
special values of standard L-functions and convolution L-functions for $GL(2)$

If one views the theory of Hilbert modular forms as but the second example (after elliptic modular forms) of much more general phenomena, then distilling the facts and finding the most economical viewpoint becomes important: a long voyage with heavy baggage is unpleasant. Partly for this reason, I have attempted to be conservative in choosing what to include: almost everything in this book is indispensable for getting to the results of Chapter 6. The new methods of 7.9–7.10 allow all these matters to fit into one volume, most of which is prerequisite for a reasonable understanding of further developments in the subject.

Given the modest prerequisites, this book ought to be accessible to graduate students and non-specialists. I also hope that it will be useful as a handbook of some standard methods in the subject.

I was introduced to this subject through Shimura's lectures at Princeton University from 1975 to 1977. The text is a rewriting of parts of courses and lectures I have given at the University of Minnesota from 1982 to 1988. While almost all these results can be found somewhere in the literature, the precise form of this text is the result of my own reflections upon the subject; therefore, I take responsibility for any imbalances, unwarranted omissions, or misattributions. The bibliography is not aimed at exhaustiveness, either in historical references or in references to current literature; such an attempt would serve no purpose. In any case, my aim has been to make the present treatment sufficiently self-contained so that it can be read fruitfully without too many auxiliary references.

The author would like to thank Professor V. Kumar Murty, University of Toronto, for his helpful suggestions in reviewing the manuscript.

This work was partially supported by a grant from the National Science Foundation.

Paul Garrett

Contents

Holomorphic Hilbert Modular Forms

1 Classical Theory of Hilbert Modular Forms

This first chapter develops most of the significant classical features of holomorphic Hilbert modular forms. Some aspects are different from the theory of elliptic modular forms. For example, Koecher's principle 1.4 holds only for fields other than $\mathbb{Q}$. On the other hand, some results from the theory of Riemann surfaces (for example, the Riemann-Hurwitz formula and the Riemann-Roch theorem) are no longer available in as usable a form, so other methods are required.

The first two sections are mainly definitions, with a few very easy lemmas. The third section is the first illustration of a novelty: The number of cusps "at level one" is the class number of the ring of integers of the field involved. Section 1.4 shows another novelty: Over fields larger than $\mathbb{Q}$, there is no need to impose growth conditions or Fourier coefficient conditions on holomorphic Hilbert modular forms, unlike the case of elliptic modular forms. Section 1.5 returns to some examples which very much resemble the elliptic modular case: holomorphic Eisenstein series of level one.

The next section, on Siegel sets, is a simple example of a very general phenomenon: It is not practical to determine a precise fundamental domain for the action of Hilbert modular groups on products of upper half-planes, but it is possible to determine in qualitatively simple terms a superset of a fundamental domain. Indeed, one may easily find examples of exact fundamental domains which are polyhedral but require a very large number of faces. Section 1.7 then provides an illustration of the fact that this approximation of a fundamental domain suffices for many purposes; the purpose at hand is to give a function-theoretic (rather than algebro-geometric) proof of the finite dimensionality of spaces of cuspforms (due to Siegel and Maass). Along the way we obtain some useful estimates on the asymptotic behavior of cuspforms and their Fourier coefficients. The next section shows that the space spanned by Eisenstein series is a complement (in the space of all Hilbert modular forms) to the space of cuspforms.

The two sections on Dirichlet series and Hecke operators associated to cuspforms are actually only first approximations; in this classical setting there are various complications arising from class numbers and units. These complications may best be eliminated by adopting an adelic viewpoint, which we do later.

To integrate functions on a quotient space, without having an explicit fundamental domain, requires some preparatory consideration of integration theory. This in hand, we compute the volume of the quotients (assuming class number one) in a quite amusing manner (after Siegel). Then the Petersson inner product on spaces of cuspforms can be defined, and properties of the Poincare series considered. In this setting we also have a reproducing kernel for cuspforms of a fixed level and weight (studied by Petersson and Selberg).

1.1 The Hilbert Modular Group

Here we define the Hilbert modular groups and certain subgroups, and examine the elementary aspects of the action of these groups on products of complex upper half-planes.

Let $GL^+(2, \mathbb{R})$ be the collection of elements of $GL(2, \mathbb{R})$ with positive determinant, where $GL(2, \mathbb{R})$ is the group of invertible two-by-two real matrices. The group $GL^+(2, \mathbb{R})$ acts on the upper complex half-plane $\mathfrak{H}$ by linear fractional transformations:

$$\begin{pmatrix} a & b \\ c & d \end{pmatrix}(z) = (az + b)/(cz + d).$$

The center of $GL^+(2, \mathbb{R})$ acts trivially. It is important to note that

$$\mathrm{Im}(gz) = (\det g)\,\mathrm{Im}(z)/|cz + d|^2.$$

The isotropy group in

$$SL(2, \mathbb{R}) = \{g \in GL(2, \mathbb{R}): \det g = 1\}$$

of the point $i \in \mathfrak{H}$ is the special orthogonal group

$$SO(2) = \{g \in SL(2, \mathbb{R}): g^T g = 1_2\},$$

and the action of $SL(2, \mathbb{R})$ on $\mathfrak{H}$ is transitive. Letting Z be the center of $GL^+(2, \mathbb{R})$, $SO(2)Z$ is the isotropy group of $i \in \mathfrak{H}$ in $GL^+(2, \mathbb{R})$.

Let F be a totally real number field of degree m over $\mathbb{Q}$, with ring of integers $\mathfrak{o}$. Let $\sigma_1, \ldots, \sigma_m$ be the real imbeddings of F; fix an ordering of them. Via the σ_i we have m imbeddings of $GL(2, F)$ into $GL(2, \mathbb{R})$, componentwise; together, we obtain an imbedding of $GL(2, F)$ into $GL(2, \mathbb{R})^m$. The image of $GL(2, \mathfrak{o})$ in $GL(2, \mathbb{R})^m$ is discrete, since the image of $\mathfrak{o}$ in $\mathbb{R}^m$ (via $\sigma_1, \ldots, \sigma_m$) is discrete, and $GL(2, \mathfrak{o}) \subset M(2, \mathfrak{o}) \approx \mathfrak{o}^4$ is discrete in $M(2, \mathbb{R})^m \approx (\mathbb{R}^m)^4$. Let $GL^+(2, F)$ (respectively, $GL^+(2, \mathfrak{o})$) be the collection of elements of $GL(2, F)$ (respectively, $GL(2, \mathfrak{o})$) with totally positive determinant. We have an action $z \to gz$ of elements g of $GL^+(2, \mathbb{R})^m$ on $z \in \mathfrak{H}^m$ componentwise, and an action of $GL^+(2, \mathfrak{o})$ on $\mathfrak{H}^m$ by our choice of ordering of the real imbeddings of F. This group $GL^+(2, \mathfrak{o})$ is the *full Hilbert modular group* (attached to the field F).

Definition For a non-zero ideal $\mathfrak{n}$ of $\mathfrak{o}$, let

$$\Gamma(\mathfrak{n}) = \{\gamma \in GL^+(2, \mathfrak{o}) : \gamma \equiv 1_2 \text{ modulo } \mathfrak{n}\}.$$

This is the *principal congruence subgroup of level* $\mathfrak{n}$. Let $Z(\mathfrak{o})$ be the center of $GL^+(2, \mathfrak{o})$. Any subgroup Γ of $GL^+(2, F)$ so that $Z(\mathfrak{o})\Gamma$ contains some $\Gamma(\mathfrak{n})$ with *finite index* is a *congruence subgroup* (of $GL^+(2, F)$).

PROPOSITION Let Γ and Γ_2 be congruence subgroups.

i. Then $\Gamma \bigcap \Gamma_2$ is also a congruence subgroup and is of finite index in both Γ and Γ_2.
ii. For $g \in GL(2, F)$, $g\Gamma g^{-1}$ is a congruence subgroup.
iii. $SL(2, \mathfrak{o})$ is a congruence subgroup.
iv. There is a congruence subgroup $\Gamma(\mathfrak{n})$ which is a normal subgroup of $Z(\mathfrak{o})\Gamma$.

Proof All these assertions are easy exercises. ∎

Remarks In the literature, there is an ambiguity about the definition of the principal congruence subgroups and some other special types of congruence subgroups. As we have defined them, they are subgroups of $GL^+(2, \mathfrak{o})$, satisfying some congruence properties. Sometimes they are defined as subgroups of $SL(2, \mathfrak{o})$, satisfying congruence properties. (Since there is no distinction for $F = \mathbb{Q}$, there is no ambiguity for elliptic modular forms). This ambiguity is essentially harmless, except for engendering commensurate ambiguities in certain subsequent definitions.

1.2 Hilbert Modular Forms

For $g = \left(\begin{smallmatrix} a & b \\ c & d \end{smallmatrix}\right) \in GL^+(2, \mathbb{R})$ and $z \in \mathfrak{h}$, put

$$\mu(g, z) = \det g^{-1/2}(cz + d).$$

For $g = (g_1, \ldots, g_m) \in GL^+(2, \mathbb{R})^m$ and $z = (z_1, \ldots, z_m) \in \mathfrak{h}^m$, and for $k = (k_1, \ldots, k_m) \in \mathbb{Z}^m$, we use a standard multi-index notation and write

$$\mu(g, z)^k = \prod_{j=1,\ldots,m} \mu(g_j, z_j)^{k_j}.$$

Definition Let f be a function on $\mathfrak{H}^m$, let $g \in GL^+(2, \mathbb{R})^m$, let $z \in \mathfrak{H}^m$, and let $k = (k_1, \ldots, k_m) \in \mathbb{Z}^m$. Then define

$$(f|_k g)(z) = f(g(z))\mu(g, z)^{-k}.$$

Definition Let Γ be a congruence subgroup, $k \in \mathbb{Z}^m$. Let $\chi : \Gamma \to \mathbb{C}^\times$ be a group homomorphism so that $\chi(\Gamma)$ is a finite group. The space of *weak Hilbert modular forms of weight k for* Γ *with character* χ is

$$\mathrm{Wfm}(\Gamma, k, \chi) = \{f \text{ holomorphic function on } \mathfrak{H}^m\colon f|_k\gamma = \chi(\gamma)f \text{ for all } \gamma \in \Gamma\}.$$

With χ the trivial character, we write

$$\mathrm{Wfm}(\Gamma, k) = \{f \text{ holomorphic function on } \mathfrak{H}^m\colon f|_k\gamma = f \text{ for all } \gamma \in \Gamma\}.$$

Let

$$\mathrm{Wfm}(k) = \bigcup_{\text{congruence subgroups } \Gamma} \mathrm{Wfm}(\Gamma, k)$$

denote the space of all weak holomorphic Hilbert modular forms of weight k with trivial character.

Remarks Very often, one is most interested in the case that the kernel of a character χ as above is a congruence subgroup Γ'. In that event, one sometimes considers $\mathrm{Wfm}(\Gamma', k)$ and neglects the character χ. However, this is not always appropriate and not always possible.

PROPOSITION Let Γ be a congruence subgroup, and let $\Lambda = \{u \in F\colon \left(\begin{smallmatrix}1 & u\\ 0 & 1\end{smallmatrix}\right) \in \Gamma\}$. Then any $f \in \mathrm{Wfm}(\Gamma, k)$ has a *Fourier expansion*

$$f(z) = \sum_{\xi \in \Lambda^*} c_\xi \exp(2\pi i \operatorname{Tr}(\xi z)),$$

where Tr is the $\mathbb{C}$-linear extension to $\mathbb{C}^m \to \mathbb{C}$ of the Galois trace $F \to \mathbb{Q}$, and where

$$\Lambda^* = \{u \in F\colon \operatorname{Tr}(u\Lambda) \subset \mathfrak{o}\}$$

is the dual $\mathbb{Z}$-module to Λ. This Fourier series is absolutely convergent, and uniformly so for z in compact subsets of $\mathfrak{H}^m$.

Proof First, note that since Γ contains some $\Gamma(\mathfrak{n})$, Λ contains some non-zero ideal $\mathfrak{n}$. Writing $z = x + iy$ with $x = (x_1, \ldots, x_m) \in \mathbb{R}^m$ and $y = (y_1, \ldots, y_m) \in \mathbb{R}^m$, $f(x + iy)$ is certainly as smooth as one could desire, as a function of $x \in \mathbb{R}^m$. Therefore, as a function of x, f has a Fourier expansion

$$f(x + iy) = \sum_{\xi \in \Lambda^*} c_\xi(y) \exp(2\pi i \operatorname{Tr}(\xi x))$$

which is absolutely convergent, and uniformly so for x in compact subsets of $\mathfrak{H}^m$. It is easy to see that the only exponentials which enter are those given by $\xi \in \Lambda^*$. For f to be holomorphic in $z = (z_1, \ldots, z_m) \in \mathfrak{H}^m$, the Cauchy-Riemann equation

$$i\partial f/\partial x_j = \partial f/\partial y_j$$

must be satisfied for $j = 1, \ldots, m$. That is, for each j, we must have

$$i \sum_{\xi \in \Lambda^*} c_\xi(y)(2\pi i \xi_j) \exp(2\pi i \operatorname{Tr}(\xi x)) = \sum_{\xi \in \Lambda^*} \partial c_\xi(y)/\partial y_j \exp(2\pi i \operatorname{Tr}(\xi x)),$$

where $\xi_j = \sigma_j(\xi)$. By uniqueness of Fourier expansions (in x) we must have

$$-2\pi \xi_j c_\xi(y) = \partial c_\xi(y)/\partial y_j.$$

Therefore, solving the indicated system of differential equations, there is a constant c_ξ so that

$$c_\xi(y) = c_\xi \exp(-2\pi \operatorname{Tr}(\xi y)).$$

Finally, to see that the Fourier series is absolutely convergent and uniformly so in compacta in $\mathfrak{H}^m$, we observe that as any component y_j increases, $\exp(-2\pi \operatorname{Tr}(\xi y))$ does not increase. As we have already ascertained the uniform absolute convergence in x, this gives the result. ■

Definition Let $f \in \mathrm{Wfm}(\Gamma, k)$. Then say that f is a *holomorphic Hilbert modular form of weight k with respect to* Γ if, for every $g \in GL^+(2, F)$, the Fourier expansion

$$(f|_k g)(z) = \sum_\xi c_\xi(g) \exp(2\pi i \operatorname{Tr}(\xi z))$$

has $c_\xi(g) = 0$ unless $\xi = 0$ or ξ is totally positive. Let $\mathrm{Mfm}(\Gamma, k)$ be the $\mathbb{C}$-vectorspace of holomorphic Hilbert modular forms of weight k with respect to Γ; let

$$\mathrm{Mfm}(k) = \bigcup_{\text{congruence subgroups } \Gamma} \mathrm{Mfm}(\Gamma, k)$$

denote the space of all *holomorphic Hilbert modular forms of weight k*.

Definition A function $f \in \mathrm{Mfm}(\Gamma, k)$ is a holomorphic Hilbert modular *cuspform* of weight k with respect to Γ if, for every $g \in GL^+(2, F)$, the Fourier expansion

$$(f|_k g)(z) = \sum_\xi c_\xi(g) \exp(2\pi i \operatorname{Tr}(\xi z))$$

has $c_\xi(g) = 0$ unless ξ is totally positive. Let $\mathrm{Cfm}(\Gamma, k)$ be the $\mathbb{C}$-vectorspace of holomorphic Hilbert modular cuspforms of weight k with respect to Γ; let

$$\mathrm{Cfm}(k) = \bigcup_{\text{congruence subgroups } \Gamma} \mathrm{Cfm}(\Gamma, k)$$

denote the space of all *holomorphic Hilbert modular cuspforms of weight k*.

PROPOSITION Let Γ be a congruence subgroup, and let $g \in GL^+(2, F)$. Take f in $\mathrm{Wfm}(\Gamma, k)$ (respectively, in $\mathrm{Mfm}(\Gamma, k)$, $\mathrm{Cfm}(\Gamma, k)$). Then $f|_k g$ is in $\mathrm{Wfm}(g^{-1}\Gamma g, k)$ (respectively, in $\mathrm{Mfm}(g^{-1}\Gamma g, k)$, $\mathrm{Cfm}(g^{-1}\Gamma g, k)$). That is, $\mathrm{Wfm}(k)$, $\mathrm{Mfm}(k)$, and $\mathrm{Cfm}(k)$ are stable under the action of $GL^+(2, F)$ given by $f \to f|_k g$.

Proof The assertion for $\mathrm{Wfm}(\Gamma, k)$ is formal: For $\gamma \in \Gamma$,

$$(f|_k g)|_k(g\gamma g^{-1}) = f|_k(g(g^{-1}\gamma g)) = f|_k g.$$

Similarly, the condition on Fourier coefficients defining modular forms and cuspforms is stable under the operation $f \to f|_k g$. ■

1.3 Class Numbers and Cusps

The main point in this section is that there is a bijection between ideal classes and certain orbits of $GL^+(2, \mathfrak{o})$ on the "boundary" of $\mathfrak{H}^m$. This result has several important consequences, which are stated in this section as corollaries after the proof of the theorem. (Furthermore, the later discussion of approximation of a fundamental domain by Siegel sets is somewhat dependent on this result.) Also, we introduce the homogeneous (projective) coordinates on $\mathfrak{H}^m \subset \mathbb{C}^m$ in order to discuss the boundary of $\mathfrak{H}^m$, the "point(s) at infinity," and the idea of a *cusp* in a precise sense.

Let $GL(2, \mathbb{C})$ act on

$$\mathbb{P}^1 = \left\{\begin{pmatrix} u \\ v \end{pmatrix} \in \mathbb{C}^2 - \{0\}\right\}\Big/\mathbb{C}^\times$$

by matrix multiplication on the left. We may identify $\mathbb{C}$ with an open subset of $\mathbb{P}^1$ by

$$\mathbb{C} = \left\{\begin{pmatrix} u \\ v \end{pmatrix} \in \mathbb{C}^2 - \{0\} : v \neq 0\right\}\Big/\mathbb{C}^\times \subset \mathbb{P}^1.$$

Let $i\infty$ denote the point represented by $\binom{1}{0}$ in $\mathbb{P}^1$; then $\mathbb{C} \cup \{i\infty\} = \mathbb{P}^1$. With this identification, we may see that the action of $GL(2, \mathbb{C})$ on $\mathbb{P}^1$ is just the linear fractional transformation action on $\mathbb{C}$. We refer to these coordinates $\binom{u}{v}$ on $\mathbb{C}$ or $\mathbb{P}^1$ as *homogeneous coordinates*; to make appropriate sense of the "point at infinity" it is useful to consider this extended action of $GL(2, \mathbb{C})$ on $\mathbb{P}^1$. Similarly, we have an action of $GL(2, \mathbb{C})^m$ on $(\mathbb{P}^1)^m$, componentwise. We have inclusions

$$\mathfrak{H}^m \subset \mathbb{C}^m \subset (\mathbb{P}^1)^m.$$

Definition The *cusps* of $GL^+(2, F)$ are the points

$$(\sigma_1(\alpha), \ldots, \sigma_m(\alpha)) \in \mathbb{R}^m \subset \partial\mathfrak{H}^m \subset \mathbb{C}^m \subset (\mathbb{P}^1)^m$$

for $\alpha \in F$, together with the point

$$i\infty = (i\infty, \ldots, i\infty) \in (\mathbb{P}^1)^m.$$

For a subgroup Γ of $GL^+(2, F)$, say that two cusps F_1 and F_2 are Γ-*inequivalent* if the orbits of F_1 and F_2 under Γ are disjoint.

Remarks It is easy to check that $GL(2, F)$ stabilizes the set of all cusps, by use of the homogeneous coordinates on $(\mathbb{P}^1)^m$. For that matter, $GL(2, \mathbb{R})^m$ stabilizes

$$(\mathbb{R} \cup \{i\infty\})^m \subset (\mathbb{P}^1)^m$$

where $\mathbb{R} \cup \{i\infty\} \subset \mathbb{C} \cup \{i\infty\} = \mathbb{P}^1$.

THEOREM Let $SL(2, \mathfrak{o}) \subset \Gamma \subset GL^+(2, \mathfrak{o})$. Then the number of Γ-inequivalent cusps is $h(\mathfrak{o})$, the absolute class number of $\mathfrak{o}$. In particular, there is a bijection between absolute ideal classes and Γ-equivalence classes of cusps, given by

$$(u\mathfrak{o} + v\mathfrak{o}) \to \begin{pmatrix} u \\ v \end{pmatrix} \mathbb{C}^\times,$$

where u, v are generators for a non-zero fractional ideal of $\mathfrak{o}$, using the homogenous coordinates on $(\mathbb{P}^1)^m$. In particular, the number of cusps is finite.

Proof First, recall that any non-zero ideal in a Dedekind domain can be generated by two elements u, v (not both 0). Two pairs (u, v) and (u', v') give the same non-zero fractional ideal if and only if

$$\begin{pmatrix} u \\ v \end{pmatrix} \in GL(2, \mathfrak{o}) \begin{pmatrix} u' \\ v' \end{pmatrix},$$

since multiplication by an element of $GL(2, \mathfrak{o})$ corresponds to a change of coordinates on the ideal. Therefore, these two pairs give the same ideal class if and only if

$$\begin{pmatrix} u \\ v \end{pmatrix} \in GL(2, \mathfrak{o}) \begin{pmatrix} u' \\ v' \end{pmatrix} F^\times.$$

Viewing this as a relation written in homogeneous coordinates on $(\mathbb{P}^1)^m$, it follows that if $\binom{u}{v}\mathbb{C}^\times$ is in the Γ-orbit of $\binom{u'}{v'}\mathbb{C}^\times$, then the two non-zero fractional ideals $u\mathfrak{o} + v\mathfrak{o}$ and $u'\mathfrak{o} + v'\mathfrak{o}$ are in the same ideal class. What remains to be shown is the converse: If the two non-zero fractional ideals $u\mathfrak{o} + v\mathfrak{o}$ and $u'\mathfrak{o} + v'\mathfrak{o}$ are in the same ideal class, then $\binom{u}{v}\mathbb{C}^\times$ is in the Γ-orbit of $\binom{u'}{v'}\mathbb{C}^\times$ in $(\mathbb{P}^1)^m$. The only complication is that Γ may be properly smaller than $GL(2, \mathfrak{o})$; it is only required to contain $SL(2, \mathfrak{o})$. It suffices to treat $\Gamma = SL(2, \mathfrak{o})$. Let $\alpha = \binom{u}{v}$, $\alpha' = \binom{u'}{v'}$; without loss of generality we may take $u\mathfrak{o} + v\mathfrak{o} = \mathfrak{a} = u'\mathfrak{o} + v'\mathfrak{o}$.

Let $\langle x, y \rangle = x_1 y_1 + x_2 y_2$ be the usual pairing on F^2. Clearly, $\langle \alpha, \mathfrak{o}^2 \rangle = \mathfrak{a} = \langle \alpha', \mathfrak{o}^2 \rangle$. From this, $\mathfrak{o} = \langle \alpha, \mathfrak{a}^{-1}\mathfrak{o}^2 \rangle = \langle \alpha', \mathfrak{a}^{-1}\mathfrak{o}^2 \rangle$. Let $\beta, \beta' \in \mathfrak{a}^{-1}\mathfrak{o}^2$ so that $\langle \alpha, \beta \rangle = 1$ and $\langle \alpha', \beta' \rangle = 1$. Since $\alpha \in \mathfrak{a}\mathfrak{o}^2$,

$$1 = \langle \alpha, \beta \rangle \in \langle \mathfrak{a}\mathfrak{o}^2, \beta \rangle \subset \langle \mathfrak{a}\mathfrak{o}^2, \mathfrak{a}^{-1}\mathfrak{o}^2 \rangle = \langle \mathfrak{o}^2, \mathfrak{o}^2 \rangle = \mathfrak{o}.$$

Therefore, $\langle \mathfrak{a}\mathfrak{d}^2, \beta\rangle = \mathfrak{o}$, since $\langle \mathfrak{a}\mathfrak{d}^2, \beta\rangle$ is an ideal of $\mathfrak{o}$ containing 1. Similarly, $\langle \mathfrak{a}\mathfrak{d}^2, \beta'\rangle = \mathfrak{o}$. Now we claim that

$$\mathfrak{a}\mathfrak{d}^2 = \mathfrak{o}\alpha \oplus M \qquad \mathfrak{a}\mathfrak{d}^2 = \mathfrak{o}\alpha' \oplus M'$$

with

$$M = \{x \in \mathfrak{a}\mathfrak{d}^2 \colon \langle x, \beta\rangle = 0\} \qquad M' = \{x \in \mathfrak{a}\mathfrak{d}^2 \colon \langle x, \beta'\rangle = 0\}.$$

That the righthand sides are contained in the lefthand sides is immediate; for the converse, note that for any $z \in \mathfrak{a}\mathfrak{d}^2$, we can write

$$z = \langle z, \beta\rangle\alpha + (z - \langle z, \beta\rangle\alpha).$$

Now it must be that

$$M = \mathfrak{m}\gamma \qquad M' = \mathfrak{m}'\gamma'$$

for some ideals $\mathfrak{m}$ and $\mathfrak{m}'$, and for some $\gamma, \gamma' \in F^2$ (from the structure theorem for finitely generated modules over Dedekind domains). Now $\bigwedge^2 F^2$ is one-dimensional, so $\alpha \wedge \gamma = x(\alpha' \wedge \gamma')$ for some $x \in F^\times$. Replace γ' by $x^{-1}\gamma'$ and $\mathfrak{m}'$ by $x\mathfrak{m}'$, so that now

$$\alpha \wedge \gamma = \alpha' \wedge \gamma' \qquad \mathfrak{a}\mathfrak{d}^2 = \mathfrak{o}\alpha \oplus \mathfrak{m}\gamma \qquad \mathfrak{a}\mathfrak{d}^2 = \mathfrak{o}\alpha' \oplus \mathfrak{m}\gamma'.$$

Take $A \in GL(2, F)$ so that $A\alpha = \alpha'$ and $A\gamma = \gamma'$. Since $\alpha \wedge \gamma = \alpha' \wedge \gamma'$, $A \in SL(2, F)$. Since A and A^{-1} send $\mathfrak{a}\mathfrak{d}^2$ to itself, A and A^{-1} preserve $\mathfrak{o}^2$. So $A \in GL(2, \mathfrak{o})$. Thus, we have $A \in SL(2, \mathfrak{o})$. ■

COROLLARY For any congruence subgroup Γ, there are only finitely many Γ-inequivalent cusps. Let $\{\kappa_1, \ldots, \kappa_v\}$ be a set of irredundant representatives for the Γ-inequivalent cusps. Let $\delta_i \in SL(2, F)$ so that $\delta_i(i\infty) = \kappa_i$. Let P be the group of upper triangular matrices in $GL^+(2, F)$. Then

$$\bigcup_i \Gamma\delta_i P = GL^+(2, F) \qquad \text{(disjoint union).}$$

Proof Let $\Gamma(\mathfrak{n})$ be a congruence subgroup contained in $Z(\mathfrak{o})\Gamma$, and let v be the index of $\Gamma(\mathfrak{n})$ in $GL^+(2, \mathfrak{o})$. Then there are at most $vh(\mathfrak{o})\Gamma(\mathfrak{n})$-inequivalent cusps of $\Gamma(\mathfrak{n})$; therefore, the number of inequivalent cusps of Γ is no greater than $vh(\mathfrak{o})$, which is finite.

It is immediate that P is exactly the isotropy group of $i\infty$ in $GL^+(2, F)$. Given $g \in GL^+(2, F)$, suppose that $g(i\infty) = \kappa$, where κ is some cusp. Let $\gamma \in \Gamma$ so that $\gamma(\kappa) = \kappa_i$, for one of the κ_i. Then $(\delta_i^{-1}\gamma g)(i\infty) = i\infty$; therefore, $\delta_i^{-1}\gamma g \in P$, so $g \in \gamma^{-1}\delta_i P$. It is clear that δ_i is uniquely determined. ■

COROLLARY Let f be a weak modular form of weight k with respect to some congruence subgroup Γ. For $g \in GL^+(2, F)$, let the Fourier expansion of $f|_k g$ be

$$(f|_k g)(z) = \sum_{\xi} c_\xi(g) \exp(2\pi i \operatorname{Tr}(\xi z)).$$

Let $\{\delta_i\}$ be a (finite) set of representatives for

$$\Gamma \backslash GL^+(2, F)/P.$$

For f to be a cuspform, it is necessary and sufficient that $c_\xi(g) = 0$ for ξ not totally positive merely for $g \in \{\delta_i\}$.

Remarks The point of this assertion is that there are only a finite number of $g \in GL^+(2, F)$ for which the condition on Fourier coefficients need be verified.

Proof For $\gamma \in \Gamma$, $f|_k \gamma = f$, by definition. On the other hand, for

$$p = \begin{pmatrix} a & b \\ 0 & d \end{pmatrix} \in P$$

and for a Hilbert modular form

$$\varphi(z) = \sum_{\xi} c(\xi) \exp(2\pi i \operatorname{Tr}(\xi z))$$

of weight k we have

$$(\varphi|_k p)(z) = (a/d)^{-k/2} \sum_{\xi} c_\xi \exp(2\pi i \operatorname{Tr}(\xi(ad^{-1}z + bd^{-1})) =$$

$$= \sum_{\xi} \{(a/d)^{-k/2} c(a^{-1}d\xi) \exp(2\pi i \operatorname{Tr}(\xi bd^{-1}))\} \exp(2\pi i \operatorname{Tr}(\xi z)).$$

That is, the Fourier coefficients of φ vanish for ξ neither 0 nor totally positive if and only if those of $\varphi|_k p$ do. For $g \in GL^+(2, F)$, by the previous corollary we can write $g = \gamma\delta_i p$ for $\gamma \in \Gamma$, $p \in P$, with some δ_i. Then

$$f|_k g = f|_k(\gamma\delta_i p) = (f|_k \delta_i)|_k p.$$

The previous remarks then yield the result. ■

1.4 Koecher's Principle

Though it would appear that there are non-trivial conditions imposed upon the Fourier coefficients of a weak Hilbert modular form in order that it be a Hilbert modular form, the following theorem shows that this is not so for $[F : \mathbb{Q}] > 1$. In the case $F = \mathbb{Q}$, it is well known that the condition on Fourier coefficients is not trivial.

THEOREM (KOECHER'S PRINCIPLE) Let Γ be a congruence subgroup of $GL^+(2, F)$, where $[F:\mathbb{Q}] = m > 1$, and let f be a weak holomorphic Hilbert modular form of weight $k = (k_1, \ldots, k_m) \in \mathbb{Z}^m$ and character χ. Then in the Fourier expansion

$$f(z) = \sum_{\xi} c(\xi) \exp(2\pi i \operatorname{Tr}(\xi z))$$

of f, we have $c(\xi) = 0$ unless $\xi = 0$ or ξ is *totally positive*. Further, unless $k_1 = k_2 = \cdots = k_m$, we must also have $c(0) = 0$.

COROLLARY When $[F:\mathbb{Q}] > 1$, every weak holomorphic Hilbert modular form is a holomorphic Hilbert modular form. Unless the weight $k = (k_1, \ldots, k_m)$ has the property that $k_1 = \cdots = k_m$, every holomorphic Hilbert modular form of weight k is a cuspform.

Proof of Theorem This is really just a corollary of the Units Theorem. Since Γ is a congruence subgroup, for some subgroup U of $\mathfrak{o}^\times$ of finite index, Γ contains a subgroup of the form

$$\left\{\tilde{y} = \begin{pmatrix} y & 0 \\ 0 & y^{-1} \end{pmatrix} : y \in U\right\}.$$

Since $\mu(\tilde{\eta}, z)^{-k} = \eta^k$ and $\tilde{\eta}(z) = \eta^2 z$ (using a multi-index notation), for $\eta \in U$ we have

$$f(\eta^2 z) = \chi(\tilde{\eta})\eta^{-k} f(z).$$

Looking at the Fourier coefficients, using the uniqueness of Fourier expansions, this implies that

$$c(\eta^{-2}\xi) = \chi(\tilde{\eta})\eta^{-k} c(\xi).$$

Already this shows (using the Units Theorem) that $c(0) = 0$ unless $k_1 = k_2 = \cdots = k_m$.

Suppose that some $\xi_j = \sigma_j(\xi)$ is negative, for some fixed $\xi \in F$. Then (using the Units Theorem) take $\eta \in U$ so that $\sigma_j(\eta)$ is a very large positive value, and so that $0 < \sigma_i(\eta) \leq 1$ for $i \neq j$. Let λ be any positive real number, and consider the subseries

$$\sum_{n \geq 0} c(\eta^{2n}\xi) \exp(2\pi i \operatorname{Tr}(\eta^{2n}\xi i\lambda)) = c(\xi) \sum_{n \geq 0} \chi(\tilde{\eta})^{-1} \eta^{nk} \exp(-2\pi\lambda \operatorname{Tr}(\eta^{2n}\xi)).$$

Since

$$\operatorname{Tr}(\eta^{2n}\xi) = \eta_j^{2n}\xi_j + \sum_{i \neq j} \eta_i^{2n}\xi_i,$$

from crude estimates we see that (no matter what weight k is) the subseries cannot be convergent unless actually $c(\xi) = 0$. The Fourier series for $f(z)$ is absolutely convergent for $z \in \mathfrak{H}^m$ (since $f(x + iy)$ is smooth as a function of x), so we find that

$c(\xi) = 0$ unless every $\sigma_j(\xi)$ is non-negative; therefore, either $\xi = 0$ or ξ is totally positive. ■

Proof of Corollary We can apply the theorem to $f|_k g$ for any $g \in GL^+(2, F)$, since we have already seen that $f|_k g$ is in Wfm(k) if f is. That is, for all $g \in GL^+(2, F)$, $f|_k g$ has non-zero Fourier coefficients $c_\xi(g)$ only for $\xi = 0$ or totally positive; f is a Hilbert modular form. A similar argument applies to the case where not all k_i are the same; all weak Hilbert modular forms of such weights are necessarily cuspforms. ■

1.5 Holomorphic Eisenstein Series of Level One

The simplest examples of Hilbert modular forms are the *holomorphic Eisenstein series*. We will consider just Eisenstein series for $GL^+(2, \mathfrak{o})$ at the moment, though an analogous construction works much more generally. Eisenstein series often appear in two guises, referred to here as "homogeneous" and "inhomogeneous." While the spaces spanned by these two versions of Eisenstein series are the same, the two sorts of Eisenstein series arise and are useful in different circumstances. So it is important to appreciate that they span the same space of functions.

In this section we will compare the space spanned by the two types of Eisenstein series and determine bases for this space. Further, we will see how an Eisenstein series may be associated to each cusp of $GL^+(2, \mathfrak{o})$, in such a manner as to determine a basis for the whole space spanned by all the Eisenstein series. Also, we will show that, with weight $2k = (2k, \ldots, 2k)$,

$$\mathrm{Mfm}(GL^+(2, \mathfrak{o}), 2k) =$$

$$= \mathrm{Cfm}(GL^+(2, \mathfrak{o}), 2k) \oplus \{\text{space spanned by Eisenstein series of weight } 2k\}.$$

That is, every modular form of such weight for $GL^+(2, \mathfrak{o})$ is a cuspform plus a linear combination of Eisenstein series. (We have already noted that if the weight has any two components unequal, then every holomorphic Hilbert modular form is a cuspform).

Definition The *inhomogeneous holomorphic Eisenstein* series of weight

$$2k = (2k, \ldots, 2k)$$

with $2 < 2k \in \mathbb{Z}$, associated to a non-zero fractional ideal $\mathfrak{m}$ of F, is defined by

$$E(z; \mathfrak{m}, 2k) = \sum_{c, d \in \mathfrak{m}}{}^{*} (cz + d)^{-2k},$$

using a multi-index notation, where $\sum^*$ indicates the sum over

$$(c, d) \in \{(\alpha, \beta) \in \mathfrak{m} \times \mathfrak{m} \colon \alpha\mathfrak{o} + \beta\mathfrak{o} = \mathfrak{m}\}/\mathfrak{o}^\times.$$

The *homogeneous holomorphic Eisenstein* series of weight $2k$, associated to a non-zero fractional ideal $\mathfrak{m}$ of F, is defined by

$$E^{\#}(z;\mathfrak{m},2k) = \sum_{c,d\in\mathfrak{m}}{}' (cz+d)^{-2k},$$

using a multi-index notation, where $\sum'$ indicates the sum over

$$(c,d) \in ((\mathfrak{m}\times\mathfrak{m}) - \{0\})/\mathfrak{o}^{\times}.$$

PROPOSITION The series defining the Eisenstein series $E^{\#}(z;\mathfrak{m},2k)$ are absolutely convergent if and only if $2k > 2$ (and then uniformly so, for z in compact subsets of $\mathfrak{H}^m$). The series defining the Eisenstein series $E(z;\mathfrak{m},2k)$ are absolutely convergent for $2k > 2$.

Proof First, note that the series for $E(z;\mathfrak{m},2k)$ is a subseries of that for $E^{\#}(z;\mathfrak{m},2k)$, so that absolute convergence of $E^{\#}(z;\mathfrak{m},2k)$ implies that of $E(z;\mathfrak{m},2k)$.

We will identify F, $\mathfrak{o}$, and so on, with their images in $\mathbb{R}^m$ via σ. Define a function φ on $F^2\otimes_{\mathbb{Q}}\mathbb{R}\approx(\mathbb{R}^m)^2$ by

$$\varphi(u,v) = \exp(-\pi(\|u\|^2+\|v\|^2))$$

where u, $v\in\mathbb{R}^m$, and $\| \ \|$ is the usual length of a vector in $\mathbb{R}^m$. Also, let

$$g = g_z = \begin{pmatrix}1 & x\\ 0 & 1\end{pmatrix}\begin{pmatrix}y & 0\\ 0 & 1\end{pmatrix} \in GL(2,\mathbb{R}^m) \approx GL(2,\mathbb{R})^m,$$

where $z = x+iy\in\mathfrak{H}^m$. It is immediate (by changing variables in the integral) that

$$\int_{(\mathbb{R}^{\times})^m} (t^2)^s \sum_{cd}{}' \varphi(t(cd)g)\, d^{\times}t_1 \ldots d^{\times}t_m = \{\pi^{-s}\Gamma(s)\}^m \sum_{cd}{}' |(cz+d)^2|^{-s}$$

where $d^{\times}\tau = d\tau/|\tau|$, and where we use a multi-index notation. It is not too hard to show that the integral is absolutely convergent for $\mathrm{Re}(s)$ sufficiently large. Thus, we have an integral representation of the series of absolute values of the terms of the series defining the Eisenstein series. We will treat this as a generalized Dirichlet series with non-negative coefficients, and apply Landau's lemma:

LEMMA Let λ_n be a monotone increasing sequence of positive numbers, and let $f(s) = \sum_n c_n/\lambda_n^s$ be a generalized Dirichlet series with $c_n \geq 0$. If this series is absolutely convergent for $\mathrm{Re}(s)\geq\sigma_0$, and if f has an analytic continuation to a function which is holomorphic on a neighborhood of $s=\sigma_0$, then there is $\delta>0$ so that the series defining $f(s)$ is absolutely convergent for $\mathrm{Re}(s) > \sigma_0-\delta$. ■

The point of this lemma is that a generalized Dirichlet series with non-negative coefficients converges absolutely exactly in the half-plane where the function it defines is holomorphic. We will show that

$$\Phi(s) = \{\pi^{-s}\Gamma(s)\}^m \sum_{(c,d)}{}' |(cz + d)^2|^{-s}$$

has an analytic continuation in s which has a pole at $s = 1$; by Landau's lemma, this implies that this series for $\Phi(s)$ is absolutely convergent exactly for $\mathrm{Re}(s) > 1$, yielding the assertion of the proposition.

We can rewrite

$$\Phi(s) = \int_{(\mathbb{R}^\times)^m/\mathfrak{o}^\times} (t^2)^s \sum_{v}{}' \varphi(tvg)\, d^\times t_1 \ldots d^\times t_m.$$

Let

$$U = \left\{t \in (\mathbb{R}^\times)^m/\mathfrak{o}^\times : \prod_j |t_j| = 1\right\}\Big/ \mathfrak{o}^\times$$

$$T = \{\tilde{t} = (t^{1/m}, \ldots, t^{1/m}) \in (\mathbb{R}^\times)^m/\mathfrak{o}^\times : t \geq 1\}$$

$$S = \{\tilde{t} = (t^{1/m}, \ldots, t^{1/m}) \in (\mathbb{R}^\times)^m/\mathfrak{o}^\times : t \leq 1\}$$

$$P = UT$$

$$Q = US.$$

By the Units Theorem for $\mathfrak{o}$, U is compact. We have

$$\Phi(s) = \int_{(\mathbb{R}^\times)^m/\mathfrak{o}^\times} (t^2)^s \sum_{v}{}' \varphi(tvg)\, d^\times t_1 \ldots d^\times t_m =$$

$$= \int_{UT} (t^2)^s \sum_{v}{}' \varphi(tvg)\, d^\times t_1 \ldots d^\times t_m + \int_{US} (t^2)^s \sum_{v}{}' \varphi(tvg)\, d^\times t_1 \ldots d^\times t_m.$$

Elementary estimates show that the integral over UT is absolutely convergent for all $s \in \mathbb{C}$. By Poisson summation, noting that φ is its own Fourier transform and using elementary properties of Fourier transforms,

$$\sum_{v} \varphi(tvg) = \mathrm{volume}(\mathbb{R}^{2m}/\mathfrak{m}^2)^{-1} \prod_j |t_j|^{-2} \prod_j y_j^{-1} \times \sum_{v} \varphi(t^{-1}vg^{T-1}).$$

Using this, letting

$$c = \mathrm{volume}(\mathbb{R}^{2m}/\mathfrak{m}^2)^{-1} \prod_j y_j^{-1}$$

the integral over US is (upon changing variables)

$$c \int_{US} (t^2)^s \prod_j |t_j|^{-2} \sum_v{}' \varphi(t^{-1} v g^{T-1})\, d^x t_1 \ldots d^x t_m -$$

$$- \int_{US} (t^2)^s \varphi(0)\, d^x t_1 \ldots d^x t_m + c \int_{US} (t^2)^s \prod_j |t_j|^{-2} \hat{\varphi}(0)\, d^x t_1 \ldots d^x t_m =$$

$$= c \int_{UT} (t^2)^{1-s} \sum_v{}' \varphi(t v g^{T-1})\, d^x t_1 \ldots d^x t_m -$$

$$- \varphi(0) \int_{US} (t^2)^s\, d^x t_1 \ldots d^x t_m + c\hat{\varphi}(0) \int_{UT} (t^2)^{1-s}\, d^x t_1 \ldots d^x t_m.$$

Again by elementary estimates, the first integral in the latter expression is an entire function of $s \in \mathbb{C}$. By an elementary change of variables,

$$\int_{US} (t^2)^s\, d^x t_1 \ldots d^x t_m = \int_U 1 \times \int_{(0,1)} t^s\, dt/t =$$

$$= (\text{non-zero constant}) \times s^{-1},$$

where we need not be too specific about what the measure on U is, except that U is compact. Likewise,

$$\int_{UT} (t^2)^{1-s}\, d^x t_1 \ldots d^x t_m = (\text{non-zero constant}) \times (s-1)^{-1}.$$

This proves that $\Phi(s)$ has a pole at $s = 1$, completing the proof by Landau's lemma. ■

PROPOSITION For $2k > 2$, the Eisenstein series $E(z; \mathfrak{m}, 2k)$ and $E^{\#}(z; \mathfrak{m}, 2k)$ are holomorphic Hilbert modular forms of weight $(2k, \ldots, 2k)$.

Proof That they are weak holomorphic modular forms is immediate from their definition, and from the convergence of the series defining them. For $[F : \mathbb{Q}] > 1$, Koecher's principle assures that they are Hilbert modular forms. (The case of $F = \mathbb{Q}$ is classical). ■

Definition Let $SL(2, \mathfrak{o}) \subset \Gamma \subset GL^+(2, \mathfrak{o})$, $2k = (2k, \ldots, 2k) \in \mathbb{Z}^m$. Define $\mathrm{Eis}(\Gamma, 2k)$ as the $\mathbb{C}$-vectorspace inside $\mathrm{Mfm}(\Gamma, 2k)$ spanned by all of the above Eisenstein series.

PROPOSITION Let $\{\mathfrak{m}_i \colon i = 1, \ldots, h\}$ be any set of ideal class representatives for $\mathfrak{o}$. Then both

$$\{E(z; \mathfrak{m}_i, 2k) \colon i = 1, \ldots, h\}$$

and

$$\{E^{\#}(z; \mathfrak{m}_i, 2k) \colon i = 1, \ldots, h\}$$

are bases for $\mathrm{Eis}(\Gamma, 2k)$ over $\mathbb{C}$.

Remarks In fact, in the course of the proof, we will obtain explicit relations between the two sorts of Eisenstein series. As a consequence of the form of the relation, we obtain the other half of the necessary and sufficient condition on the weight for convergence of the Eisenstein series of the form $E(z; \mathfrak{m}, k)$.

COROLLARY OF PROOF The Eisenstein series $E(z; \mathfrak{m}, 2k)$ is absolutely convergent only if $2k > 2$.

Proof Abbreviate

$$E(\mathfrak{m}) = E(z; \mathfrak{m}, 2k) \qquad E^{\#}(\mathfrak{m}) = E^{\#}(z; \mathfrak{m}, 2k).$$

First, one has the obvious

$$E^{\#}(\mathfrak{m}) = \sum_{\mathfrak{n} \subset \mathfrak{m}} E(\mathfrak{n})$$

where $\mathfrak{n}$ runs over non-zero ideals (contained in $\mathfrak{m}$). Also, for $\alpha \in F^{\times}$, $\mathfrak{a}$ a fractional ideal of F,

$$E(\alpha\mathfrak{a}) = N\alpha^{-2k} E(\mathfrak{a}), \qquad E^{\#}(\alpha\mathfrak{a}) = N\alpha^{-2k} E^{\#}(\mathfrak{a}).$$

Let $\{\rho_i\}$ be ideal class representatives for $\mathfrak{o}$. From these observations, it follows that $\{E(z; \rho_i, 2k)\}$ spans the space spanned by

$$\{E(z; \mathfrak{m}, 2k) \colon \mathfrak{m} \text{ is a non-zero ideal of } \mathfrak{o}\}.$$

A similar statement holds for $\{E^{\#}(z; \rho_i, 2k)\}$.

Then

$$E^{\#}(\mathfrak{a}) = \sum_{\mathfrak{m} \subset \mathfrak{a}} E(\mathfrak{m}) = \sum_i \sum_{\alpha \in \mathfrak{a}\rho_i^{-1}/\mathfrak{o}^{\times}} E(\alpha\rho_i) =$$

$$= \sum_i \sum_\alpha N\alpha^{-2k} E(\rho_i) = \sum_i E(\rho_i^{-1}) \left\{ \sum_\alpha N(\alpha)^{-2k} \right\} = \sum_i E(\rho_i) \zeta(2k; \mathfrak{a}\rho_i^{-1}),$$

where $\zeta(*; \mathfrak{n})$ is the partial zeta function

$$\zeta(s; \mathfrak{n}) = \sum_{\alpha \in \mathfrak{n}/\mathfrak{o}^{\times}} |N(\alpha)|^{-s}.$$

From this it follows that if the series defining each $E(\rho_i)$ is absolutely convergent and if the series defining $\zeta(2k; \mathfrak{a}\rho_i^{-1})$ is absolutely convergent, then the series for

$E^{\#}(\mathfrak{a})$ must be absolutely convergent. The latter can happen only for $2k > 2$; this gives the convergence assertion of the corollary.

Let X be the set of absolute ideal class characters. Let $h = {}^{\#}X$ be the class number of $\mathfrak{o}$. It is elementary that

$$\zeta(s;\mathfrak{n}) = h^{-1} \sum_{\omega \in X} \bar{\omega}(\mathfrak{n}) N\mathfrak{n}^{-s} L(s,\omega),$$

where $L(s,\omega)$ is the L-function over F with character ω:

$$L(s,\omega) = \prod_{\mathfrak{p}} (1 - \omega(\mathfrak{p}) N\mathfrak{p}^{-s})^{-1} = \sum_{\mathfrak{m} \subset \mathfrak{o}} \omega(\mathfrak{m}) N\mathfrak{m}^{-s}.$$

Therefore,

$$N\mathfrak{a}^{2k} E^{\#}(\mathfrak{a}) = h^{-1} \sum_{i,\omega} E(\rho_i) \bar{\omega}(\mathfrak{a}\rho_i^{-1}) N\rho_i^{2k} L(2k,\omega).$$

Now let $\tilde{E}^{\#}$ be the vector with jth component $N\rho_j^{2k} E^{\#}(\rho_j)$, $\tilde{E}$ the vector with jth component $N\rho_i^{2k} E(\rho_j)$, $\Lambda(2k)$ the diagonal matrix with ω-th entry $L(2k,\omega)$, Ω the matrix with (i,ω)-th component $\bar{\omega}(\rho_i)^{-1}$. Note that $\Omega\bar{\Omega}^T = h$. Then the last equation gives *the fundamental relation between the two sorts of Eisenstein series*:

$$\tilde{E}^{\#} = \Omega\Lambda(2k)\Omega^{-1}\tilde{E}.$$

The determinant of Ω is non-zero, as is that of $\Lambda(2k)$ for all $1 < k \in \mathbb{Z}$. We also have

$$\tilde{E} = \Omega\Lambda(2k)^{-1}\Omega^{-1}\tilde{E}^{\#}.$$

We have already noted that Ω is non-singular; $\Lambda(2k)$ is certainly non-singular, since the L-functions do not vanish in the region of convergence. Thus, we have shown that either of the two sets of Eisenstein series *spans* $\mathrm{Eis}(\Gamma, 2k)$.

For the linear independence, we need:

LEMMA Let $E = E(\ ;\mathfrak{m}, 2k)$. Take

$$\delta = \begin{pmatrix} A & B \\ C & D \end{pmatrix} \in GL^+(2,F).$$

Let

$$(E|_{2k}\delta)(z) = \sum_{\xi} c_\xi \exp(2\pi i \operatorname{Tr}\xi z).$$

Then $c_0 = 0$ unless $\mathfrak{m}$ is in the same ideal class as $A\mathfrak{o} + C\mathfrak{o}$, in which case it is non-zero.

Proof We have, by definition,

$$(E|_{2k}\delta)(z) = \sum_{c\mathfrak{o}+d\mathfrak{o}=\mathfrak{m}}^{*} N(c(Az+B)(Cz+D)^{-1}+d)^{-2k}(Cz+D)^{-2k}N(\det\delta)^k =$$

$$= N(\det\delta)^k \sum N(c(Az+B)+d(Cz+D))^{-2k} =$$

$$= N(\det\delta)^k \sum N((cA+dC)z+(cB+dD))^{-2k}.$$

Now $c_\xi = 0$ unless $\xi = 0$ or ξ is totally positive. If $z = (i\lambda, \dots, i\lambda)$ with $\lambda > 0$, then it is elementary that

$$\lim_{\lambda\to\infty} (E|_{2k}\delta)(z) = c_0.$$

Looking at the other expression, we note that

$$\lim_{\lambda\to\infty} N((cA+dC)i\lambda+(cB+dD)^{-2k} = \begin{cases} 0 & (cA+dC)\neq 0 \\ N(cB+dD)^{-2k}>0 & (cA+dC)=0. \end{cases}$$

It is straightforward to check that $cA + dC = 0$ if and only if $c\mathfrak{o} + d\mathfrak{o} = \mathfrak{m}$ and $A\mathfrak{o} + C\mathfrak{o}$ are in the same ideal class. Therefore, if this condition does not hold,

$$c_0 = \lim_{\lambda\to\infty} (E|_{2k}\delta)(z) = 0.$$

If $\mathfrak{m}$ and $A\mathfrak{o} + C\mathfrak{o}$ are in the same ideal class, then

$$c_0 = \lim_{\lambda\to\infty} (E|_{2k}\delta)(z) > 0. \quad \blacksquare$$

To complete the proof of the proposition, we observe that this lemma provides us with linear functionals which separate the Eisenstein series. This gives linear independence. ■

COROLLARY OF PROOF Let $SL(2,\mathfrak{o}) \subset \Gamma \subset GL^+(2,\mathfrak{o})$. Given $f \in \mathrm{Mfm}(\Gamma, 2k)$, there is a linear combination E of Eisenstein series of weight $2k$ so that $f - E$ is a cuspform.

Proof As in the proof of the previous proposition, for $g \in GL(2,F)^+$ we have a linear functional L_g defined on $f \in \mathrm{Mfm}(\Gamma, 2k)$, defined by

$$L_g f = \lim_{\lambda\to\infty} (f|_{2k}g)(i\lambda) = \text{0-th Fourier coefficient of } f|_{2k}g.$$

As we saw in our discussion of cusps, we need consider only $g \in \{\delta_i\}$, where $\{\delta_i\}$ is a finite subset of $GL(2, F)^+$ so that $\{\delta_j(i\infty)\}$ is the collection of Γ-inequivalent cusps. Let δ_j have entries A_j, B_j, C_j, D_j (in the usual form); then $\delta_j(i\infty) = A_j/C_j \in (\mathbb{P}^1)^m$, and $\{\mathfrak{m}_j = A_j\mathfrak{o} + C_j\mathfrak{o}\}$ is a collection of representatives for the ideal classes. We saw in the proof of the proposition that $L_{\delta_j}E(z; \mathfrak{m}_i, 2k) = \delta_{ij}$ (Kronecker delta). This shows that subtraction of a suitable linear combination E of Eisenstein series from f makes $L_g(f - E) = 0$ for all g. ∎

Last, we compute the Fourier expansion ("at infinity") of the holomorphic Eisenstein series $E^\#(z; \mathfrak{m}, 2k)$. It should be noted that all the numbers appearing are actually quite elementary objects.

PROPOSITION

$E^\#(z; \mathfrak{m}, 2k) =$

$$= \sum_{\alpha \in \mathfrak{m}/\mathfrak{o}^\times} N\alpha^{-2k} + D^{-1/2} N\mathfrak{m}^{-1} [(-2\pi i)^{2k}/\Gamma(2k)]^m \sum_{0 \ll \xi \in \mathfrak{o}^*} \sigma_{2k-1}(\xi; \mathfrak{m}) \exp(2\pi i \operatorname{Tr}(\xi z)),$$

where D is the discriminant of $\mathfrak{o}$ over $\mathbb{Z}$, $N\mathfrak{m}$ is the ideal norm of $\mathfrak{m}$,

$$\mathfrak{m}^* = \{\alpha \in F : \operatorname{Trace}_{F/\mathbb{Q}} \alpha\mathfrak{m} \subset \mathbb{Z}\}$$

and

$$\sigma_{2k-1}(\xi; \mathfrak{m}) = \sum_{\delta \in \mathfrak{m}^*/\mathfrak{o}^\times :\, \xi/\delta \in \mathfrak{m}} |\operatorname{Norm}_{F/\mathbb{Q}} \delta|^{2k-1}.$$

Proof Let N denote both the ideal norm and the Galois norm from F to $\mathbb{Q}$. Note that $D^{1/2}$ is the measure of $\mathbb{R}^m/\mathfrak{o}$. We have

$$E^\#(z; \mathfrak{m}, k) = \sum_{c, d \in \mathfrak{m}}' N(cz + d)^{-2k} = \sum_{c=0} + \sum_{c \neq 0} =$$

$$= \sum_{\alpha \in \mathfrak{m}/\mathfrak{o}^\times} N\alpha^{-2k} + \sum_{c \in (\mathfrak{m}-0)/\mathfrak{o}^\times} Nc^{-2k} \left\{ \sum_{d \in \mathfrak{m}} N(z + d/c)^{-2k} \right\}.$$

We may index the Fourier coefficients by $\xi \in \mathfrak{o}^*$, the absolute inverse different of $\mathfrak{o}$. The ξ-th Fourier coefficient of the sum in curly brackets is

$$D^{-1/2} \int_{\mathbb{R}^m/\mathfrak{o}} \exp(-2\pi i \operatorname{Tr}(\xi x)) \left\{ \sum_{d \in \mathfrak{m}} N(z + d/c)^{-2k} \right\} dx =$$

$$= D^{-1/2} \int_{\mathbb{R}^m} \exp(-2\pi i \operatorname{Tr}(\xi x)) \left\{ \sum_{d \in \mathfrak{m}/c\mathfrak{o}} N(z + d/c)^{-2k} \right\} dx =$$

$$= D^{-1/2} \sum_{d \in \mathfrak{m}/c\mathfrak{o}} \exp(2\pi i \operatorname{Tr}(\xi d/c)) \times \int_{\mathbb{R}^m} \exp(-2\pi i \operatorname{Tr}(\xi x)) N(z)^{-2k}\, dx.$$

By the usual arguments, the sum is

$$\sum_{d\in\mathfrak{m}/c\mathfrak{o}} \exp(2\pi i\,\mathrm{Tr}(\xi d/c)) = 0 \qquad (\xi/c \notin \mathfrak{m}^* = \mathfrak{m}^{-1}\mathfrak{o}^*),$$

$$= |Nc|N\mathfrak{m}^{-1} \quad (\xi/c \in \mathfrak{m}^* = \mathfrak{m}^{-1}\mathfrak{o}^*).$$

By residues, the integral is 0 unless ξ is *totally positive*, in which case it is

$$\prod_j \int_{\mathbb{R}} \exp(-2\pi i\,\mathrm{Tr}(\xi_j x_j)) z_j^{2k}\,dx = \prod_j \{(-2\pi i)^{2k}\Gamma(2k)^{-1}\xi_j^{2k-1}\exp(-2\pi\xi_j y_j)\} =$$

$$= [(-2\pi i)^{2k}/\Gamma(2k)]^m N\xi^{2k-1}\exp(-2\pi\,\mathrm{Tr}(\xi y)).$$

So, for $\xi \in \mathfrak{o}^*$ totally positive, the ξ-th coefficient is

$$N\mathfrak{m}^{-1}D^{-1/2}[(-2\pi i)^{2k}/\Gamma(2k)]^m N\xi^{2k-1}\exp(-2\pi\,\mathrm{Tr}(\xi y))\left\{\sum_{0\neq c\in\mathfrak{m}/\mathfrak{o}^\times:\ \xi\in c\mathfrak{m}^*} |Nc|^{1-2k}\right\} =$$

$$= N\mathfrak{m}^{-1}D^{-1/2}[(-2\pi i)^{2k}/\Gamma(2k)]^m \exp(-2\pi\,\mathrm{Tr}(\xi y))\left\{\sum_{0\neq c\in\mathfrak{m}^*/\mathfrak{o}^\times:\ \xi\in c\mathfrak{m}} |Nc|^{2k-1}\right\} =$$

(note the interchange of $\mathfrak{m}$ and $\mathfrak{m}^*$ in the last two lines)

$$= N\mathfrak{m}^{-1}D^{-1/2}[(-2\pi i)^{2k}/\Gamma(2k)]^m \exp(-2\pi\,\mathrm{Tr}(\xi y))\sigma_{2k-1}(\xi;\mathfrak{m}).$$

Thus, indeed,

$$E^{\#}(z;\mathfrak{m},2k) =$$

$$= \sum_{\alpha\in\mathfrak{m}/\mathfrak{o}^\times} N\alpha^{-2k} + D^{-1/2}N\mathfrak{m}^{-1}[(-2\pi i)^{2k}/\Gamma(2k)]^m \sum_{0\ll\xi\in\mathfrak{o}^*} \sigma_{2k-1}(\xi;\mathfrak{m})\exp(2\pi i\,\mathrm{Tr}(\xi z)). \quad \blacksquare$$

1.6 Siegel Sets: An Approximate Fundamental Domain

For many subsequent discussions, we will need to know that there is some "nice, small" set S in $\mathfrak{H}^m$ so that $\Gamma S = \mathfrak{H}^m$. In general, it is not practical to try to determine a precise *fundamental domain*—that is, a measurable set Φ so that $\Gamma\Phi = \mathfrak{H}^m$ and $\Phi \cap \gamma\Phi$ is of measure 0 for all γ acting non-trivially on $\mathfrak{H}^m$. Of course, for $F = \mathbb{Q}$, there is a well-known, very nice fundamental domain for the action of $SL(2,\mathbb{Z})$ on $\mathfrak{H}$:

$$\{z \in \mathfrak{H}: |z| \geq 1 \text{ and } |\mathrm{Re}(z)| \leq \tfrac{1}{2}\}.$$

Though in principle an analogous-general result is possible, the actual general result is very much more complicated and is commensurately less useful in application. Further, some other discussions, which seem to depend upon the *existence* of a fundamental domain, in reality depend only upon the possibility of making sense of integrals on quotient spaces.

Definition Let R be a compact set in $\mathbb{R}^m$. For $B > 0$, define a *standard Siegel set* to be a subset of $\mathfrak{H}^m$ of the form

$$S = \{z \in \mathfrak{H}^m \colon \mathrm{Re}(z) \in R \text{ and } \mathrm{Im}(z_j) \geq B \text{ for } j = 1, \ldots, m\}.$$

THEOREM Let Γ be a congruence subgroup. Let $\{\kappa_1, \ldots, \kappa_\nu\}$ be representatives for the Γ-inequivalent cusps. Let $\{\delta_1, \ldots, \delta_\nu\}$ be a set of elements of $SL(2, F)$ so that $\delta_j(i\infty) = \kappa_j$. Then there is a standard Siegel set S so that

$$\mathfrak{H}^m = \bigcup_{\gamma \in \Gamma} \gamma\left(\bigcup_i \delta_i S\right).$$

Proof Without loss of generality, we may as well suppose that Γ contains $Z(\mathfrak{o}) \approx \mathfrak{o}^\times$, since $Z(\mathfrak{o})$ acts trivially on $\mathfrak{H}^m$. Note that from our prior discussion of cusps, we do know that there are only finitely many Γ-inequivalent cusps. Let P be the group of upper triangular matrices in $GL^+(2, F)$. We have seen that

$$\bigcup_i \Gamma\delta_i P = GL^+(2, F) \qquad \text{(disjoint union)}.$$

By taking inverses, this yields

$$\bigcup_i P\delta_i^{-1}\Gamma = GL^+(2, F) \qquad \text{(disjoint union)}.$$

Let c_i, d_i be the bottom two entries of δ_i^{-1}.
We recall a well-known result of Minkowski:

LEMMA Let Λ be a lattice in $\mathbb{R}^N$. Let U be an open set so that $\Lambda + U = \mathbb{R}^N$. Let V be a convex open set containing 0 so that $-V = V$. Then if the measure of V is greater than 2^N times the measure of U, it must be that there is some non-zero element of Λ lying in V. ■

Given $z \in \mathfrak{H}^m$, consider the lattice $\Lambda = \mathfrak{o}z + \mathfrak{o} \subset \mathbb{C}^m$ (using a multi-index notation). It is not hard to check that the volume of $\mathbb{C}^m/\Lambda$ is $(y_1 \ldots y_m)D$, where $y_j = \mathrm{Im}(z_j)$ and D is the discriminant of $\mathfrak{o}$ over $\mathbb{Z}$. For a constant K, the volume of the convex open set

$$V = \{u \in \mathbb{C}^m \colon |u_j|^2 < Ky_j\}$$

is seen to be $\pi^m K^m (y_1 \ldots y_m)$. Therefore, by Minkowski's result, for K sufficiently large there are $c, d \in \mathfrak{o}$ (not both 0) so that

$$|\sigma_j(c)z_j + \sigma_j(d)|^2 < Ky_j \qquad (j = 1, \ldots, m).$$

Take $a, b \in F$ so that $g = \left(\begin{smallmatrix} a & b \\ c & d \end{smallmatrix}\right)$ is in $SL(2, F)$. Then

$$\operatorname{Im}(g(z)_j) = \operatorname{Im}(z_j)/|\sigma_j(c)z_j + \sigma_j(d)|^2 > y_j/Ky_j = K^{-1}.$$

Select

$$p = \begin{pmatrix} p_{11} & p_{12} \\ 0 & p_{22} \end{pmatrix} \in P \qquad \gamma \in \Gamma$$

so that $g = p\delta_i^{-1}\gamma$, for some index i. That is,

$$(c \quad d) = p_{22}(c_i d_i)\gamma.$$

LEMMA There is some non-zero fractional ideal $\mathfrak{n} \subset \mathfrak{o}$, depending upon Γ, so that entries of $\gamma \in \Gamma$ lie in $\mathfrak{n}^{-1}$.

Proof There is an ideal so that Γ contains $\Gamma(\mathfrak{m})$ with finite index. Certainly all the entries of elements of $\Gamma(\mathfrak{m})$ are in $\mathfrak{o}$. Let $\{\gamma_j\}$ be some finite collection of coset representatives for $\Gamma/\Gamma(\mathfrak{m})$, and let $\mathfrak{n}$ be a non-zero ideal in $\mathfrak{o}$ so that all the entries of every γ_j lie in $\mathfrak{n}^{-1}$; this $\mathfrak{n}$ satisfies the requirement of the lemma. ■

The equality $(cd)\gamma^{-1} = p_{22}(c_i d_i)$, together with the lemma, implies that

$$p_{22}(c_i\mathfrak{o} + d_i\mathfrak{o}) \subset \mathfrak{n}^{-1}(c\mathfrak{o} + d\mathfrak{o}).$$

Therefore, upon taking ideal norms,

$$N\mathfrak{n}^{-1}N(c\mathfrak{o} + d\mathfrak{o}) \geq |\operatorname{Norm}_{F/\mathbb{Q}} p_{22}| \times N(c_i\mathfrak{o} + d_i\mathfrak{o}).$$

Since $c\mathfrak{o} + d\mathfrak{o} \subset \mathfrak{o}$, we find that

$$|\operatorname{Norm}_{F/\mathbb{Q}} p_{22}|^{-1} \leq N\mathfrak{n}^{-1}N(c\mathfrak{o} + d\mathfrak{o})/N(c_i\mathfrak{o} + d_i\mathfrak{o}) \leq$$

$$\leq N\mathfrak{n}^{-1}/N(c_i\mathfrak{o} + d_i\mathfrak{o}).$$

Let $A = \max_i N\mathfrak{n}^{-1}/N(c_i\mathfrak{o} + d_i\mathfrak{o})$. For $\eta \in \mathfrak{o}^\times$, write

$$\tilde{\eta} = \begin{pmatrix} \eta & 0 \\ 0 & \eta \end{pmatrix} \in Z(\mathfrak{o}).$$

As an application of the Units Theorem, there exists $\eta \in \mathfrak{o}^\times$ so that

$$C'|\operatorname{Norm}_{F/\mathbb{Q}} p_{22}|^{1/m} \geq |\sigma_j(p_{22}\eta)| \geq C|\operatorname{Norm}_{F/\mathbb{Q}} p_{22}|^{1/m} \qquad (j = 1, \ldots, m)$$

for some constants C, C' depending only on $\mathfrak{o}$. From this it follows that

$$|\sigma_j(p_{22}\eta)|^{-1} \leq C^{-1}|\operatorname{Norm}_{F/\mathbb{Q}} p_{22}|^{-1/m} \leq C^{-1}A^{1/m}.$$

We certainly may replace p by $p\tilde{\eta}$ and γ by $\gamma\tilde{\eta}^{-1}$ without affecting the relation $g = p\delta_i^{-1}\gamma$; thus, we may assume that

$$|\sigma_j(p_{22})|^{-1} \leq C^{-1}A^{1/m}.$$

For $\eta \in \mathfrak{o}^\times$, put

$$\eta' = \begin{pmatrix} \eta & 0 \\ 0 & 1 \end{pmatrix} \in P$$

There is $\mathfrak{m} \subset \mathfrak{o}$ so that if

$$\eta \in U = \{\varepsilon \in \mathfrak{o}^\times : \varepsilon \equiv 1 \text{ modulo } \mathfrak{m}\},$$

then

$$\delta_i \eta' \delta_i^{-1} \in \Gamma$$

for all indices i. Therefore, we may replace p by $p\eta'^{-1}$ and γ by $\delta_i\eta'\delta_i^{-1}\gamma$, preserving the relation $g = p\delta_i^{-1}\gamma$. This allows us to adjust p_{11} by elements of U. Since

$$p_{11}p_{22} = \det p = (\det g)(\det \delta_i)(\det \gamma) = (\det \gamma),$$

we have

$$p_{11} \in (1/p_{22})\mathfrak{o}^\times/U,$$

which is a finite set. Let X be a finite set of representatives for $\mathfrak{o}^\times/U$, and put

$$B = \max_j \max_{\varepsilon \in \chi} |\sigma_j(\varepsilon)|.$$

Combining these results, we have

$$\begin{aligned} K^{-1} < \mathrm{Im}(g(z)_j) &= \mathrm{Im}((p\delta_i^{-1}\gamma)(z)_j) = \\ &= |\sigma_j(p_{11})/\sigma_j(p_{22})| \times \mathrm{Im}((\delta_i^{-1}\gamma)(z)_j) = \\ &= |\sigma_j(\varepsilon)|/\sigma_j(p_{22})^2 \times \mathrm{Im}((\delta_i^{-1}\gamma)(z)_j) \end{aligned}$$

for some $\varepsilon \in X$. Then we have

$$BK^{-1} < \sigma_j(p_{22})^{-2}\,\mathrm{Im}((\delta_i^{-1}\gamma)(z))_j \qquad (j = 1, \ldots, m).$$

Also, recall that we have

$$|\sigma_j(p_{22})|^{-1} \leq C^{-1}A^{1/m}.$$

Therefore,

$$BK^{-1}C^2A^{-2/m} < \operatorname{Im}((\delta_i^{-1}\gamma)(z))_j \qquad (j = 1, \ldots, m).$$

Let $\mathfrak{n}$ be an ideal in $\mathfrak{o}$ so that $\Gamma(\mathfrak{n}) \subset \Gamma$, and let $\mathfrak{n}$ be sufficiently small so that for $x \in \mathfrak{n}$,

$$\Gamma(\mathfrak{n})\delta_i^{-1} \subset \delta_i^{-1}\Gamma.$$

Then, in particular, for $u \in \mathfrak{n}$ the matrix $\tilde{u} = \left(\begin{smallmatrix} 1 & u \\ 0 & 1 \end{smallmatrix}\right)$ lies in $\Gamma(\mathfrak{n})$. Let R be an open subset of $\mathbb{R}^m$ with compact closure, so that $R + \mathfrak{n} = \mathbb{R}^m$, viewing $\mathfrak{o}$ as imbedded in $\mathbb{R}^m$. The action of such an element $\tilde{u}$ on $(\delta_i^{-1}\gamma)(z)$ does not affect $\operatorname{Im}((\delta_i^{-1}\gamma)(z))$. Therefore, given $z \in \mathfrak{H}^m$, we can find δ_i, $\gamma \in \Gamma$, and $\tilde{u}$ with $u \in \mathfrak{n}$, so that

$$(\tilde{u}\delta_i^{-1}\gamma)(z) \in S = \{z \in \mathfrak{H}^m \colon \operatorname{Im}(z)_j > BK^{-1}C^2A^{-2/m}, \operatorname{Re}(z) \in R\}.$$

By choice of $\mathfrak{n}$, $\tilde{u}\delta_i^{-1} = \delta_i^{-1}u'$ for some $u' \in \Gamma$; therefore, $(\delta_i^{-1}u'\gamma)(z)$ is in S. That is,

$$(u'\gamma)(z) \in \delta_i S.$$

At last, this shows that $\mathfrak{H}^m = \bigcup_{\gamma \in \Gamma} \gamma(\bigcup_i \delta_i S)$. ■

1.7 Finite Dimensionality of Spaces of Cuspforms, Estimates on Cuspforms

The result of the previous section, which may seem rather rough and unhelpful, is all that is needed to obtain a fundamental structural result: the finite dimensionality of spaces of cuspforms. In the course of the proof, we will also point out some useful facts concerning the behavior of cuspforms on Siegel sets and obtain an estimate (following Hecke) on the asymptotic size of Fourier coefficients of cuspforms. Indeed, these properties are two of the important points in the proof. Also, better than mere finite dimensionality of spaces of cuspforms, a careful reading of the proof yields a crude estimate on this dimension.

In the proof, we use the fact that there exist cuspforms (not identically zero) for some sufficiently small congruence subgroup; this is proven in the next section. Since these are holomorphic functions, there are non-empty open subsets of $\mathfrak{H}^m$ on which they are non-zero.

THEOREM Fix a congruence subgroup Γ with v inequivalent cusps. Let $k = (k_1, \ldots, k_m)$ with all $k_i \in \mathbb{Z}$. Then

$$\dim_{\mathbb{C}} \operatorname{Cfm}(\Gamma, k) < \infty.$$

Further, if any $k_j < 0$, then $\mathrm{Mfm}(\Gamma, k) = \{0\}$; if some $k_i = 0$ and some $k_j > 0$, then $\mathrm{Mfm}(\Gamma, k) = \{0\}$, and $\mathrm{Mfm}(\Gamma, 0) = \mathbb{C}$.

COROLLARY OF PROOF For a fixed weight k and congruence subgroup Γ, there is a constant ω so that for any $f(z) = \sum_\xi c_\xi \exp(2\pi i \operatorname{Tr} \xi z)$ in $\mathrm{Cfm}(\Gamma, k)$, we have an estimate

$$|c_\xi| \leq \omega \xi^{k/2}$$

(using a multi-index notation).

COROLLARY OF PROOF Given a cuspform f of weight k, there is a constant $C < \infty$ so that, for any $z = x + iy$ in $\mathfrak{H}^m$,

$$y^{k/2}|f(z)| < C.$$

COROLLARY OF PROOF Given a Siegel set S, and given a cuspform f of weight k, there is a constant $c > 0$ so that for $z = x + iy$ in S

$$y^{k/2}|f(z)| \leq \exp(-cy^{1/m}).$$

Proof of Theorem To illustrate the idea without peripheral technical obfuscation, we first give the proof for $SL(2, \mathbb{Z})$, granting that

$$S = \{z = x + iy \in \mathfrak{H}: |x| \leq \tfrac{1}{2} \text{ and } y \geq \tfrac{1}{2}\}$$

is a Siegel domain so that $SL(2, \mathbb{Z})S = \mathfrak{H}$. Let $f \in \mathrm{Cfm}(SL(2, \mathbb{Z}), k)$ with $k \geq 0$, with Fourier coefficients c_n $(0 < n \in \mathbb{Z})$, and suppose that $c_n = 0$ for $n \leq N$. The function $y^{k/2}|f(z)|$ is $SL(2, \mathbb{Z})$-invariant; therefore,

$$\sup_{\mathfrak{H}} y^{k/2}|f(z)| = \sup_{S} y^{k/2}|f(z)|.$$

Since f is of exponential decay on S, the latter supremum is attained, say at $z_0 = x_0 + iy_0 \in S$. On the other hand, we also have an estimate

$$C = \sup_S \exp(\pi N y)|f(z)| < \infty,$$

and this supremum is also attained, say at $z_1 = x_1 + iy_1$.

For $k < 0$, for any $z = x + iy$ in $\mathfrak{H}$ we have

$$|f(z)| \leq y^{-k/2} y_0^{k/2} |f(z_0)| = y^{-k/2} y_0^{k/2} \exp(-\pi N y_0) \exp(\pi N y_0)|f(z_0)| \leq$$

$$\leq y^{-k/2} y_0^{k/2} \exp(-\pi N y_0) C.$$

As $k < 0$, the last expression is decreasing as $y_0 \to \infty$ and so is dominated by the expression obtained by replacing y_0 by $\frac{1}{2}$:

$$|f(z)| \leq y^{-k/2}(\tfrac{1}{2})^{k/2}\exp(-\pi N/2)C.$$

Then, for all $y > 0$, we have

$$|c_n| = \exp(2\pi ny)\left|\int_{[0,1]} f(x+iy)\exp(-2\pi inx)\,dx\right| \leq$$

$$\leq \exp(2\pi ny)\sup_{0\leq x\leq 1}|f(x+iy)| \leq \exp(2\pi ny)y^{-k/2}C(\tfrac{1}{2})^{k/2}\exp(-\pi N/2).$$

Since this holds for all $y > 0$, we may let $y \to 0^+$ and find that $c_n = 0$: $\mathrm{Cfm}(\Gamma, k) = \{0\}$. If there is $\varphi \in \mathrm{Mfm}(\Gamma, k)$ with $k < 0$, then let ψ be any (not identically zero) cuspform (of positive weight) with respect to any congruence subgroup. Then $\varphi^n\psi$ is a cuspform of negative weight, for some sufficiently large n. Therefore, the holomorphic function φ must be zero on some non-empty open set and so must be identically zero.

Now consider $k = 0$: Let $\varphi \in \mathrm{Mfm}(\Gamma, 0)$. Note that φ^n is in $\mathrm{Mfm}(\Gamma, 0)$ for every $0 \leq n \in \mathbb{Z}$. Granting the assertion of the theorem for $k > 0$, for any non-zero cuspform $f \in \mathrm{Cfm}(\Gamma', k')$, the elements $\varphi^n f$ are in $\mathrm{Cfm}(\Gamma \cap \Gamma', k')$. These cannot be linearly independent, so there is a relation

$$\sum_n \alpha_n \varphi^n f = 0.$$

This implies vanishing of $\sum_n \alpha_n\varphi^n$ on a non-empty open subset of $\mathfrak{H}^m$, which implies that $\sum_n \alpha_n\varphi^n = 0$. Therefore, φ is algebraic over $\mathbb{C}$ and so is a constant.

For $k > 0$, for any $z = x + iy$ in $\mathfrak{H}$, we have

$$|f(z)| \leq y^{-k/2}y_0^{k/2}|f(z_0)| = y^{-k/2}y_0^{k/2}\exp(-\pi Ny_0)\exp(\pi Ny_0)|f(z_0)| \leq$$

$$\leq y^{-k/2}y_0^{k/2}\exp(-\pi Ny_0)C \leq y^{-k/2}C(k/2\pi eN)^{k/2},$$

where the last constant arises from maximizing $y^{k/2}\exp(-\pi Ny)$ on $[0, \infty)$. Then, for all $y > 0$, we have

$$|c_n| = \exp(2\pi ny)\left|\int_{[0,1]} f(x+iy)\exp(-2\pi inx)\,dx\right| \leq$$

$$\leq \exp(2\pi ny)\sup_{0\leq x\leq 1}|f(x+iy)| \leq \exp(2\pi ny)y^{-k/2}C(k/2\pi eN)^{k/2}.$$

It is here that we obtain the estimate on Fourier coefficients: For $N \geq 1$, minimizing $\exp(2\pi ny)y^{-k/2}$ on $(0, \infty)$ yields

$$|c_n| \leq (4\pi ne/k)^{k/2}C(k/2\pi eN)^{k/2} = C(2n/N)^{k/2}.$$

Now we have

$$C = \exp(\pi N y_1)|f(z_1)| \le \exp(\pi N y_1) \sum_{n>N} |c_n| \exp(-2\pi n y_1) \le$$

$$\le \exp(\pi N y_1) \sum_{n>N} C(2n/N)^{k/2} \exp(-2\pi n y_1) =$$

$$= CN^{-k/2} \sum_{n>N} (2n)^{k/2} \exp(-2\pi(n - \tfrac{1}{2}N)y_1).$$

By dividing both sides by C, we have

$$N^{k/2} \le \sum_{n>N} (2n)^{k/2} \exp(-2\pi(n - \tfrac{1}{2}N)y_1).$$

By calculus, we have

$$(2n)^{k/2} \le (2k)^{k/2} \exp(\pi n/2).$$

Therefore,

$$N^{k/2} \le (2k)^{k/2} \sum_{n>N} \exp(\pi n/2) \exp(-2\pi(n - \tfrac{1}{2}N)y_1).$$

It is clear that the sum is a decreasing function of y_1 as $y_1 \to \infty$, so its maximum occurs for $y_1 = \frac{1}{2}$; therefore,

$$N^{k/2} \le (2k)^{k/2} \sum_{n>N} \exp(\pi n/2) \exp(-\pi(n - \tfrac{1}{2}N)) =$$

$$= (2k)^{k/2} \sum_{n>N} \exp(-\tfrac{1}{2}\pi(n - N)) = (2k)^{k/2} \sum_{n\ge 1} \exp(-\tfrac{1}{2}\pi n).$$

The righthand side is bounded independently of N, so N must be bounded.

Now if $\mathrm{Cfm}(SL(2,\mathbb{Z}), k)$ were not finite-dimensional, we could find a cuspform with arbitrarily many Fourier coefficients vanishing, contradicting the previous result. This gives the result for $SL(2,\mathbb{Z})$. ■

For the general case, we must deal with various complications, most of which are purely technical rather than conceptual. An example of this is provided by the third assertion of the lemma below, which has a trivial analogue for $m = 1$.

First consider $k = 0$: Let $\varphi \in \mathrm{Mfm}(\Gamma, 0)$. Note that φ^n in $\mathrm{Mfm}(\Gamma, 0)$ for every $0 \le n \in \mathbb{Z}$. Granting the assertion of the theorem for $k > 0$, for any not identically zero cuspform $f \in \mathrm{Cfm}(\Gamma', k')$, the elements $\varphi^n f$ are in $\mathrm{Cfm}(\Gamma \cap \Gamma', k')$. These cannot be linearly independent, so there is a relation

$$\sum_n \alpha_n \varphi^n f = 0.$$

This implies vanishing of $\sum_n \alpha_n \varphi^n$ on a non-empty open set, which implies that $\sum_n \alpha_n \varphi^n = 0$. Therefore, φ is algebraic over $\mathbb{C}$ and so is a constant.

We collect some elementary estimates:

LEMMA Take $\alpha > 0$.

i. $\sup_{y>0} y^\alpha \exp(-c\alpha) = (\alpha/ec)^\alpha$.
ii. For $y > 0$, for all $c > 0$, $y^\alpha \leq (\alpha/ec)^\alpha \exp(cy)$.
iii. For $y_1, \ldots, y_m > 0$,

$$\inf\left\{\sum_i \xi_i y_i \colon \xi_1, \ldots, \xi_m > 0, \prod_i \xi_i \geq n > 0\right\} = mn^{1/m}\left(\prod_i y_i\right)^{1/m}.$$

Proof The first assertion follows from calculus; the second is merely a paraphrase of the first. The third assertion follows from application of the "Lagrange multiplier method": It is clear that the infimum must occur on the hypersurface $\prod_i \xi_i = n$. The infimum will occur at a point $(\xi_1, \ldots, \xi_m)$ so that (for some $\lambda \in \mathbb{R}$)

$$\nabla \prod_i \xi_i = \lambda \nabla \sum_i \xi_i y_i.$$

That is,

$$\left(\prod_i \xi_i\right)(\xi_1^{-1}, \ldots, \xi_m^{-1}) = \lambda(y_1, \ldots, y_m).$$

From this, $\lambda = n/\xi_i y_i$ for all indices i; we find that

$$\lambda^m = n^{m-1} \Big/ \prod_i y_i.$$

Therefore, at the infimum,

$$\xi_i = n/\lambda y_i = n \Big/ \left\{\left[n^{m-1} \Big/ \prod_i y_i\right]^{1/m} y_i\right\} = n^{1/m}\left(\prod_i y_i\right)^{1/m} \Big/ y_i,$$

$$y_i \xi_i = n^{1/m}\left(\prod_i y_i\right)^{1/m}.$$

■

Let $\{\kappa_i \colon i = 1, \ldots, v\}$ be a (finite) collection of representatives for the Γ-equivalence classes of cusps, and take $\delta_i \in SL(2, F)$ so that $\delta_i(i\infty) = \kappa_i$. We have seen that $\Gamma(\bigcup_i \delta_i S) = \mathfrak{H}^m$ for some Siegel set

$$S = \{z \in \mathfrak{H}^m \colon \operatorname{Im}(z_i) \geq \eta > 0, \operatorname{Re}(z) \text{ in a fixed compact set}\}.$$

For each index i, and for $f \in \operatorname{Cfm}(\Gamma, k)$, let

$$(f|_k\delta_i)(z) = \sum_{\xi} c_i(\xi)\exp(2\pi i\,\mathrm{Tr}(\xi z)).$$

Also, write $\delta_1 = 1 \in SL(2, F)$. For each i, let Λ_i be the lattice of translations $z \to z + x_0$, leaving $f|_k\delta_i$ invariant; let Λ_i^* be the dual lattice (with respect to Tr); and let Λ_i^+ be the subset of Λ_i^* consisting of totally positive elements.

Now take $f \in \mathrm{Cfm}(\Gamma, k)$ so that, for all indices i,

$$c_i(\xi) = 0 \quad \text{for} \quad N\xi \le n,$$

where we write $N\xi = \prod_i \xi_i$. Let $\varphi(z) = y^{k/2}|f(z)|$; then φ is Γ-invariant. Therefore,

$$\sup\{\varphi(z)\colon z \in \mathfrak{H}^m\} = \sup\left\{\varphi(z)\colon z \in \bigcup_i \delta_i S\right\}.$$

PROPOSITION We have

$$\sup\left\{\varphi(z)\colon z \in \bigcup_i \delta_i S\right\} < \infty,$$

and this supremum is attained.

Proof Let $d = \sup\{1, \ldots, k_i, \ldots\}$. From the lemma, for any $\xi = (\xi_1, \ldots, \xi_m)$ with all $\xi_i > 0$ and $N\xi > n$,

$$\begin{aligned}\prod_i y_i^{k_i/2} &= \eta^{-\sum_i k_i/2} \prod_i (y_i/\eta)^{k_i/2} \le \eta^{-\sum_i k_i/2} \prod_i (y_i/\eta)^{d/2} \le \\ &\le \eta^{-\sum_i k_i/2} \prod_i (k_i\eta/2\pi e\xi_i)^{d/2} \exp(\pi\,\mathrm{Tr}\,\xi y/\eta) \le \\ &\le \eta^{-\sum_i k_i/2} k^{k/2} \prod_i (k_i\eta/2\pi e)^{d/2} \prod_i \xi_i^{-d/2} \times \exp(\pi\,\mathrm{Tr}\,\xi y) \le \\ &\le \eta^{-\sum_i k_i/2} k^{k/2} \prod_i (k_i\eta/2\pi e)^{d/2} n^{-d/2} \times \exp(\pi\,\mathrm{Tr}\,\xi y)\end{aligned}$$

where η is the constant occurring in the defining inequalities of the Siegel set S. Therefore, taking $z \in S$,

$$\begin{aligned}\mathrm{Im}(\delta_i z)^{k/2}|f(\delta_i z)| &= y^{k/2}|(f|_k\delta_i)(z)| \le \\ &\le \eta^{-\sum_i k_i/2} k^{k/2} \prod_i (k_i\eta/2\pi e)^{d/2} n^{-d/2} \exp(\pi\,\mathrm{Tr}\,\xi y) \sum_{\xi} |c_i(\xi)| \exp(-2\pi\,\mathrm{Tr}\,\xi y) \le \\ &\le \eta^{-\sum_i k_i/2} k^{k/2} \prod_i (k_i\eta/2\pi e)^{d/2} n^{-d/2} \sum_{\xi} |c_i(\xi)| \exp(-\pi\,\mathrm{Tr}\,\xi y).\end{aligned}$$

The latter series is the series of absolute values of the terms of $(f|_k\delta_i)(\frac{1}{2}z))$ and so is indeed absolutely convergent. For $y_i \ge \eta > 0$, as any $y_i \to \infty$ this series goes to 0 termwise; by the dominated convergence theorem, the series must likewise go to 0. Therefore, the supremum is finite and is attained. ∎

Suppose that the supremum occurs at $\delta_i z_0 = \delta_i(x_0 + iy_0) \in \delta_i S$; then for all $z \in \mathfrak{H}^m$

$$|f(z)| = y^{-k/2}\varphi(z) \leq y^{-k/2}\varphi(\delta_i z_0).$$

This yields the second of the corollaries of the proof. Now

$$\varphi(\delta_i z) = \operatorname{Im}(\delta_i z)^{k/2}|f(\delta_i z)| = y^{k/2}(f|_k \delta_i)(z).$$

Therefore, by replacing f by $f|_k \delta_i^{-1}$ we may suppose that the supremum actually occurs in S itself—that is, for $\delta_i = 1$.

PROPOSITION In the situation above

$$C = \sup\left\{\exp\left(\pi m n^{1/m} \prod_i y_i^{1/m}\right) \times |f(z)|: z = x + iy \in S\right\} < \infty.$$

Proof We have the convergent Fourier expansion

$$f(z) = \sum_{\xi} c_1(\xi)\exp(2\pi i \operatorname{Tr}(\xi z)),$$

where only ξ with $N\xi > n$ occurs. Since also $\xi \in \Lambda_j^*$, there is $\varepsilon > 0$ so that $N\xi \geq n + \varepsilon$ whenever $N\xi > n$. From the lemma above, for any such ξ,

$$mn^{1/m}\prod_i y_i^{1/m} = (n/N\xi)^{1/m} \times m(N\xi)^{1/m}\prod_i y_i^{1/m} \leq (n/N\xi)^{1/m} \times \operatorname{Tr}\xi y \leq$$

$$\leq (n/n + \varepsilon)^{1/m} \times \operatorname{Tr}\xi y.$$

Therefore,

$$\exp\left(-2\pi m n^{1/m}\prod_i y_i^{1/m}\right)|f(z)| =$$

$$= \exp\left(-2\pi m n^{1/m}\prod_i y_i^{1/m}\right)\left|\sum_{\xi} c_1(\xi)\exp(2\pi i \operatorname{Tr}(\xi z))\right| \leq$$

$$\leq \sum_{\xi}|c_1(\xi)|\exp\left(2\pi m n^{1/m}\prod_i y_i^{1/m}\right)\exp(-2\pi \operatorname{Tr}\xi y) \leq$$

$$\leq \sum_{\xi}|c_1(\xi)|\exp(\pi(n/n + \varepsilon)^{1/m}\operatorname{Tr}\xi y) \times \exp(-2\pi \operatorname{Tr}\xi y) =$$

$$= \sum_{\xi}|c_1(\xi)|\exp(-\pi\{2 - (n/n + \varepsilon)^{1/m}\}\operatorname{Tr}\xi y),$$

and the latter series is necessarily absolutely convergent, since it is the series of absolute values of the terms of the Fourier series at

$$z' = \tfrac{1}{2}\{2 - (n/n + \varepsilon)^{1/m}\} \times z \in \mathfrak{H}^m.$$

In particular, in the region $\{y_i \geq \eta > 0: i = 1, \ldots, m\}$, as any $y_i \to \infty$ the series goes to 0 termwise; by the dominated convergence theorem, the whole expression goes to 0. Therefore, the supremum exists (is finite) and is attained. ■

An easy variant of this proposition yields the third corollary of the proof.

Therefore, returning to our previous calculations, with $y_i \geq \eta$ for all i, for any $z = x + iy$ lying in $\mathfrak{H}^m$,

$$|f(z)| \leq y^{-k/2} y_0^{k/2} |f(z_0)| \leq y^{-k/2} y_0^{k/2} C \exp\left(-\pi m n^{1/m} \prod_i y_{0i}^{1/m}\right) \leq$$

$$\leq y^{-k/2} y_0^{k/2} C \exp\left(-\pi n^{1/m} \eta^{(m-1)/m} \sum_i y_{0i}^{1/m}\right).$$

If some $k_i \leq 0$, then

$$y^{-k/2} y_0^{k/2} C \exp\left(-\pi n^{1/m} \eta^{(m-1)/m} \sum_i y_{0i}^{1/m}\right)$$

is a decreasing function of $y_{0i} \to \infty$ and so is dominated by its value at $y_{0i} = \eta$:

$$|f(z)| \leq y^{-k/2} \eta^{k/2} C \exp(-\pi m n^{1/m} \eta).$$

Therefore,

$$|c_1(\xi)| = \exp(2\pi \operatorname{Tr} \xi y) \left| (\operatorname{vol}(\mathbb{R}^m/\Lambda_1)^{-1} \int_{\mathbb{R}^m/\Lambda_1} \exp(-2\pi i \operatorname{Tr} \xi x) f(x+iy)\, dx \right| \leq$$

$$\leq \exp(2\pi \operatorname{Tr} \xi y) \sup\{|f(x+iy)|: x \in \mathbb{R}^m/\Lambda_1\} \leq$$

$$\leq \exp(2\pi \operatorname{Tr} \xi y) y^{-k/2} \eta^{k/2} C \exp(-\pi m n^{1/m} \eta).$$

This holds for every $y_i > 0$; therefore, if $k_i < 0$, then by letting $y_i \to 0^+$ we see that $c_1(\xi) = 0$ for every ξ. That is, $\operatorname{Cfm}(\Gamma, k) = \{0\}$ when some $k_i < 0$. If $\varphi \in \operatorname{Mfm}(\Gamma, k)$ with some $k_i < 0$, let f be some cuspform not identically zero; then $\varphi^n f$ is a cuspform with some negative weight component for large enough n and so is 0. Therefore, $\varphi = 0$. If some $k_i = 0$ (and some $k_j \neq 0$), then this argument shows that $c_1(\xi)$ are bounded in absolute value (and we already have seen, in the discussion of Koecher's principle, that $\operatorname{Mfm}(\Gamma, k) = \operatorname{Cfm}(\Gamma, k)$ in this case). But the relation

$$c_1(\varepsilon^2 \xi) = c_1(\xi) \varepsilon^{-k}$$

for ε in some finite-index subgroup of $\mathfrak{o}^\times$ shows (as in the discussion of Koecher's principle) that $c_1(\xi) = 0$, by invocation of the Units Theorem. Thus, $\mathrm{Cfm}(\Gamma, k) = \{0\}$ if some $k_i < 0$ and some $k_j \neq 0$.

If every $k_i > 0$, then

$$y^{-k/2} y_0^{k/2} C \exp\left(-\pi n^{1/m} \eta^{(m1)/m} \sum_i y_{0i}^{1/m} \right) \leq$$

$$\leq C y^{-k/2} \prod_i \sup\{ t^{k_i/2} \exp(-\pi n^{1/m} \eta^{(m-1)/m} t^{1/m}) \colon t > 0\} =$$

$$= C y^{-k/2} \prod_i \sup\{ t^{mk_i/2} \exp(-\pi n^{1/m} \eta^{(m-1)/m} t) \colon t > 0\} =$$

$$= C y^{-k/2} \prod_i \{ mk_i / [2\pi e n^{1/m} \eta^{(m-1)/m}] \}^{mk_i/2}$$

by the lemma regarding estimates. Therefore,

$$|c_1(\xi)| = \exp(2\pi \operatorname{Tr} \xi y) \left| (\mathrm{vol}(\mathbb{R}^m/\Lambda_1)^{-1} \int_{\mathbb{R}^m/\Lambda_1} \exp(-2\pi i \operatorname{Tr} \xi x) f(x + iy)\, dx \right| \leq$$

$$\leq \exp(2\pi \operatorname{Tr} \xi y) \sup\{ |f(x + iy)| \colon x \in \mathbb{R}^m/\Lambda_1 \} \leq$$

$$\leq \exp(2\pi \operatorname{Tr} \xi y) C y^{-k/2} \prod_i \{ mk_i / [2\pi e n^{1/m} \eta^{(m-1)/m}] \}^{mk_i/2}.$$

Minimizing this over $\{y_1, \ldots, y_m > 0\}$ yields (again by the lemma on estimates)

$$|c_1(\xi)| \leq C \prod_i (4\pi \xi_i e / k_i)^{k_i/2} \times \prod_i \{ mk_i / [2\pi e n^{1/m} \eta^{(m-1)/m}] \}^{mk_i/2}.$$

As a corollary, for fixed $k = (k_1, \ldots, k_m)$ and Γ, with

$$n \geq \inf\{ N\xi \colon \xi \in \Lambda_i^+, i = 1, \ldots, v \} > 0,$$

this gives the estimate asserted in the first of the corollaries of the proof:

$$|c_1(\xi)| = 0\left(\prod_i \xi_i^{k_i/2} \right)$$

depending only on Γ and k.

Returning to the main proof, let $w = u + iv \in \mathfrak{H}^m$ be the point in S where the supremum

$$C = \sup\left\{ \exp\left(\pi m n^{1/m} \prod_i y_i^{1/m} \right) \times |f(z)| \colon z = x + iy \in S \right\}$$

is attained. Using the estimate on the Fourier coefficients, we have

$$C\exp\left(-\pi m n^{1/m}\prod_i v_i^{1/m}\right) =$$

$$= |f(w)| \leq \sum_\xi |c_1(\xi)| \exp(-2\pi \operatorname{Tr}\xi v) \leq$$

$$\leq \sum_\xi \left\{C\prod_i (k_i/4\pi\xi_i e)^{-k_i/2}\{mk_i/[2\pi e n^{1/m}\eta^{(m-1)/m}]\}^{mk_i/2}\right\}\exp(-2\pi\operatorname{Tr}\xi v) =$$

$$= C\prod_i (k_i/4\pi e)^{-k_i/2}\{mk_i/[2\pi e n^{1/m}\eta^{(m-1)/m}]\}^{mk_i/2}\sum_\xi\left(\prod_i \xi_i^{k_i/2}\right)\exp(-2\pi\operatorname{Tr}\xi v).$$

Now we use an estimate (again from the lemma):

$$\prod_i \xi_i^{k_i/2} \leq \left\{\prod_i (k_i/\pi e v)^{k_i/2}\right\}\exp(\tfrac{1}{2}\pi\operatorname{Tr} v\xi).$$

Also, recall that in the course of the proof of the second proposition, we found that for $\xi \in \Lambda_1^+$ with $N\xi > n$, there is $\varepsilon > 0$ so that

$$mn^{1/m}\prod_i y_i^{1/m} \leq (n/n + \varepsilon)^{1/m} \times \operatorname{Tr}\xi y.$$

Therefore,

$$\exp\left(-\pi m n^{1/m}\prod_i v_i^{1/m}\right) \geq \exp(-\pi(n/n+\varepsilon)^{1/m}\operatorname{Tr}\xi y).$$

Putting this together, we have

$$C\exp(-\pi(n/n+\varepsilon)^{1/m}\operatorname{Tr}\xi y) \leq$$

$$\leq C\prod_i \{(k_i/4\pi e)^{-k_i/2}\{mk_i/[2\pi e n^{1/m}\eta^{(m-1)/m}]\}^{mk_i/2}\} \times$$

$$\times \sum_\xi \left\{\prod_i (k_i/\pi e v)^{k_i/2}\right\}\exp(\tfrac{1}{2}\pi\operatorname{Tr} v\xi)\exp(-2\pi\operatorname{Tr}\xi v).$$

Dividing by C and rearranging, this gives

$$1 \leq \prod_i \{(k_i/4\pi e)^{-k_i/2}\{mk_i/[2\pi e n^{1/m}\eta^{(m-1)/m}]\}^{mk_i/2}(k_i/\pi e v)^{k_i/2}\} \times$$

$$\times \sum_\xi \exp(-\pi(n/n+\varepsilon)^{1/m}\operatorname{Tr}\xi v)\exp(\tfrac{1}{2}\pi\operatorname{Tr} v\xi)\exp(-2\pi\operatorname{Tr}\xi v) =$$

$$= \prod_i \{(k_i/4\pi e)^{-k_i/2}\{mk_i/[2\pi e n^{1/m}\eta^{(m-1)/m}]\}^{mk_i/2}(k_i/\pi e v)^{k_i/2}\} \times$$

$$\times \sum_\xi \exp(-\pi\{-(n/n+\varepsilon)^{1/m} - \tfrac{1}{2} + 2\}\operatorname{Tr}\xi v).$$

Since $v_i \geq \eta > 0$, we have

$$\sum_\xi \exp(-\pi\{-(n/n+\varepsilon)^{1/m} - \tfrac{1}{2} + 2\} \operatorname{Tr} \xi v) \leq$$

$$\leq \sum_\xi \exp(-\pi\{-(n/n+\varepsilon)^{1/m} - \tfrac{1}{2} + 2\}\eta \operatorname{Tr} \xi) \leq$$

$$\leq \sum_\xi \exp(-\tfrac{1}{2}\pi\eta \operatorname{Tr} \xi) = \sigma < \infty$$

with σ depending only upon Γ. Therefore, upon simplifying, we obtain

$$1 \leq \sigma \prod_i \{\{mk_i/[2\pi e n^{1/m}\eta^{(m-1)/m}]\}^{mk_i/2}(2/v)^{k_i/2}\} \leq$$

$$\leq \sigma \prod_i \{\{mk_i/[2\pi e n^{1/m}\eta^{(m-1)/m}]\}^{mk_i/2}(2/\eta)^{k_i/2}\}.$$

From this, we have

$$n^{\sum_i k_i/2} \leq \sigma\{m/[2\pi e\eta^{(m-1)/m}]\}^{m\sum_i k_i/2}(2/\eta)^{\sum_i k_i/2} \times \prod_i k_i^{k_i/2}.$$

The point is that the righthand side does not depend on n: There is an upper bound on n which does not depend on f.

Therefore, we have found that there is a constant n depending only on Γ and on the weight so that, if $c_i(\xi) = 0$ for every index i and for every $\xi \in \Lambda_i^+$ with $N\xi \leq n$, then $f \equiv 0$. There is only a finite number of indices i to be considered; also, the coefficients $c_i(\xi)$ depend only on ξ modulo a subgroup U of $\mathfrak{o}^\times$ of finite index, since f is a modular form with respect to some congruence subgroup. Elementary considerations show that each quotient

$$\{\xi \in \Lambda_i^+ : N\xi \leq n\}/U$$

must be finite. That is, there are only finitely many vanishing conditions on f which assure that f is identically zero. Yet if the space of cuspforms were infinite-dimensional, we could make as many $c_i(\xi)$ zero as desired, by taking finite linear combinations. This proves the finite dimensionality of the space of cuspforms. ■

1.8 Holomorphic Eisenstein Series and Cuspforms

Now we give the general version of a previous result we proved only for $SL(2,\mathfrak{o}) \subset \Gamma \subset GL(2,\mathfrak{o})^+$. We will show that any space $\mathrm{Mfm}(\Gamma, k)$ is a direct sum of $\mathrm{Cfm}(\Gamma, k)$ with a subspace $\mathrm{Eis}(\Gamma, k) \subset \mathrm{Mfm}(\Gamma, k)$ consisting of explicitly constructed modular forms. Of course, for weights $k = (k_1, \ldots, k_m)$ with some $k_i \neq k_j$, we already know that $\mathrm{Cfm}(\Gamma, k) = \mathrm{Mfm}(\Gamma, k)$. Incidentally, here we show that *there exist non-zero cuspforms.*

Definition Let Γ be a congruence subgroup, and let $k > 2$. For $\alpha \in GL(2, F)^+$ define an Eisenstein series of weight $(k, \ldots, k)$ by

$$E_\alpha(z) = \sum_\gamma \mu(\alpha^{-1}\gamma, z)^{-k}$$

where γ is summed over $(\alpha P \alpha^{-1} \cap \Gamma)\backslash\Gamma$. One may readily verify that these series are well-defined, if absolutely convergent. It is immediate and formal that, if absolutely convergent, these Eisenstein series are in $\mathrm{Mfm}(\Gamma, k)$ (invoking Koecher's principle for $[F : \mathbb{Q}] > 1$). Let $\mathrm{Eis}(\Gamma, k)$ denote the subspace of $\mathrm{Mfm}(\Gamma, k)$ spanned by these Eisenstein series.

PROPOSITION These Eisenstein series are absolutely convergent for weight $k > 2$.

Proof Now

$$\{\mu(\alpha^{-1}\gamma, z)^{-k} : \gamma \in (\alpha P \alpha^{-1} \cap \Gamma)\backslash\Gamma\} = \{\mu(\gamma\alpha^{-1}, z)^{-k} : \gamma \in (P \cap \alpha\Gamma\alpha^{-1})\backslash\alpha\Gamma\alpha^{-1}\}.$$

Thus, by changing the congruence subgroup Γ, we may suppose that $\alpha = 1$. Also, since Γ contains some $\Gamma(\mathfrak{n})$ with finite index (at least up to the center), Eisenstein series for Γ are just finite sums of those for $\Gamma(\mathfrak{n})$. So we may assume that $\Gamma = \Gamma(\mathfrak{n})$ for discussions of convergence. Then

$$\{\mu(\gamma, z)^{-k} : \gamma \in (P \cap \Gamma)\backslash\Gamma\} \subset \{\mu(\gamma, z)^{-k} : \gamma \in (P \cap GL^+(2, \mathfrak{o}))\backslash GL^+(2, \mathfrak{o})\}.$$

Thus, the convergence of the Eisenstein series for $GL^+(2, \mathfrak{o})$ implies the convergence in general. ■

PROPOSITION Take $\delta \in GL^+(2, F)$. Let

$$(E_\alpha|_k\delta)(z) = \sum_\xi c_\xi \exp(2\pi i \operatorname{Tr} \xi z).$$

Then $c_0 = 0$ unless $\Gamma\alpha P = \Gamma\delta P$, in which case $c_0 \neq 0$. Further, if $\Gamma\alpha P = \Gamma\beta P$, then $E_\alpha = E_\beta$. In particular, no one of these Eisenstein series is identically zero.

COROLLARY There exist cuspforms which are not identically zero. In particular, take a sufficiently small Γ so that $\Gamma\backslash GL^+(2, F)/P$ has cardinality > 1. (We claim that $\Gamma = \Gamma(\mathfrak{n})$ suffices, for a proper ideal $\mathfrak{n}$ of $\mathfrak{o}$.) Let E_α and E_β be two Eisenstein series of weight (for example), $(4, \ldots, 4)$ for Γ, attached to distinct double cosets $\Gamma\alpha P$ and $\Gamma\beta P$ (that is, to distinct cusps). Then $f = E_\alpha E_\beta$ is a cuspform for Γ of weight $(8, \ldots, 8)$.

Proof of Proposition We have, by definition,

$$(E_\alpha|_k\delta)(z) = \sum_\gamma \mu(\alpha^{-1}\gamma, \delta z)^{-k}\mu(\delta, z)^{-k} = \sum_\gamma \mu(\alpha^{-1}\gamma\delta, z)^{-k}.$$

Now $c_\xi = 0$ unless $\xi = 0$ or ξ is totally positive; and, if $z = (i\lambda, \ldots, i\lambda)$ with $\lambda > 0$, then

$$\lim_{\lambda\to\infty} (E_\alpha|_k\delta)(z) = c_0.$$

We note that

$$\lim_{\lambda\to\infty} \mu\left(\begin{pmatrix} a & b \\ c & d \end{pmatrix}, z\right)^{-k} = \begin{cases} 0 & \text{for } c \neq 0 \\ d^{-k} & \text{for } c = 0. \end{cases}$$

That is, this limit is 0 unless $\left(\begin{smallmatrix} a & b \\ c & d \end{smallmatrix}\right)$ lies in P. Therefore, $c_0 = 0$ unless $\alpha^{-1}\gamma\delta \in P$ for some γ; that is, $c_0 = 0$ unless $\delta \in \Gamma\alpha P$, as claimed. On the other hand, if $\delta \in \gamma_1^{-1}\alpha P$ and $\delta \in \gamma_2^{-1}\alpha P$ with $\gamma_1, \gamma_2 \in \Gamma$, then for some $p \in P$ we have

$$\gamma_1^{-1}\alpha p = \gamma_2^{-1}\alpha$$

or

$$\gamma_1 \in \alpha P \alpha^{-1}\gamma_1.$$

That is, γ_1 and γ_2 are in the same coset in $(\alpha P\alpha^{-1} \cap \Gamma)\backslash\Gamma$. Thus, when $\Gamma\delta P = \Gamma\alpha P$, there is precisely one term which does not vanish as $\lambda \to \infty$. Therefore, in this case, $c_0 \neq 0$. This proves the first assertion.

Now suppose that $\Gamma\alpha P = \Gamma\beta P$; take $\delta \in \Gamma$ and $p \in P$ so that $\delta\alpha p = \beta$. By definition,

$$E_\beta(z) = \sum_\gamma \mu(\beta^{-1}\gamma, z)^{-k} \qquad (\gamma \in (\beta P\beta^{-1} \cap \Gamma)\backslash\Gamma.$$

By hypothesis,

$$(\beta P\beta^{-1} \cap \Gamma)\backslash\Gamma \approx (\alpha P\alpha^{-1} \cap \Gamma)\backslash\Gamma$$

insofar as these two coset spaces have a common set of representatives. This proves the second assertion. ■

Proof of Corollary Take a sufficiently small Γ so that $\Gamma\backslash GL^+(2, F)/P$ has cardinality > 1. Let E_α and E_β be two Eisenstein series of weight (for example), $(4, \ldots, 4)$ for Γ, attached to distinct double cosets $\Gamma\alpha P$ and $\Gamma\beta P$ (that is, to distinct cusps). Then the argument in the proof of the previous proposition shows that $f = E_\alpha E_\beta$ is a cuspform for Γ of weight $(8, \ldots, 8)$. If $\Gamma = \Gamma(\mathfrak{n})$ with a proper ideal $\mathfrak{n}$ of $\mathfrak{o}$, then 0 and $i\infty$ are readily checked to be inequivalent cusps. ■

THEOREM Let Γ be a congruence subgroup, and let $k > 2$. For any choice $\{g_i\}$ of representatives for $\Gamma\backslash GL(2, F)^+/P$, the Eisenstein series defined by

$$E_i(z) = \sum_\gamma \mu(g_i^{-1}\gamma, z)^{-k} \qquad (\text{with } \gamma \in (g_i P g_i^{-1} \cap \Gamma)\backslash\Gamma)$$

form a basis for $\mathrm{Eis}(\Gamma, k)$. Hence, $\mathrm{Eis}(\Gamma, k)$ is finite-dimensional. Further, given $f \in \mathrm{Mfm}(\Gamma, k)$, there is $E \in \mathrm{Eis}(\Gamma, k)$ so that $f - E \in \mathrm{Cfm}(\Gamma, k)$. That is,

$$\mathrm{Mfm}(\Gamma, k) = \mathrm{Eis}(\Gamma, k) \oplus \mathrm{Cfm}(\Gamma, k).$$

Hence, $\mathrm{Mfm}(\Gamma, k)$ itself is finite-dimensional.

Proof The first assertion follows from the second assertion of the previous proposition. For each $\beta \in GL(2, F)^+$ we have a linear functional L_β defined on $f \in \mathrm{Mfm}(\Gamma, k)$, defined by

$$L_\beta f = \lim_{\lambda \to \infty} (f|_k \beta)(i\lambda) = \text{0-th Fourier coefficient of } f|_k \beta.$$

By the definitions, $f \in \mathrm{Mfm}(\Gamma, k)$ is a cuspform if and only if $L_\beta f = 0$ for all $\beta \in GL(2, F)^+$. An elementary calculation shows that $L_\beta f = 0$ if and only if $L_{\gamma\beta p} f = 0$ for all $\gamma \in \Gamma$, $p \in P$. Therefore, it suffices merely to show that $L_{gi} f = 0$ for all g_i in order to show that f is a cuspform. Given $f \in \mathrm{Mfm}(\Gamma, k)$, consider

$$E = \sum_i (L_{gi} f)(L_{gi} E_i)^{-1} E_i.$$

Using the previous proposition, we have $L_{gi} E_j = 0$ unless $i = j$, in which case $L_{gi} E_j \neq 0$. Therefore,

$$L_{gj}(f - E) = 0$$

for all indices i. That is, $f - E$ is a cuspform.

Since we already have demonstrated the finite dimensionality of $\mathrm{Cfm}(\Gamma, k)$, this finishes the proof. ■

1.9 Dirichlet Series Associated to Cuspforms

Just as in the case of elliptic modular forms, from a Hilbert modular form one may make a (generalized) Dirichlet series whose coefficients are essentially those of the modular form. (We do this here in a somewhat crude form; this is unavoidable in a classical treatment over fields F larger than $\mathbb{Q}$). Because such Dirichlet series may be represented as integrals against the cuspform, these Dirichlet series have analytic continuations and functional equations; this is the principal result of this section. When $[F : \mathbb{Q}] > 1$, there arises a new phenomenon: There are infinitely many Dirichlet series naturally attached to each Hilbert modular cuspform, since there exist infinitely many grossencharacters with given ramification.

Throughout this section, fix a non-zero ideal $\mathfrak{n}$ of $\mathfrak{o}$, let $\Gamma = \Gamma(\mathfrak{n})$, and let

$$f(z) = \sum_\xi c_\xi \exp(2\pi i \operatorname{Tr} \xi z)$$

be in $\mathrm{Cfm}(\Gamma, k)$, with $k = (k_1, \ldots, k_m)$.

PROPOSITION Let $J = \begin{pmatrix} 0 & -1 \\ 1 & 0 \end{pmatrix}$. Then the map $f \to f^{\natural}$ defined on Mfm(k) by

$$f^{\natural}(z) = f|_k J$$

maps Mfm(Γ, k) and Cfm(Γ, k) to themselves. Further, both Mfm(Γ, k) and Cfm(Γ, k) are decomposable into eigenspaces with respect to $f \to f^{\natural}$, with eigenvalues in $\{\pm 1, \pm i\}$.

Proof From general principles, Mfm(k) and Cfm(k) are stabilized by this map. Since J normalizes Γ, it follows immediately that Mfm(Γ, k) is stabilized. This gives the first assertion. Since $J^4 = 1_2$, the second assertion follows. ∎

Definition Let $\mathbb{R}^+$ denote the positive real numbers, and let $T = (\mathbb{R}^+)^m$. Let $s \in \mathbb{C}$, and let $\chi: T \to \mathbb{C}^\times$ be a continuous group homomorphism which is trivial on the two subgroups

$$\Delta = \{(y, \ldots, y) \in (\mathbb{R}^+)^m : y > 0\} \qquad U = \{\eta \in T : \eta \in \mathfrak{o}^\times, \eta \equiv 1 \bmod \mathfrak{n}\}.$$

Then define an L-function attached to f by

$$L(s, \chi, f) = \sum_{\xi \bmod U} c_\xi \chi(\xi) \xi^{-(s+k/2-1/2)}$$

where we use a multi-index notation. One may check that the presence of the $k/2$ in the exponent assures that this sum is well-defined, if convergent. For each such χ, write

$$\chi(y) = \chi(y_1, \ldots, y_m) = \prod_j y_j^{\pi i \alpha_j}$$

for suitable complex numbers α_j.

Remarks The choice of the exponents α_j is clearly constrained by the logarithms of the real imbeddings of the units in U. In any case, each α_j must be real, and $\sum_j \alpha_j = 0$. If $F = \mathbb{Q}$, the only such χ is $\chi \equiv 1$.

THEOREM The series defining $L(s, \chi, f)$ is absolutely convergent for Re(s) > 3/2. The integral

$$\tilde{L}(s, \chi, f) = \int_{T/U} y^{s+k/2-1/2} \bar{\chi}(y) f(iy) \, d^\times y$$

is absolutely convergent for Re(s) > $\frac{1}{2}$, where

$$d^\times y = (y_1 \ldots y_m)^{-1} \, dy_1 \ldots dy_m.$$

For Re(s) > 3/2,

$$\tilde{L}(s,\chi,f) = \left\{\prod_j (2\pi)^{-(s+1/2k_j-1/2+\pi i\alpha_j)}\Gamma(s+\tfrac{1}{2}k_j-\tfrac{1}{2}+\pi i\alpha_j)\right\} L(s,\chi,f).$$

The function $\tilde{L}(s,\chi,f)$ has an analytic continuation to an entire function of $s \in \mathbb{C}$. Further, if f is an eigenfunction for the map $f \to f^{\natural}$, with eigenvalue λ, then there is the functional equation

$$\tilde{L}(s,\chi,f) = \lambda i^N \tilde{L}(1-s,\bar{\chi},f)$$

where $N = \sum_j k_j$, and $\lambda i^N = \pm 1$.

Proof We leave the necessary estimates to the end of the proof.

Granting absolute convergence, we have

$$\tilde{L}(s,\chi,f) = \int_{T/U} y^{s+k/2-1/2}\chi(y)f(iy)\,d^\times y =$$

$$= \int_{T/U} \sum_{\xi} c_\xi\, y^{s+k/2-1/2}\chi(y)\exp(-2\pi \operatorname{Tr}\xi y)\,d^\times y =$$

$$= \int_T \sum_{\xi/U} c_\xi y^{s+k/2-1/2}\chi(y)\exp(-2\pi \operatorname{Tr}\xi y)\,d^\times y =$$

$$= \sum_{\xi/U} c_\xi \int_T y^{s+k/2-1/2}\chi(y)\exp(-2\pi \operatorname{Tr}\xi y)\,d^\times y.$$

Replacing y by $y\xi^{-1}$ gives

$$\tilde{L}(s,\chi,f) = \sum_{\xi/U} c_\xi \chi(\xi)\xi^{-(s+k/2-1/2)} \int_T y^{s+k/2-1/2}\chi(y)\exp(-2\pi \operatorname{Tr} y)\,d^\times y.$$

The integral over T breaks up into a product of m integrals over $\mathbb{R}^+$; the j-th is

$$\int_{\mathbb{R}^+} y_j^{s+k_j/2-1/2+\pi i\alpha_j}\chi(y_j)\exp(-2\pi y_j)\,dy_j/y_j =$$

$$= (2\pi)^{-(s+1/2k_j-1/2+\pi i\alpha_j)}\Gamma(s+\tfrac{1}{2}k_j-\tfrac{1}{2}+\pi i\alpha_j).$$

This proves the equality indicated.

Now for the proof of the analytic continuation. Let $T(+)$ be the subgroup of T where $\prod_i y_i \geq 1$, and let $T(-)$ be the subgroup where $\prod_i y_i \leq 1$. Then

$$\tilde{L}(s,\chi,f) = \int_{T(+)/U} y^{s+k/2-1/2}\chi(y)f(iy)\,d^\times y + \int_{T(-)/U} y^{s+k/2-1/2}\chi(y)f(iy)\,d^\times y.$$

We will see below that the first integral is entire in $s \in \mathbb{C}$. In the second integral, replace y by y^{-1} to obtain:

$$\int_{T(-)/U} y^{s+k/2-1/2}\chi(y)f(iy)\,d^{\times}y = \int_{T(+)/U} y^{-(s+k/2-1/2)}\bar{\chi}(y)f(-1/iy)\,d^{\times}y.$$

Since f is a cuspform of weight k, we have

$$f(-1/iy) = (iy)^k f^{\natural}(iy).$$

Therefore,

$$\int_{T(-)/U} y^{s+k/2-1/2}\chi(y)f(iy)\,d^{\times}y = i^N \int_{T(+)/U} y^{1-s+k/2-1/2}\bar{\chi}(y)f^{\natural}(iy)\,d^{\times}y.$$

The latter integral is entire in $s \in \mathbb{C}$, since $f^{\natural}$ is also a cuspform; again, this is demonstrated below. This proves that $\tilde{L}(s, \chi, f)$ has an analytic continuation to an entire function of $s \in \mathbb{C}$.

Now suppose that $f^{\natural} = \lambda f$. By the previous discussion, we have

$$\tilde{L}(s, \chi, f) = \int_{T(+)/U} y^{s+k/2-1/2}\chi(y)f(iy)\,d^{\times}y + \lambda i^N \int_{T(+)/U} y^{1-s+k/2-1/2}\bar{\chi}(y)f(iy)\,d^{\times}y.$$

On the other hand (at first for $\mathrm{Re}(1-s) > 1$, and then by analytic continuation for $s \in \mathbb{C}$),

$$\tilde{L}(1-s, \bar{\chi}, f) = \int_{T(+)/U} y^{1-s+k/2-1/2}\bar{\chi}(y)f(iy)\,d^{\times}y + \lambda i^N \int_{T(+)/U} y^{s+k/2-1/2}\chi(y)f(iy)\,d^{\times}y.$$

Therefore, if we can show that $\lambda i^N = \pm 1$, we are done. It is immediately apparent that $\mu(J^2, *)^{-k} = (-1)^N$. Therefore,

$$\lambda^2 f = f|_k J^2 = (-1)^N f.$$

Therefore,

$$(\lambda i^N)^2 = \lambda^2 i^{2N} = \lambda^2(-1)^N = (-1)^{2N} = 1,$$

which proves what we wanted.

For the convergence and entireness assertions regarding the integrals, we use the corollaries from the proof of the finite dimensionality of spaces of cuspforms, giving estimates on the growth of cuspforms in Siegel sets, etc. First, observe that $T(+)/U$ has representatives in some Siegel set S. We had shown that on S

$$y^{k/2}|f(x+iy)| \le C\exp(-c(Ny)^{1/m})$$

for some positive constants c and C. We can write

$$T(+)/U \approx \left\{(y_1,\ldots,y_m)\colon \prod_i y_i = 1\right\}\Big/U \times \{(\eta^{1/m},\ldots,\eta^{1/m})\colon \eta \geq 1\}.$$

By the Units Theorem, the first factor is compact; choose a compact set X in T of representatives for the first factor. With $y' \in X$, $\eta \geq 1$, in these coordinates we have

$$y^{k/2}|f(iy)| \leq \exp(-c\eta).$$

Therefore, letting $\sigma = \operatorname{Re}(s)$ and noting that $|\chi(y)| = 1$,

$$\int_{T(+)/U} y^{s+1/2k-1/2}\chi(y)f(iy)\,d^{\times}y$$

is dominated by

$$\int_X y'^{\sigma+1/2k-1/2} \times \int_{\eta\geq 1} \eta^{\sigma-1/2}\exp(-c\eta)\,d\eta/\eta$$

with a suitable measure in the first integral. The first integral is that of a continuous function over a compact set, and the second is clearly absolutely convergent for all $\sigma \in \mathbb{R}$. This proves that the indicated integrals over $T(+)/U$ are absolutely convergent and are entire functions of $s \in \mathbb{C}$.

Regarding the integral over $T(-)/U$, we use the same coordinates. We had the estimate

$$|y^{k/2}f(iy)| < C$$

for some finite constant C. Thus, the indicated integral over $T(-)/U$ is dominated by

$$\int_X y'^{\sigma-1/2} \times \int_{0<\eta\leq 1} \eta^{\sigma-1/2}\,d\eta/\eta = \int_X 1 \times \int_{0<\eta\leq 1} \eta^{\sigma-1/2}\,d\eta/\eta$$

with a suitable measure in the first integral. The first integral is that of a continuous function over a compact set and so converges absolutely. The second is clearly absolutely convergent for $\sigma - \frac{1}{2} > 0$, as desired. This result, together with the result on the integrals on $T(+)/U$, proves the absolute convergence of the integral defining $\tilde{L}(s,\chi,f)$ for $\operatorname{Re}(s) > \frac{1}{2}$.

For absolute convergence of the series, recall that we had

$$c_\xi = O(\xi^{k/2}).$$

Also, $c_\xi \xi^{-k/2}$ is well-defined for ξ modulo U, since f is a modular form. Therefore, letting $\sigma = \operatorname{Re}(s)$,

$$\sum_{\xi \bmod U} c_\xi \xi^{-k/2} N\xi^{-(s-1/2)}$$

is dominated by

$$\sum_{\xi \bmod U} N\xi^{-(\sigma-1/2)}$$

which is dominated by the series defining the zeta function of F, evaluated at $\sigma - 1/2$. Therefore, we have absolute convergence for $\sigma > 3/2$. ■

1.10 Some Integration Theory

As it is not so appropriate to try to describe a "fundamental domain" for the action of congruence subgroups Γ on $\mathfrak{H}^m$, integration theory on $\Gamma\backslash\mathfrak{H}^m$ needs a little consideration. In the following, we define an abstract integral on $\Gamma\backslash\mathfrak{H}^m$. In the end, we reach the conclusion that we need not have worried: Things work as well as one could hope. See Appendix A.1 for a more detailed discussion of the background material on integration on quotients.

Let G be a locally compact Hausdorff topological group, with a left Haar measure μ. There is a continuous positive real-valued homomorphism $\Delta\colon G \to \mathbb{R}$ so that, for a measurable subset E of G, and for $g \in G$,

$$\mu(Eg) = \Delta(g)\mu(E).$$

This function Δ is the *modular function* (for G and μ). Let H be a closed (not necessarily normal) topological subgroup of G, with left Haar measure ν and modular function δ. Then G/H is a locally compact Hausdorff space with a continuous action of G. A fundamental result here is

PROPOSITION There is a G-invariant (positive regular Borel) measure $\bar{\mu}$ on G/H if and only if

$$\Delta|_H = \delta.$$

If this condition is met, then we have a uniqueness result: Given $\bar{\mu}$, there is a positive constant C such that for every continuous compactly supported function f on G,

$$\int_G f(g)\,d\mu(g) = C\int_{G/H}\int_H f(gh)\,d\nu(h)\,d\bar{\mu}(gH).$$

Given a Haar measure dh on H and given a compactly supported continuous function F on G/H, there is a compactly supported continuous function f on G so that for all $g \in G$ we have

$$F(g) = \int_H f(gh)\,dh.$$ ■

COROLLARY If G is unimodular (i.e., if $\Delta \equiv 1$) and if H is discrete or in the center of G, or if H is compact, then there is always a G-invariant (positive regular Borel) measure on G/H, and this measure is unique up to multiplication by a positive real constant.

Proof If G is unimodular and H is discrete, then the measure giving each point of H mass 1 is a Haar measure on H, so the modular function on H is identically 1, and the condition of the proposition is met. If G is unimodular and H is in its center, then the condition is also met, since H is unimodular. If H is compact, then its modular function δ must be identically 1, since $\delta(H)$ is a compact subgroup of the positive real numbers; likewise, $\Delta(H) = 1$. ■

Now let G be a unimodular locally compact Hausdorff topological group with (left and right) Haar measure μ, with a compact subgroup K, and with a discrete subgroup Γ. We wish to establish a natural integration theory on the double coset space

$$(\Gamma\backslash G)/K \approx \Gamma\backslash(G/K).$$

Definition Let f be a continuous left Γ-invariant complex-valued function on G/K so that the obvious induced function on $\Gamma\backslash G/K$ is compactly supported. Let $\tilde{f}$ be the function on $\Gamma\backslash G$ defined by

$$\tilde{f}(\Gamma g) = f(\Gamma g K).$$

Then define

$$\int_{\Gamma\backslash G/K} f = {}^{\#}(\Gamma \cap K) \int_{\Gamma\backslash G} \tilde{f}(g)\, d\bar{\mu}(g)$$

where $\bar{\mu}$ is a right G-invariant positive regular Borel measure on $\Gamma\backslash G$ (as in the previous corollary).

PROPOSITION The previous definition uniquely determines a positive regular Borel measure v on the double coset space $\Gamma\backslash G/\mathrm{K}$ so that, in the above notation,

$$\int_{\Gamma\backslash G/K} f\, dv = \int_{\Gamma\backslash G} \tilde{f}(g)\, d\bar{\mu}(g).$$

Proof By the Riesz representation theorem, we merely need to show that the functional given by the righthand side of the asserted equality gives a (well-defined) continuous linear functional on continuous compactly supported functions, with respect to the semi-norms σ_C defined by

$$\sigma_C(f) = \sup_{z \in C} |f(z)|$$

for compact subsets C of $\Gamma\backslash G/K$. It is certainly well-defined, as for each f the associated function $\tilde{f}$ is still left Γ-invariant and compactly supported on $\Gamma\backslash G$, since K is compact. Since $\bar{\mu}$ is a positive regular Borel measure on the locally compact Hausdorff space $\Gamma\backslash G$, the map

$$\tilde{f} \to \int_{\Gamma\backslash G} \tilde{f}(g)\, d\bar{\mu}(g)$$

is certainly continuous with respect to the analogous semi-norms $\tilde{\sigma}_C$ on $\Gamma\backslash G$. But the values $\sigma_C(f)$ are just a subset of the values $\tilde{\sigma}_C(\tilde{f})$. This gives the continuity, proving the proposition. ■

PROPOSITION Suppose that there is a closed subgroup P of G so that $G = PK \approx P \times K$; let dp denote a left Haar measure on P, and let dk denote a Haar measure on K. Then for a given Haar measure dg on G, there is a non-zero positive constant c so that

$$dp \otimes dk = c\, dg.$$

Proof We can write

$$d(p, k) = \varphi(p, k)\, dp \otimes dk$$

for some continuous function φ on $P \times K$. Since the Haar measure on G is left P-invariant, $\varphi(p, k)$ must be constant in $p \in P$; likewise, it cannot depend on K. ■

PROPOSITION In the situation of the previous discussion, suppose that K has total measure 1 (with respect to its Haar measure). Let Γ_0 be a subgroup of $P \cap \Gamma$, and let φ be a left Γ_0-invariant right K-invariant function on G. Then (assuming absolute convergence of the integrals)

$$\int_{\Gamma\backslash G} \left(\sum_{\gamma \in \Gamma_0\backslash\Gamma} \varphi(\gamma g) \right) dg = \int_{\Gamma_0\backslash P} \varphi(p)\, dp$$

where the dg and dp refer to the uniquely determined measures induced on the quotients.

Proof By the Riesz representation theorem, we need consider only compactly supported continuous functions. By the remarks after the first proposition, there is a compactly supported continuous function ψ on G so that

$$\varphi(g) = \sum_{\gamma \in \Gamma_0} \psi(\gamma g).$$

Then

$$\int_{\Gamma\backslash G}\left(\sum_{\gamma\in\Gamma_0\backslash\Gamma}\varphi(\gamma g)\right)dg = \int_{\Gamma\backslash G}\left(\sum_{\gamma\in\Gamma_0\backslash\Gamma}\sum_{\delta\in\Gamma_0}\psi(\delta\gamma g)\right)dg =$$

$$= \int_{\Gamma\backslash G}\left(\sum_{\gamma\in\Gamma}\psi(\gamma g)\right)dg = \int_G \psi(g)\,dg$$

where the last two equalities follow from the definitions of the induced measures. By the previous proposition, we have

$$\int_G \psi(g)\,dg = \int_{P\times K}\psi(pk)\,dp\,dk = \int_P \psi(p)\,dp \times \int_K 1\,dk = \int_P \psi(p)\,dp.$$

Going backward, we have

$$\int_P \psi(p)\,dp = \int_{\Gamma_0\backslash P}\sum_{\gamma\in\Gamma_0}\psi(\gamma p)\,dp = \int_{\Gamma_0\backslash P}\varphi(\gamma p)\,dp. \qquad \blacksquare$$

Now we apply these general considerations to the concrete situation of interest. First, we claim that the measure μ on $SL(2,\mathbb{R})$ given by

$$d\mu\left(\begin{pmatrix}1 & x\\ 0 & 1\end{pmatrix}\begin{pmatrix}y^{1/2} & 0\\ 0 & y^{-1/2}\end{pmatrix}\begin{pmatrix}\cos\theta & \sin\theta\\ -\sin\theta & \cos\theta\end{pmatrix}\right) = (2\pi)^{-1}y^{-2}\,dx\,dy\,d\theta$$

is a left and right Haar measure, where $x\in\mathbb{R}$, $y>0$, $\theta\in\mathbb{R}$. (It is not hard to check that every element of $SL(2,\mathbb{R})$ can be written in this form, and uniquely so). Let

$$K = \left\{\begin{pmatrix}\cos\theta & \sin\theta\\ -\sin\theta & \cos\theta\end{pmatrix} : \theta\in\mathbb{R}\right\} \qquad P = \left\{\begin{pmatrix}1 & x\\ 0 & 1\end{pmatrix}\begin{pmatrix}y & 0\\ 0 & y^{-1}\end{pmatrix} : x\in\mathbb{R}, y>0\right\}.$$

Then we have $SL(2,\mathbb{R}) = PK \approx P\times K$, as above. It is easy to check that

$$dp = d\left(\begin{pmatrix}1 & x\\ 0 & 1\end{pmatrix}\begin{pmatrix}y^{1/2} & 0\\ 0 & y^{-1/2}\end{pmatrix}\right) = y^{-2}\,dx\,dy$$

is a left Haar measure on P, and that $(2\pi)^{-1}\,d\theta$ is a Haar measure on K with total measure 1. Therefore, the previous discussion applies, and this is a Haar measure on $SL(2,\mathbb{R})$.

Next, we note the identification

$$P \approx \mathfrak{H}$$

via

$$\begin{pmatrix}1 & x\\ 0 & 1\end{pmatrix}\begin{pmatrix}y^{1/2} & 0\\ 0 & y^{-1/2}\end{pmatrix} \to \begin{pmatrix}1 & x\\ 0 & 1\end{pmatrix}\begin{pmatrix}y^{1/2} & 0\\ 0 & y^{-1/2}\end{pmatrix}(i) = x+iy.$$

Then the measure

$$dv(x + iy) = y^{-2}\,dx\,dy$$

on $\mathfrak{H}$ is a $SL(2, \mathbb{R})$-invariant measure on $\mathfrak{H}$, obtained via

$$SL(2, \mathbb{R})/SO(2) \approx P \times K/K \approx P \approx \mathfrak{H}.$$

Further,

$$\int_{SL(2,\mathbb{R})} f(g(i))\,d\mu(g) = \int_{\mathfrak{H}} f(x + iy)\,dv(x + iy)$$

for a function f on $\mathfrak{H}$. Clearly a similar discussion applies to products $\mathfrak{H}^m$ and $SL(2, \mathbb{R})^m$.

Now let Γ be a discrete subgroup of $SL(2, \mathbb{R})^m$, and let $\Gamma_0 = \Gamma \cap P^m$. Let φ be a Γ_0-invariant function on $\mathfrak{H}^m$. The general considerations above show that

$$\int_{\Gamma\backslash\mathfrak{H}^m} \sum_{\gamma\in\Gamma_0\backslash\Gamma} \varphi(\gamma z)\,dv(z) = \int_{\Gamma_0\backslash\mathfrak{H}^m} \varphi(z)\,dv(z) =$$

$$= \int_{\Gamma_0\backslash P^m} \varphi(p(i))\,dp$$

where now dp and dv denote the obvious product measures. In particular, if $\Gamma_0 = \{1\}$, then

$$\int_{\Gamma\backslash\mathfrak{H}^m} \sum_{\gamma\in\Gamma} \varphi(\gamma z)\,dv(z) = \int_{\mathfrak{H}^m} \varphi(z)\,dv(z).$$

Now we adapt these considerations to discrete subgroups of $GL^+(2, \mathbb{R})^m$. Let Z be the center of $GL^+(2, \mathbb{R})^m$. We note that

$$GL^+(2, \mathbb{R})^m/Z \approx SL(2, \mathbb{R})^m/\{\pm 1\}^m.$$

If Γ is a discrete subgroup of $GL^+(2, \mathbb{R})^m$, then it is elementary that $\Gamma/\Gamma \cap Z$ is discrete in $GL^+(2, \mathbb{R})^m/Z$, and the inverse image Γ' in $SL(2, \mathbb{R})^m$ is still discrete.

Remarks In light of the previous discussion, for a discrete subgroup Γ of $GL^+(2, \mathbb{R})^m$ we will often abuse notation by writing

$$\int_{\Gamma\backslash\mathfrak{H}^m} f(x_1 + iy_1, \ldots, x_m + iy_m)y^{-2}\,dx\,dy$$

though it should be understood that this must be construed as an integral in $\Gamma\backslash\mathfrak{H}^m$

as an abstract space, being a continuous image of

$$Z\Gamma\backslash GL^+(2,\mathbb{R})^m \approx \Gamma'\backslash SL(2,\mathbb{R})^m/\{\pm 1\}^m.$$

1.11 A Volume Computation

The ideas of the previous section allow computation of the volume of $\Gamma\backslash\mathfrak{H}^m$ for $\Gamma = SL(2,\mathfrak{o})$ under the hypothesis that $\mathfrak{o}$ is a principal ideal domain.

CLAIM With measures normalized as below (in an eminently reasonable way),

$$\text{volume}(\Gamma\backslash\mathfrak{H}^m) = \pi^{-m}\zeta_F(2)D^{3/2}.$$

Proof By our conventions, this is the same as the computation of the volume of

$$\Gamma\backslash SL(2,\mathbb{R})^m.$$

Let $G = SL(2,\mathbb{R})^m$, let

$$U = \left\{u(x) = \begin{pmatrix} 1 & x \\ 0 & 1 \end{pmatrix} : x \in \mathbb{R}^m\right\} \subset G$$

$$T = \left\{t(y) = \begin{pmatrix} y^{1/2} & 0 \\ 0 & y^{-1/2} \end{pmatrix} : y \in \{\eta > 0\}^m\right\} \subset G$$

$$K = \left\{k(\theta) = \begin{pmatrix} \cos\theta & \sin\theta \\ -\sin\theta & \cos\theta \end{pmatrix} : \theta \in \mathbb{R}^m\right\} \subset G$$

and write $P^+ = UT = TU$. Then

$$d(u(x)) = dx_1 \ldots dx_m$$

gives a two-sided Haar measure on U,

$$d(t(y)) = \left(\prod_i y_i\right)^{-1} dy_1 \ldots dy_m$$

gives a two-sided Haar measure on T,

$$d(u(x)t(y)) = \left(\prod_i y_i\right)^{-2} dx_1 \ldots dx_m\, dy_1 \ldots dy_m$$

gives a left Haar measure on P^+,

$$d(k(\theta)) = (2\pi)^{-m}\, d\theta_1 \ldots d\theta_m$$

gives a two-sided Haar measure on K, and

$$d(u(x)t(y)k(\theta)) = d(u(x)t(y))d(k(\theta))$$

gives a two-sided Haar measure on G. We let G act on row vectors v in $(\mathbb{R}^m)^2$ by right multiplication.

The first observation is that we can identify $U\backslash G$ with $V = (\mathbb{R}^m)^2 - \{0\}$ via the map

$$Ug \to (0, 1)g.$$

Since G is unimodular and U is unimodular, there is a G-invariant (positive regular Borel) measure η on $U\backslash G \approx (\mathbb{R}^m)^2 - \{0\}$ so that for a continuous function f on G with compact support we have

$$\int_G f(g)\, dg = \int_V \int_U f(u(x)g)\, dx\, d\eta(Ug).$$

Since the usual (Haar) measure dv on $(\mathbb{R}^m)^2$ is another G-invariant measure, it must be that

$$d\eta(v) = c\, dv$$

for some positive constant c.

Now let f be a Schwartz function on $\mathbb{R}^{2m}$, and consider

$$I(f) = \int_{\Gamma\backslash G} \sum_{v \in \mathfrak{o}^2} f(vg)\, dg$$

$$J(f) = \int_{\Gamma\backslash G} \sum_{v \in \mathfrak{o}^{*2}} f(vg)\, dg$$

where dg denotes the Haar measure on G and also the invariant measure on $\Gamma\backslash G$, and $\mathfrak{o}^*$ is the inverse different of $\mathfrak{o}$. On one hand, by the Poisson summation formula, this is

$$I(f) = D^{-1} \int_{\Gamma\backslash G} \sum_{v \in \mathfrak{o}^{*2}} \hat{f}(vg^{T-1})\, dg$$

where D is the discriminant of D over $\mathbb{Q}$ (and also is the natural volume of $\mathbb{R}^{2m}/\mathfrak{o}^2$). Since $g \to g^{T-1}$ is an automorphism of order 2 of G, it must preserve Haar measure, so we may replace g by g^{T-1} in the latter integral to find

$$I(f) = D^{-1}J(\hat{f}).$$

On the other hand, using the fact that $\mathfrak{o}$ is a principal ideal domain, and using facts about integration on homogeneous spaces,

$$I(f) - f(0)\,\mathrm{vol}(\Gamma\backslash G) = I(f) - \int_{\Gamma\backslash G} f(0)\,dg = \int_{\Gamma\backslash G} \sum_{v \in \mathfrak{o}^2 - 0} f(vg)\,dg =$$

$$= \int_{U\cap\Gamma\backslash G} \sum_{\alpha \in (\mathfrak{o}-0)/\mathfrak{o}^\times} f(\alpha(0,1)g)\,dg =$$

$$= \int_{U\backslash G} \sum_{\alpha \in (\mathfrak{o}-0)/\mathfrak{o}^\times} f(\alpha(0,1)g)\,dg \times \int_{U\cap\Gamma\backslash U} 1\,du =$$

$$= D^{1/2} \int_{U\backslash G} \sum_{\alpha \in (\mathfrak{o}-0)/\mathfrak{o}^\times} f(\alpha(0,1)g)\,dg =$$

$$= D^{1/2} c \int_{\mathbb{R}^{2m}} \sum_{\alpha \in (\mathfrak{o}-0)/\mathfrak{o}^\times} f(\alpha v)\,dv =$$

$$= D^{1/2} c \left\{ \sum_{\alpha \in (\mathfrak{o}-0)/\mathfrak{o}^\times} N\alpha^{-2} \right\} \int_{\mathbb{R}^{2m}} f(v)\,dv = D^{1/2} c\zeta_F(2)\hat{f}(0).$$

Similarly, regarding $J(\hat{f})$, letting $\mathfrak{o}^* = \delta\mathfrak{o}$, we have

$$J(\hat{f}) - \hat{f}(0)\,\mathrm{vol}(\Gamma\backslash G) = J(\hat{f}) - \int_{\Gamma\backslash G} \hat{f}(0)\,dg = \int_{\Gamma\backslash G} \sum_{v \in \mathfrak{o}^{*2} - 0} \hat{f}(vg)\,dg =$$

$$= \int_{U\cap\Gamma\backslash G} \sum_{\alpha \in (\mathfrak{o}-0)/\mathfrak{o}^\times} \hat{f}(\delta\alpha(0,1)g)\,dg =$$

$$= \int_{U\backslash G} \sum_{\alpha \in (\mathfrak{o}-0)/\mathfrak{o}^\times} \hat{f}(\delta\alpha(0,1)g)\,dg \times \int_{U\cap\Gamma\backslash U} 1\,du =$$

$$= D^{1/2} \int_{U\backslash G} \sum_{\alpha \in (\mathfrak{o}-0)/\mathfrak{o}^\times} \hat{f}(\delta\alpha(0,1)g)\,dg =$$

$$= D^{1/2} c \int_{\mathbb{R}^{2m}} \sum_{\alpha \in (\mathfrak{o}-0)/\mathfrak{o}^\times} \hat{f}(\delta\alpha v)\,dv =$$

$$= D^{1/2} c\,\mathrm{Norm}_{F/\mathbb{Q}}\,\delta^{-2} \left\{ \sum_{\alpha \in (\mathfrak{o}-0)/\mathfrak{o}^\times} \mathrm{Norm}_{F/\mathbb{Q}}\,\alpha^{-2} \right\} \int_{\mathbb{R}^{2m}} \hat{f}(v)\,dv =$$

$$= D^{1/2} c D^2 \zeta_F(2) f(0) = D^{5/2} c\zeta_F(2) f(0).$$

Combining these two results, we have

$$D^{1/2} c\zeta_F(2)\hat{f}(0) + f(0)\,\mathrm{vol}(\Gamma\backslash G) = D^{-1}\{D^{5/2} c\zeta_F(2) f(0) + \hat{f}(0)\,\mathrm{vol}(\Gamma\backslash G)\}$$

or

$$\mathrm{vol}(\Gamma\backslash G)\{f(0) - D^{-1}\hat{f}(0)\} = c\zeta_F(2)D^{3/2}\{f(0) - D^{-1}\hat{f}(0)\}.$$

Therefore, as we may certainly find f so that $f(0) - D^{-1}\hat{f}(0)$ is non-zero, we have

$$\mathrm{vol}(\Gamma\backslash G) = c\zeta_F(2)D^{3/2}.$$

To determine the constant c, take $f(v) = \exp(-\pi\|v\|^2)$, and consider the relation

$$\int_{U\backslash G} f((0,1)g)\,dg = c\int_V f(v)\,dv.$$

The righthand side is just c. Since $f((0,1)g)$ is right K-invariant as a function of $g \in G$, the definitions of the measures involved allow us to transform the lefthand side to

$$\int_T f((0,1)t(y))\prod_i y_i^{-2}\,dy_1\ldots dy_m =$$

$$= \int_T \exp(\pi(y_1^{-1} + \cdots y_m^{-1}))\prod_i y_i^{-2}\,dy_1\ldots dy_m =$$

$$= \int_T \exp(\pi(y_1 + \cdots y_m))\,dy_1\ldots dy_m = \pi^{-m}.$$

Thus, $c = \pi^{-m}$, and we have

$$\mathrm{vol}(\Gamma\backslash G) = \pi^{-m}\zeta_F(2)D^{3/2}. \qquad \blacksquare$$

1.12 The Petersson Inner Product

Now we are in a position to discuss some integrals which (for example) give inner products $\langle\ ,\ \rangle$ on spaces of cuspforms. Further, we will see that the integral defining $\langle f, E\rangle$ exists for a cuspform f and an Eisenstein series E, and $\langle f, E\rangle = 0$.

Definition Let f_1, f_2 be two elements of $\mathrm{Mfm}(\Gamma, k)$ for some $k \in \mathbb{Z}^m$, Γ a congruence subgroup. Then define (the weight k *Petersson inner product*)

$$\langle f_1, f_2\rangle = \int_{\Gamma\backslash\mathfrak{H}^m} f_1(z)\overline{f_2}(z)y^k y^{-2}\,dx\,dy$$

where this integral is in the sense of the previous section.

PROPOSITION If either f_1 or f_2 is a cuspform, then the integral defining $\langle f_1, f_2 \rangle$ is absolutely convergent. This inner product gives $\mathrm{Cfm}(\Gamma, k)$ a structure of (finite-dimensional) Hilbert space.

Proof First, by virtue of the transformation properties of y^k under $GL^+(2, \mathbb{R})^m$, and because the f_i are of modular forms with respect to Γ of weight k, the expression

$$\varphi(z) = f_1(z)\bar{f}_2(z)y^k$$

is a Γ-invariant function on $\mathfrak{H}^m$. Therefore, from section 1.10, we have available a definition of such an integral.

To prove the absolute convergence of the integral, let S be a standard Siegel set, and $\{\delta_i\}$ a finite set of elements of $GL^+(2, F)$ so that

$$\mathfrak{H}^m = \Gamma\left(\bigcup_i \delta_i S\right).$$

Let ψ be the characteristic function of S, and put

$$\Psi(z) = \sum_i \sum_{\gamma \in \Gamma} \psi(\delta_i \gamma z).$$

Then $\Psi \geq 1$ on $\mathfrak{H}^m$. Therefore, the integral for $\langle f_1, f_2 \rangle$ is dominated by

$$\int_{\Gamma \backslash \mathfrak{H}^m} \Psi(z)|f_1(z)f_2(z)|y^k y^{-2}\,dx\,dy.$$

By general integration theory, and by the definition of the integral defining $\langle f_1, f_2 \rangle$, we have

$$\int_{\Gamma \backslash \mathfrak{H}^m} \Psi(z)|f_1(z)f_2(z)|y^k y^{-2}\,dx\,dy = \sum_i \int_{\mathfrak{H}^m} \psi(\delta_i z)|f_1(z)f_2(z)|y^k y^{-2}\,dx\,dy =$$

$$= \sum_i \int_{\mathfrak{H}^m} \psi(z)|f_1(\delta_i^{-1}z)f_2(\delta_i^{-1}z)|\,\mathrm{Im}(\delta_i^{-1}z)^k y^{-2}\,dx\,dy =$$

$$= \sum_i \int_S |(f_1|_k\delta_i^{-1})(z)(f_2|_k\delta_i^{-1})(z)|y^k y^{-2}\,dx\,dy.$$

Therefore, it suffices to prove convergence of such integrals over standard Siegel sets.

Suppose that f_1 is a cuspform. Previous general estimates on the behavior of cuspforms showed that on a standard Siegel set S we have

$$|f_1(z) \leq C\exp(-cy^{1/m})$$

for some finite positive constants C and c. On the other hand, a modular form f_2

has a convergent Fourier series on S; let c_0 be the 0-th Fourier coefficient. By the same calculus estimates we have used before, we also have

$$|f_2(z) - c_0| \leq C\exp(-cy^{1/m})$$

by adjusting C and c if necessary. Then the convergence follows by crude estimates. ∎

PROPOSITION Let $f \in \mathrm{Cfm}(\Gamma, k)$ and $E \in \mathrm{Eis}(\Gamma, k)$. Then $\langle f, E\rangle = 0$.

Proof Since $\mathrm{Eis}(\Gamma, k)$ is spanned by the Eisenstein series denoted by E_α in a previous section, it suffices to prove the assertion for $E = E_\alpha$. Further, one may readily reduce to the case that $\alpha = 1$; that is, take

$$E(z) = \sum_{\gamma \in P\cap\Gamma\backslash\Gamma} \mu(\gamma, z)^{-k}$$

where P is the group of upper triangular matrices in $GL^+(2, F)$. Also, without loss of generality we may simply take $\Gamma = \Gamma(\mathfrak{n})$; this multiplies integrals $\langle f_1, f_2\rangle$ by the uniform constant

$$\mathrm{vol}(\Gamma(\mathfrak{n})\backslash\mathfrak{H}^m) \times \mathrm{vol}(\Gamma\backslash\mathfrak{H}^m)^{-1}.$$

Let U be the units congruent to 1 modulo $\mathfrak{n}$, and let $T = (0, \infty)^m$. Then, by unwinding the integral,

$$\langle f, E\rangle = \int_{\Gamma\backslash\mathfrak{H}^m} f(z) \sum_{\gamma \in P\cap\Gamma\backslash\Gamma} \mu(\gamma, z)^{-k} y^k y^{-2}\, dx\, dy =$$

$$= \int_{\Gamma\cap P\backslash\mathfrak{H}^m} f(z) y^k y^{-2}\, dx\, dy = \int_{T/U} \left(\int_{\mathbb{R}^m/\mathfrak{n}} f(z)\, dx\right) y^k y^{-2}\, dy.$$

In the inner integral we are merely integrating exponential functions (in the Fourier expansion of f), and such an integral is zero unless it is the trivial exponential function. Since f is a cuspform, it has 0-th Fourier coefficient 0, so the inner integral is 0. ∎

1.13 Poincare Series

Definition Let Γ be a congruence subgroup containing $\Gamma(\mathfrak{n})$ for some non-zero ideal $\mathfrak{n}$. For a totally positive element ν of $\mathfrak{n}^*$, and for weight $k = (k_1, \ldots, k_m)$, define the *ν-th Poincare series of weight k with respect to* Γ to be

$$P(z; k, \nu) = \sum_{\gamma} \mu(\gamma, z)^{-k} \exp(2\pi i\, \mathrm{Tr}(\nu(\gamma z)))$$

where γ is summed over

$$\left\{\gamma \text{ of the form } \begin{pmatrix} 1 & x \\ 0 & 1 \end{pmatrix} \text{ in } \Gamma\right\}\Big\backslash\Gamma.$$

PROPOSITION If each $k_j > 2$, then this series is absolutely convergent, and uniformly so for z in compact subsets of $\mathfrak{H}^m$. In this circumstance, such a Poincare series is in $\mathrm{Cfm}(\Gamma, k)$.

Proof Let $f(z) = P(z; k, v)$, and let

$$\Gamma_\infty = \left\{\gamma \text{ of the form } \begin{pmatrix} 1 & x \\ 0 & 1 \end{pmatrix} \text{ in } \Gamma\right\}.$$

We use a general fact:

LEMMA For a sequence $\{f_i\}$ of holomorphic functions on an open subset U of $\mathbb{C}^n$, uniform pointwise convergence of $\{f_i\}$ on a compact subset C of U is equivalent to $L^1(C)$-convergence of $\{f_i\}$ (with Lebesgue measure). ∎

We claim that, in fact,

$$\int_{\Gamma\backslash\mathfrak{H}^m} y^{k/2} \sum_\gamma |\mu(\gamma, z)^{-k} \exp(2\pi i \operatorname{Tr}(v(\gamma z)))| y^{-2}\, dx\, dy < \infty.$$

This would surely imply local integrability, which would yield the desired convergence. Observe that

$$y^{k/2} \sum_\gamma |\mu(\gamma, z)^{-k} \exp(2\pi i \operatorname{Tr}(v(\gamma z)))|$$

is indeed Γ-invariant. Then

$$\int_{\Gamma\backslash\mathfrak{H}^m} y^{k/2} \sum_\gamma |\mu(\gamma, z)^{-k} \exp(2\pi i \operatorname{Tr}(v(\gamma z)))| y^{-2}\, dx\, dy = \int_{\Gamma_\infty\backslash\mathfrak{H}^m} y^{k/2} |\exp(2\pi i \operatorname{Tr}(vz))| y^{-2}\, dx\, dy$$

by general properties of homogeneous space integration, noting that the integrand is non-negative, so that a generalization of Fubini's theorem can be applied without knowing absolute convergence. This is

$$\int_{\Gamma_\infty\backslash\mathfrak{H}^m} y^{k/2} |\exp(-2\pi \operatorname{Tr} vy)| y^{-2}\, dx\, dy = \int_{\mathbb{R}^m/\Lambda \times (\mathbb{R}^+)^m} y^{k/2} |\exp(-2\pi \operatorname{Tr} vy)| y^{-2}\, dx\, dy =$$

$$= \int_{\mathbb{R}^m/\Lambda} 1\, dx \times \int_{(0,\infty)^m} y^{k/2} |\exp(-2\pi \operatorname{Tr} vy)| y^{-2}\, dy$$

where

$$\Lambda = \left\{ x \in \mathbb{R}^m : \begin{pmatrix} 1 & x \\ 0 & 1 \end{pmatrix} \in \Gamma_\infty \right\}.$$

Since Γ is a congruence subgroup, the integral in x is over a compact set, so this integral is dominated by

$$\prod_{j=1,\ldots,m} \int_{(0,\infty)} y_j^{k_j/2} \exp(-2\pi \operatorname{Tr} v_j y_j) y_j^{-2}\, dy_j$$

where we write v_j for the j-th imbedded image of v. The j-th integral is absolutely convergent if and only if $\frac{1}{2}k_j - 1 > 0$, i.e., if and only if $k_j > 2$. This proves the asserted convergence.

Assuming convergence, it is purely formal that these Poincare series are in $\mathrm{Mfm}(\Gamma, k)$. What remains to be shown is that the Poincare series are cuspforms. Applying a criterion we have used before in discussion of Eisenstein series, we need show only that

$$\lim_{\lambda \to \infty} (f|_k g)(i\lambda) = 0$$

for every $g \in GL^+(2, F)$. From the definition,

$$(f|_k g)(i\lambda) = \sum_\gamma \mu(\gamma, g(i\lambda))^{-k} \exp(2\pi i \operatorname{Tr}(v(\gamma g(i\lambda))))\mu(g, i\lambda)^{-k} =$$

$$= \sum_\gamma \mu(\gamma g, i\lambda)^{-k} \exp(2\pi i \operatorname{Tr}(v(\gamma g(i\lambda)))).$$

As $\lambda \to \infty$, $\mu(\gamma g, i\lambda)^{-k}$ goes to 0 unless γg is upper triangular. In case γg is upper triangular, the exponential term goes to 0 as $\lambda \to \infty$. In either case the decay is monotone, so by the Monotone Convergence Theorem the sum goes to 0. That is, the Poincare series f is indeed a cuspform. ∎

PROPOSITION Let

$$f(z) = \sum_\xi c_\xi \exp(2\pi i \operatorname{Tr}(\xi z))$$

be in $\mathrm{Cfm}(\Gamma, k)$, with every $k_j > 2$, and let $P(z; k, v)$ the v-th Poincare series of weight k for Γ. Then

$$\langle f, P(\ ; k, v)\rangle = c_v v^{1-k} \times \operatorname{vol}(\Lambda \backslash \mathbb{R}^m) \prod_{j=1,\ldots,m} (4\pi)^{1-k_j} \Gamma(k_j - 1)$$

using a multi-index notation, where

$$\Lambda = \left\{ x \in \mathbb{R}^m : \begin{pmatrix} 1 & x \\ 0 & 1 \end{pmatrix} \in \Gamma \right\}$$

and where we write v_j for the j-th component of v in $\mathbb{R}^m$.

Proof This follows from "unwinding" the integral for the Petersson inner product:

$$\langle f, P(\ ;k,v)\rangle = \int_{\Gamma\backslash\mathfrak{H}^m} f(z)P(z;k,v)^- y^k y^{-2}\,dx\,dy =$$

$$= \int_{\Gamma_\infty\backslash\mathfrak{H}^m} \sum_\xi c_\xi \exp(2\pi i\,\mathrm{Tr}(\xi z))\exp(2\pi i\,\mathrm{Tr}(vz))y^k y^{-2}\,dx\,dy.$$

Integrating first in x, the orthogonality of distinct exponential functions gives

$$\int_{\Gamma_\infty\backslash\mathfrak{H}^m} \sum_\xi c_\xi \exp(2\pi i\,\mathrm{Tr}(\xi z))\exp(2\pi i\,\mathrm{Tr}(vz))^- y^k y^{-2}\,dx\,dy =$$

$$= c_v \operatorname{vol}(\Lambda\backslash\mathbb{R}^m) \prod_{j=1,\ldots,m} \int_{(0,\infty)} y_j^{k_j-1} e^{-4\pi v_j y_j}\,dy_j/y_j =$$

$$= c_v \operatorname{vol}(\Lambda\backslash\mathbb{R}^m) \prod_{j=1,\ldots,m} (4\pi v_j)^{1-k_j}\Gamma(k_j-1). \qquad \blacksquare$$

COROLLARY The Poincare series span spaces of cuspforms.

Proof If $\langle f, P\rangle = 0$ for a cuspform f, for every Poincare series P, then every Fourier coefficient of f is 0, so f is 0. That is, the orthogonal complement of the space spanned by the Poincare series is just $\{0\}$. ■

Remarks There is a certain error possible in considering the integral for $\langle E, P_v\rangle$, where E is in $\mathrm{Eis}(\Gamma, k)$ and P_v is the v-th Poincare series. In the previous section, we claimed to prove that this must be 0, since P is a cuspform. On the other hand, the discussion just above might lead one to believe that this integral essentially computes the v-th Fourier coefficient. Note well that it is not true that the only non-zero Fourier coefficient of an Eisenstein series is the 0-th one. The reconciliation of this seeming paradox is that *the "unwinding" in the proof that integrals of (cuspforms) f against Poincare series compute Fourier coefficients is not justifiable unless the modular form f is a cuspform.* Indeed, termwise the series defining the Poincare series is *not* very rapidly decreasing in Siegel sets, though the function is so: There is evidently a great deal of cancellation. This cancellation is disrupted during such an unwinding. What is needed to justify such an unwinding is that the other modular form occurring in the integral be of rapid decay (e.g., be a cuspform). In contrast, Eisenstein series have (typically) a finite non-zero limit at $i\infty$, and the y^k occurring in the definition of the integral therefore disallows an unwinding of the Poincare series.

This item may be one of the more important examples of the non-triviality of Fubini-type theorems. The compactness of simpler (e.g., abelian) sorts of quotients (e.g., $\mathbb{R}^n/\mathbb{Z}^n$) prevents construction of analogous examples in such situations.

1.14 A Reproducing Kernel for Cuspforms

Definition Let Γ be a congruence subgroup. For $k = (k_1, \ldots, k_m)$, and for z, $w \in \mathfrak{H}^m$, define

$$Q(z, w) = \sum_{\gamma \in \Gamma} \mu(\gamma, z)^{-k}(\gamma z + w)^{-k}.$$

THEOREM If each $k_j > 2$, this series is absolutely convergent, and uniformly so for (z, w) in compact subsets of $\mathfrak{H}^m \times \mathfrak{H}^m$. As a function of z, $Q(z, w) \in \mathrm{Cfm}(\Gamma, k)$ for each fixed $w \in \mathfrak{H}^m$, and is a cuspform in w, with respect to the congruence subgroup $\Gamma^{\natural}$, where

$$\begin{pmatrix} a & b \\ c & d \end{pmatrix}^{\natural} = \begin{pmatrix} d & b \\ c & a \end{pmatrix}.$$

Further, for $f \in \mathrm{Cfm}(\Gamma, k)$,

$$\langle f, Q(*, -\bar{w})\rangle = \prod_{j=1,\ldots,m} \{(2\pi i)(-4i)^{1-k_j}\Gamma(k_j - 1)\} \times f(w).$$

Proof Again we use the idea that L^1 convergence (on compacta) of a sequence of holomorphic functions implies uniform pointwise convergence (on compacta). Note that

$$y^{k/2} \sum_{\gamma \in \Gamma} |\mu(\gamma, z)^{-k}(\gamma z + w)^{-k}|$$

is Γ-invariant. It suffices to show that

$$\int_{\Gamma\backslash\mathfrak{H}^m} y^{k/2} \sum_{\gamma \in \Gamma} |\mu(\gamma, z)^{-k}(\gamma z + w)^{-k}| y^{-2}\, dx\, dy < \infty.$$

By unwinding this integral (using properties of integration on homogeneous spaces), it becomes

$$\int_{\mathfrak{H}^m} y^{k/2}|(z + w)^{-k}| y^{-2}\, dx\, dy.$$

Since everything is non-negative, this unwinding is justified. This integral is a product of the m integrals

$$\int_{\mathfrak{H}} y^{k_j/2}|(z + w_j)^{-k_j}| y^{-2}\, dx\, dy.$$

Let $w_j = u + iv \in \mathfrak{H}$. Replacing x by $x - u$, and then replacing x by $x(y + v)$, this integral is

$$\int_{\mathbb{R}} (x^2 + 1)^{-k_j/2}\, dx \times \int_{(0,\infty)} y^{k_j/2}(y + v)^{1-k_j/2} y^{-2}\, dy.$$

Elementary and well-known estimates give the convergence of these integrals for $k_j > 2$.

For the reproducing property, a similar unwinding gives

$$\langle f, Q(\ , -\overline{w})\rangle = \int_{\mathfrak{H}^m} f(z) y^k (\overline{z} - w)^{-k} y^{-2}\, dx\, dy.$$

One may justify integrating termwise in the Fourier expansion of f; the effect on an exponential is

$$\exp(2\pi i \operatorname{Tr} \xi z) \to \int_{\mathfrak{H}^m} \exp(2\pi i \operatorname{Tr} \xi z) y^k (\overline{z} - w)^{-k} y^{-2}\, dx\, dy,$$

which factors into m integrals

$$\int_{\mathfrak{H}} \exp(2\pi i \xi_j z) y^k (x - iy - w_j)^{-k} y^{-2}\, dx\, dy.$$

Integrating first in $x \in \mathbb{R}$ by residues, this becomes

$$2\pi i \int_{(0,\infty)} (2\pi i \xi_j)^{k_j - 1} \exp(2\pi i \xi_j (2iy + w_j)) y^{k-1}\, dy/y =$$

$$= (2\pi i)(2\pi i \xi_j)^{k_j-1} \exp(2\pi i \xi_j w_j) \int_{(0,\infty)} \exp(-4\pi \xi_j y) y^{k_j - 1}\, dy/y =$$

$$= (2\pi i)(2\pi i \xi_j)^{k_j-1} \exp(2\pi i \xi_j w_j)(4\pi \xi_j)^{1-k_j}\Gamma(k_j - 1) =$$

$$= (2\pi i)(-4i)^{1-k_j}\Gamma(k_j - 1) \exp(2\pi i \xi_j w_j).$$

This proves the reproducing property.

To show that $Q(z, w)$ is a cuspform in z, rewrite $Q(z, w)$ as

$$Q(z, w) = \sum_{\gamma \in \Gamma_\infty \backslash \Gamma} \sum_{\delta \in \Gamma_\infty} \mu(\delta\gamma, z)^{-k} (\delta\gamma z + w)^{-k}$$

where

$$\Gamma_\infty = \left\{ \begin{pmatrix} 1 & x \\ 0 & 1 \end{pmatrix} : x \text{ in a lattice } \Lambda \text{ in } F \subset \mathbb{R}^m \text{ depending on } \Gamma \right\}$$

is the subgroup of Γ consisting of upper triangular unipotent matrices. As

$$\mu(\delta\gamma, z) = \mu(\delta, \gamma z)\mu(\gamma, z) = \mu(\gamma, z),$$

this is

$$Q(z, w) = \sum_{\gamma \in \Gamma_\infty \backslash \Gamma} \mu(\gamma, z)^{-k} \sum_{\delta \in \Gamma_\infty} (\delta\gamma z + w)^{-k} =$$

$$= \sum_{\gamma \in \Gamma_\infty \backslash \Gamma} \mu(\gamma, z)^{-k} \sum_{\lambda \in \Lambda} (\gamma z + w + \lambda)^{-k} =$$

$$= \sum_{\gamma \in \Gamma_\infty \backslash \Gamma} \mu(\gamma, z)^{-k} \sum_{\xi} \prod_{j} \{(-2\pi i)^{k_j - 1} \Gamma(k_j - 1)\} \xi^{k-1} \exp(2\pi i \operatorname{Tr} \xi(\gamma z + w))$$

by residues, where ξ runs over totally positive elements of the dual lattice to Λ with respect to the trace (as in computation of Fourier coefficients of Eisenstein series). Taking out the uniform constant, this is

$$\sum_{\gamma \in \Gamma_\infty \backslash \Gamma} \mu(\gamma, z)^{-k} \sum_{\xi} \xi^{k-1} \exp(2\pi i \operatorname{Tr} \xi(\gamma z + w)) =$$

$$= \sum_{\xi} \xi^{k-1} \exp(2\pi i \operatorname{Tr} \xi w) \left\{ \sum_{\gamma \in \Gamma_\infty \backslash \Gamma} \mu(\gamma, z)^{-k} \exp(2\pi i \operatorname{Tr} \xi(\gamma z)) \right\} =$$

$$= \sum_{\xi} \xi^{k-1} \exp(2\pi i \operatorname{Tr} \xi w) P(z; k, \xi)$$

with the ξ-th Poincare series in z appearing. Thus, as we already are assured of convergence, $Q(z, w)$ is a linear combination of cuspforms, so is a cuspform (in z).

On the other hand, for $\gamma = \left(\begin{smallmatrix} a & b \\ c & d \end{smallmatrix}\right)$,

$$\mu(\gamma z)(\gamma z + w) = (cz + d)[(az + b)(cz + d)^{-1} + w] =$$

$$= (dw + b) + (cw + a)z = (cw + a)^{-1}[z + (dw + b)(cw + a)^{-1}] =$$

$$= \mu(\gamma^{\natural}, z)(z + \gamma^{\natural} w).$$

Therefore, by an argument symmetrical to the above, $Q(z, w)$ is a cuspform in w, with respect to $\Gamma^{\natural}$. ■

1.15 Hecke Operators in a Special Case

For this section, we require that $\mathfrak{o}$ be a principal ideal domain and that the totally positive units are squares of units. The latter condition implies, for example, that every ideal has a totally positive generator. Also, we only consider $\Gamma = SL(2, \mathfrak{o})$. In this situation we will be able to give a very simple classical treatment of Hecke operators, and obtain the standard arithmetic corollaries on the Fourier coefficients of cuspforms and on the Euler product factorization of associated Dirichlet series.

To consider any other case than this it is nearly imperative to use an adelic formulation, as will be done later.

Definition Let $\mathfrak{n} = \eta\mathfrak{o}$ be an ideal in $\mathfrak{o}$, with η totally positive. Let

$$\Delta(\mathfrak{n}) = \left\{\delta = \begin{pmatrix} a & b \\ c & d \end{pmatrix} \in GL(2, F)\text{: } a, b, c, d \in \mathfrak{o}, \eta^{-1} \det \delta = \text{totally positive unit}\right\}.$$

Let $Z(\mathfrak{o})$ be the center of $GL^+(2, \mathfrak{o})$. Define an operator $T_\mathfrak{n}$ on $\mathrm{Mfm}(\Gamma, k)$ by

$$(T_\mathfrak{n} f)(z) = \sum_{\delta \in \Gamma Z(\mathfrak{o}) \backslash \Delta(\mathfrak{n})} (f|_k \delta)(z).$$

This is the $\mathfrak{n}$-th Hecke operator. This is well-defined (if convergent) since for $\gamma \in \Gamma$ and $\delta \in \Delta(\mathfrak{n})$ we have

$$f|_k \gamma\delta = (f|_k \gamma)|_k \delta = f|_k \delta.$$

PROPOSITION The Hecke operators map $\mathrm{Mfm}(\Gamma, k)$ to $\mathrm{Mfm}(\Gamma, k)$ and $\mathrm{Cfm}(\Gamma, k)$ to $\mathrm{Cfm}(\Gamma, k)$. In particular, for

$$f(z) = \sum_{\xi \in \mathfrak{o}^*} c(\xi) \exp(2\pi i \operatorname{Tr} \xi z)$$

in $\mathrm{Mfm}(\Gamma, k)$ (where $\mathfrak{o}^*$ is the inverse different), we have

$$(T_\mathfrak{n} f)(z) = \sum_{\xi} c'(\xi) \exp(2\pi i \operatorname{Tr} \xi z)$$

where

$$c'(\xi) = \eta^{1-k/2} \sum_{d} d^{k-1} c(\xi\eta/d^2).$$

where d runs over divisors of η modulo $\mathfrak{o}^\times$ so that $\xi/d \in \mathfrak{o}^*$.

Proof Let $\mathfrak{n} = \eta\mathfrak{o}$ with η totally positive. Since $\mathfrak{o}$ is a principal ideal domain, for $\delta \in \Delta(\mathfrak{n})$ there are $\gamma \in \Gamma$ and $\zeta \in Z(\mathfrak{o})$ so that

$$\zeta\gamma\delta = \begin{pmatrix} a & b \\ 0 & d \end{pmatrix}$$

where $d \in \mathfrak{o}/\mathfrak{o}^\times$, $ad\mathfrak{o} = \eta\mathfrak{o}$, $b \in \mathfrak{o}$ modulo d, $a \in \mathfrak{o}/(\mathfrak{o}^\times)^2$, and da is totally positive. That is, we may choose (finitely many) representatives for $\Gamma Z(\mathfrak{o}) \backslash \Delta(\mathfrak{n})$ of this form; this has the virtue that we can compute the effect of the $\mathfrak{n}$-th Hecke operator termwise on the Fourier expansion.

Since the sum in the definition of the Hecke operator is finite, it is formal to see that $T_{\mathfrak{n}}$ maps $\mathrm{Wfm}(\Gamma, k)$ to $\mathrm{Wfm}(\Gamma, k)$; for $\gamma \in \Gamma$ we have

$$(T_{\mathfrak{n}}f)|_k\gamma = \sum_{\delta \in \Gamma\backslash\Delta(\mathfrak{n})} (f|_k\delta)|_k\gamma = \sum_{\delta \in \Gamma\backslash\Delta(\mathfrak{n})} f|_k\delta\gamma = \sum_{\delta \in \Gamma\backslash\Delta(\mathfrak{n})} f|_k\delta$$

since $\Gamma\backslash\Delta(\mathfrak{n})$ is stable under right multiplication by Γ. On the other hand, each translate $f|_k\delta$ is a modular form for some congruence subgroup; therefore, $f \in \mathrm{Mfm}(\Gamma, k)$.

Termwise in the Fourier expansion, we have

$$\exp(2\pi i \operatorname{Tr} \xi z)|_k\begin{pmatrix} a & b \\ 0 & d \end{pmatrix} = \eta^{k/2}d^{-k}\exp(2\pi i \operatorname{Tr} \xi az/d)\exp(2\pi i \operatorname{Tr} \xi b/d).$$

The sum over b modulo d is 0 unless ξ/d is in $\mathfrak{o}^*$, in which case it is the ideal norm Nd. Therefore,

$$(T_{\mathfrak{n}}f)(z) = \sum_{\xi} c(\xi)\eta^{k/2} \sum_{a,d} Nd\, d^{-k}\exp(2\pi i \operatorname{Tr} \xi az/d)$$

where d runs over divisors of η modulo $\mathfrak{o}^\times$, Nd is the ideal norm of $d\mathfrak{o}$, $a \in \mathfrak{o}/(\mathfrak{o}^\times)^2$, ad is totally positive, $\xi/d \in \mathfrak{o}^*$, and $ad\mathfrak{o} = \eta\mathfrak{o}$. Our hypothesis on units implies that this is the same as

$$(T_{\mathfrak{n}}f)(z) = \sum_{\xi} c(\xi)\eta^{k/2} \sum_{d} d^{1-k}\exp(2\pi i \operatorname{Tr} \xi\eta z/d^2)$$

where d runs over totally positive divisors of η modulo $\mathfrak{o}^\times$ so that $\xi/d \in \mathfrak{o}^*$. Replacing d by η/d, this gives the ξ'-th Fourier coefficient of $T_{\mathfrak{n}}f$ as

$$c'(\xi') = \eta^{1-k/2} \sum_{d} d^{k-1}c(\xi'\eta/d^2)$$

where d runs over divisors of η modulo $\mathfrak{o}^\times$ so that $\xi'/d \in \mathfrak{o}^*$.

Since the class number of $\mathfrak{o}$ is 1, there is only one cusp, and the determination of whether or not a weak modular form is a cuspform depends only on the 0-th Fourier coefficient. The previous formula shows that $c'(0) = 0$ if $c(0) = 0$. ■

PROPOSITION The effect of $T_{\mathfrak{n}}$ on the ν-th Poincare series is described by

$$T_{\mathfrak{n}}P(*; k, \nu) = \eta^{k/2} \sum_{d} d^{1-k}P(z; k, \nu\eta/d^2)$$

where the sum is over totally positive divisors d of η modulo $\mathfrak{o}^\times$ so that $\nu/d \in \mathfrak{o}^*$.

Proof Let Γ_∞ be the upper triangular unipotent matrices in Γ, and let X be a collection of upper triangular representatives for $\Gamma\backslash\Delta(\mathfrak{n})$ as above. From the definitions,

$$T_\mathfrak{n}P(*;k,v) = \sum_{\delta\in\Gamma Z(\mathfrak{o})\backslash\Delta(\mathfrak{n})}\ \sum_{\gamma\in\Gamma_\infty\backslash\Gamma}(\exp(2\pi i\operatorname{Tr} vz))|_k\gamma|_k\delta =$$

$$= \sum_{\delta\in\Gamma Z(\mathfrak{o})\backslash\Delta(\mathfrak{n})}\ \sum_{\gamma\in\Gamma_\infty\backslash\Gamma}(\exp(2\pi i\operatorname{Tr} vz))|_k(\gamma\delta) =$$

$$= \sum_{\delta\in\Gamma_\infty Z(\mathfrak{o})\backslash\Delta(\mathfrak{n})}(\exp(2\pi i\operatorname{Tr} vz))|_k\delta.$$

Fix a set X of representatives for $\Gamma Z(\mathfrak{o})\backslash\Delta(\mathfrak{n})$ of the form $\left(\begin{smallmatrix} a & b \\ 0 & d\end{smallmatrix}\right)$ with d lying in a fixed set of representatives for $\mathfrak{o}/\mathfrak{o}^\times$. Fix a set of representatives Y for $\Gamma_\infty\backslash\Gamma$. We claim that

$$\{\delta\gamma\colon \gamma\in Y \text{ and } \delta\in X\}$$

is also a set of representatives for $\Gamma_\infty Z(\mathfrak{o})\backslash\Delta(\mathfrak{n})$. Given $\delta\in\Delta(\mathfrak{n})$, there is $\gamma\in\Gamma$ so that $\delta\gamma^{-1}$ is of the form $\left(\begin{smallmatrix} a & b \\ 0 & d\end{smallmatrix}\right)$ with d in a given set of representatives for $\mathfrak{o}/\mathfrak{o}^\times$; it is elementary that this condition determines the bottom row of γ uniquely, since $\gamma\in SL(2,\mathfrak{o})$ (i.e., the bottom two entries of γ must be relatively prime). But a coset in $\Gamma_\infty\backslash\Gamma$ is determined precisely by the bottom row of one of its representatives. Allowing the further adjustment by $Z(\mathfrak{o})$ gives the desired conclusion.

Therefore, we have

$$T_\mathfrak{n}P(*;k,v) = \sum_{\gamma\in\Gamma_\infty\backslash\Gamma}\ \sum_{\delta\in X}(\exp(2\pi i\operatorname{Tr} vz))|_k\delta|_k\gamma =$$

$$= \sum_{\gamma\in\Gamma_\infty\backslash\Gamma}\ \sum_{\delta}\eta^{k/2}d^{-k}\exp(2\pi i\operatorname{Tr} vb/d)(\exp(2\pi i\operatorname{Tr} vaz/d))|_k\gamma$$

where for each δ we write $\delta = \left(\begin{smallmatrix} a & b \\ 0 & d\end{smallmatrix}\right)$. Now b is summed over $\mathfrak{o}$ modulo d, so the sum of the terms $\exp(2\pi i\operatorname{Tr} vb/d)$ is 0 unless $v/d\in\mathfrak{o}^*$, in which case it is Nd. Therefore,

$$T_\mathfrak{n}P(*;k,v) = \sum_{\gamma\in\Gamma_\infty\backslash\Gamma}\ \sum_{\delta \text{ with } v/d\in\mathfrak{o}^*}\eta^{k/2}Nd\,d^{-k}(\exp(2\pi i\operatorname{Tr} vaz/d))|_k\gamma =$$

$$= \eta^{k/2}\sum_{\delta \text{ with } v/d\in\mathfrak{o}^*} Nd\,d^{-k}P(z;k,va/d).$$

The hypothesis on units allows us to take $a=\eta/d$ and d totally positive, yielding the desired result. ■

The following corollary is a weak version of a much more general result; in particular, the hypotheses on the weight are completely unnecessary, being required only to give the proof in its present special form.

COROLLARY Assume that the weight is $k=(k_1,\dots,k_m)$ with each $k_j>2$. Then the Hecke operators are self-adjoint with respect to the Petersson inner product.

Proof Since we have seen that Poincare series span spaces of cuspforms, it suffices to show that

$$\langle T_{\mathfrak{n}} f, P(*; k, \nu)\rangle = \langle f, T_{\mathfrak{n}} P(*; k, \nu)\rangle$$

for every $\nu \in \mathfrak{o}^*$. To this end we use three things: that inner products against Poincare series are Fourier coefficients (up to some explicit constants), that we know the Fourier coefficients of $T_{\mathfrak{n}} f$ in terms of those of f, and that we can express $T_{\mathfrak{n}} P(*; k, \nu)$ as a linear combination of Poincare series. Let $c(\xi)$ be the ξ-th Fourier coefficient of f, let $\mathfrak{n} = \eta\mathfrak{o}$, and let

$$\omega = \mathrm{vol}(\mathfrak{o}\backslash\mathbb{R}^m) \prod_{j=1,\ldots,m} (4\pi)^{1-k_j}\Gamma(k_j - 1).$$

We have

$$\begin{aligned}\langle T_{\mathfrak{n}} f, P(*; k, \nu)\rangle &= \omega\nu^{1-k} \times (\nu\text{-th Fourier coefficient of } T_{\mathfrak{n}} f) = \\ &= \omega\nu^{1-k}\eta^{1-k/2} \sum_d d^{k-1} c(\eta\nu/d^2)\end{aligned}$$

where d runs over divisors of η modulo $\mathfrak{o}^\times$ so that $\nu/d \in \mathfrak{o}^*$. This is

$$\begin{aligned}\eta^{k/2} \sum_d d^{1-k}(\omega(\eta\nu/d^2)^{1-k} c(\eta\nu/d^2)) &= \eta^{k/2} \sum_{a,d} d^{1-k}\langle f, P(*; k, \eta\nu/d^2)\rangle = \\ &= \left\langle f, \eta^{k/2} \sum_d d^{1-k} P(*; k, \eta\nu/d^2)\right\rangle.\end{aligned}$$

Now

$$T_{\mathfrak{n}} P(*; k, \nu) = \eta^{k/2} \sum_d d^{1-k} P(*; k, \nu\eta/d^2).$$

From this, finally,

$$\langle T_{\mathfrak{n}} f, P(*; k, \nu)\rangle = \langle f, T_{\mathfrak{n}} P(*; k, \nu)\rangle.$$

■

PROPOSITION The ring H of operators on $\mathrm{Cfm}(\Gamma, k)$ generated by the Hecke operators is commutative. In particular, $T_{\mathfrak{mn}} = T_{\mathfrak{m}} T_{\mathfrak{n}}$ for $\mathfrak{m} + \mathfrak{n} = \mathfrak{o}$, and for $\mathfrak{p}$ prime we have

$$T_{\mathfrak{p}} T_{\mathfrak{p}^m} = T_{\mathfrak{p}^{m+1}} + N\mathfrak{p}\, T_{\mathfrak{p}^{m-1}}.$$

Proof This result is a consequence of the existence of the explicit upper triangular representatives for $\Gamma\backslash\Delta(\mathfrak{n})$, which is a consequence of the hypotheses on class number and units in $\mathfrak{o}$. The commutativity is a consequence of the explicit relations. Again, if $\mathfrak{n} = \eta\mathfrak{o}$ with η totally positive, there is a set of representatives

$$X(\mathfrak{n}) = \left\{\begin{pmatrix} a & b \\ 0 & d\end{pmatrix} : d \text{ is a totally positive divisor of } \eta \text{ modulo } \mathfrak{o}^\times,\ a = \eta/d,\ b \bmod d\right\}$$

for $\Gamma\backslash\Delta(\mathfrak{n})$. It is a consequence of the Chinese Remainder Theorem that if $\mathfrak{m} + \mathfrak{n} = \mathfrak{o}$

we have

$$X(\mathfrak{m})X(\mathfrak{n}) = X(\mathfrak{mn}).$$

For the last assertion, let π be a totally positive generator of $\mathfrak{p}$; the Chinese Remainder Theorem gives

$$X(\mathfrak{p})X(\mathfrak{p}^m) = X(\mathfrak{p}^{m+1}) \cup (N\mathfrak{p} \text{ copies of } \{\pi\delta\colon \delta \in \Delta(\pi^{m-1})\}).$$

Since $\pi\delta$ has the same action on modular forms as does δ, this gives the second explicit formula of the proposition. ∎

COROLLARY Suppose that the weight is of the form $k = (k_1, \ldots, k_m)$ with each $k_j > 2$. Then there is an orthogonal basis for $\mathrm{Cfm}(\Gamma, k)$ consisting of simultaneous eigenvectors for all the Hecke operators. Let δ be a totally positive generator for the inverse different $\mathfrak{o}^*$. For a simultaneous eigenvector

$$f(z) = \sum_{\xi \in \mathfrak{o}^*} c(\xi)\exp(2\pi i \operatorname{Tr} \xi z)$$

so that $T_\mathfrak{n} f = \lambda(\mathfrak{n}) f$, let η be a totally positive generator for $\mathfrak{n}$; then

$$\lambda(\mathfrak{n})c(\delta) = \eta^{1-k/2}c(\delta\eta).$$

Proof The first assertion follows from the previous results by finite-dimensional linear algebra: A finite-dimensional Hilbert space with a commutative ring of self-adjoint operators has an orthogonal basis consisting of simultaneous eigenvectors. We saw above that the effect of $T_\mathfrak{n}$ on Fourier coefficients is given by

$$\xi\text{-th Fourier coefficient of } T_\mathfrak{n} f = \eta^{1-k/2} \sum_d d^{k-1} c(\xi\eta/d^2)$$

where d runs over divisors of η modulo $\mathfrak{o}^\times$ so that $\xi/d \in \mathfrak{o}^*$. Taking $\xi = \delta$ gives

$$\delta\text{-th Fourier coefficient of } T_\mathfrak{n} f = \eta^{1-k/2}c(\delta\eta).$$

Therefore,

$$\lambda(\mathfrak{n})c(\delta) = \eta^{1-k/2}c(\delta\eta).$$

∎

COROLLARY Let $f \in \mathrm{Cfm}(\Gamma, k)$ be an eigenfunction for the Hecke operators normalized so that $c(\delta) = 1$, where δ is a totally positive generator for $\mathfrak{o}^*$. For α and β totally positive elements of $\mathfrak{o}$ so that $\mathfrak{o} = \alpha\mathfrak{o} + \beta\mathfrak{o}$,

$$c(\alpha\beta\delta) = c(\alpha\delta)c(\beta\delta).$$

If π is a totally positive prime element of $\mathfrak{o}$, then

$$c(\pi\delta)c(\delta\pi^m) = c(\delta\pi^{m+1}) + \pi^{k-1}c(\delta\pi^{m-1}).$$

Therefore, the Dirichlet series associated to f and to an absolutely unramified finite-order Hecke character χ has an Euler product factorization

$$\sum_{\alpha} \chi(\alpha)c(\alpha\delta)\alpha^{-(s+k/2-1/2)} = \prod_{\mathfrak{p}} (1 - \chi(\pi)c(\pi\delta)\pi^{-s-1/2k+1/2} + \chi(\pi)^2\pi^{-2s})^{-1}$$

where the sum is over totally positive elements α of $\mathfrak{o}$ modulo $\mathfrak{o}^\times$, and the product is over primes of $\mathfrak{p} = \pi\mathfrak{o}$ of $\mathfrak{o}$ where π is a totally positive generator.

Proof The first two assertions are combinations of the last two corollaries. The third assertion follows by (for example) some elementary linear algebra from the first two, noting again that $c(\alpha\delta)(\alpha\delta)^{-k/2}$ is well-defined for α modulo totally positive units. ∎

2 Automorphic Forms on $GL(2, \mathbb{A})$

The purpose of this chapter is to develop a formalism and appropriate supporting results regarding automorphic forms on $GL(2, \mathbb{A})$. We will see in the next chapter that holomorphic Hilbert modular forms and their accoutrements may be turned into objects and structures on $GL(2, \mathbb{A})$ whose nature is thereby made more accessible; therefore, we develop the adele-group analogues of classical things in as much generality and formality as possible. Though we do not discuss such matters here, the adelic notion of automorphic form includes not only holomorphic modular forms, but many other types of automorphic forms as well.

The first section is quite elementary, giving some notation and some elementary results on the structure of $GL(2)$ over local fields and over the adeles. The second section returns to the issue of computing the volume of quotients, though now we consider $GL(2, F)\backslash GL(2, \mathbb{A})$, which is not at all a manifold and has no convenient geometric picture. The point is that such volume computations do not require these things.

The third section defines the spherical Hecke algebras at finite primes as right-convolution algebras. The main result is the quite formal and straightforward fact that these convolution algebras are commutative; the proof is given in a manner which illustrates that such a result holds in great generality. The fourth secton is an analogous discussion for infinite primes; here we consider invariant differential operators rather than convolution operators, though there is a close connection between the two. The determination of some such operators in coordinates is in Appendix A.4, since the precise form is of no particular moment in the general definitions.

The section on adelic Fourier expansions starts to illustrate some other increased simplicity of the adelic version. In some sense there is only one Fourier coefficient of an adelic cuspform. In this section we also consider some special Hecke operators at "bad" primes, i.e., those "dividing the level." Then we show that the Hecke eigenvalues essentially determine the Fourier coefficients.

Next, we combine these elements into a very general definition of automorphic form. In this context, via an adelic Mellin transform, for pleasing formal reasons, cuspforms (eigenfunctions of Hecke operators) give rise to L-functions with Euler products, with Euler factors of the expected form. This "trivializes" the classical theory in the same way that an adelic approach "trivializes" treatment of zeta and L-functions of number fields. (We neglect issues of convergence, as these matters reduce to the classical ones for automorphic forms arising from holomorphic Hilbert modular forms.)

2.1 Structure of $GL(2, \mathbb{A})$

For this chapter, let F be *any* number field, with ring of integers $\mathfrak{o}$, adeles $\mathbb{A}$, ideles $\mathbb{J}$, etc. For any commutative $\mathfrak{o}$-algebra R, let

$$GL(2, R) = \{2 \times 2 \text{ matrices over } R \text{ with determinant in } R^{\times}\}$$

$$P(R) = \{\text{upper triangular matrices in } GL(2, R)\}$$

$$U(R) = \left\{u(x) = \begin{pmatrix} 1 & x \\ 0 & 1 \end{pmatrix} : x \in R\right\} \qquad T(R) = \left\{\ell(y) = \begin{pmatrix} y & 0 \\ 0 & 1 \end{pmatrix} : y \in R^{\times}\right\}$$

$$Z(R) = \text{center of } GL(2, R) = \left\{\begin{pmatrix} t & 0 \\ 0 & t \end{pmatrix} : t \in R^{\times}\right\}.$$

Thus, $GL(2, \mathbb{A})$ makes sense as an abstract group; we note that, as a group, $\mathbb{A}^{\times} = \mathbb{J}$, so that

$$GL(2, \mathbb{A}) = \{2 \times 2 \text{ matrices over } \mathbb{A} \text{ with determinant in } \mathbb{J}\}.$$

For a finite prime $\mathfrak{p}$ of F, let

$$K_{\mathfrak{p}} = GL(2, \mathfrak{o}_{\mathfrak{p}}).$$

If $\mathfrak{p}$ is a real prime of F, identify $GL(2, F_{\mathfrak{p}}) \approx GL(2, \mathbb{R})$, and let

$$K_{\mathfrak{p}} = SO(2).$$

If $\mathfrak{p}$ is a complex prime, identify $GL(2, F_{\mathfrak{p}}) \approx GL(2, \mathbb{C})$, and let

$$K_{\mathfrak{p}} = SU(2).$$

It is easy to check that

$$GL(2, \mathbb{A}) = \left\{(\ldots, g_{\mathfrak{p}}, \ldots) \in \prod_{\mathfrak{p}} GL(2, F_{\mathfrak{p}}) : \text{for almost all finite primes } \mathfrak{p},\ g_{\mathfrak{p}} \in K_{\mathfrak{p}}\right\}.$$

We give $GL(2, \mathbb{A})$ the topology with basis at 1

$$\left\{U = \prod_{\mathfrak{p}} U_{\mathfrak{p}} \colon U_{\mathfrak{p}} \subset GL(2, F_{\mathfrak{p}}) \text{ is open, and, for almost all finite primes } \mathfrak{p},\ U_{\mathfrak{p}} = K_{\mathfrak{p}}\right\}.$$

It is not hard to verify that, for every finite or infinite prime, we have an *Iwasawa decomposition*

$$GL(2, F_{\mathfrak{p}}) = P(F_{\mathfrak{p}})K_{\mathfrak{p}} = U(F_{\mathfrak{p}})(Z(F_{\mathfrak{p}})T(F_{\mathfrak{p}}))K_{\mathfrak{p}}.$$

In the case that $\mathfrak{p}$ is infinite, this is exactly the "classical" Iwasawa decomposition, provable by elementary linear algebra over $\mathbb{R}$ or $\mathbb{C}$. When $\mathfrak{p}$ is finite, this decomposition follows from the fact that $\mathfrak{o}_{\mathfrak{p}}$ is a principal ideal domain. There is also the *Cartan decomposition*

$$GL(2, F_{\mathfrak{p}}) = K_{\mathfrak{p}}(Z(F_{\mathfrak{p}})T(F_{\mathfrak{p}}))K_{\mathfrak{p}}.$$

For $\mathfrak{p}$ infinite this assertion may be strengthened by replacing $Z(F_{\mathfrak{p}})T(F_{\mathfrak{p}})$ by the group of *diagonal matrices with positive real entries*; proof of this assertion may be reduced to the spectral theorem for finite-dimensional self-adjoint operators, for $\mathfrak{p}$ infinite. For $\mathfrak{p}$ finite, this is essentially equivalent to the structure theory of finitely generated torsion $\mathfrak{o}_{\mathfrak{p}}$-modules.

2.2 A Volume Computation

We now give the adelic version of the volume computation of 1.11, still restricting our attention to totally real number fields. The point is that now class numbers and units receive no special attention. This method applies equally well to arbitrary number fields.

THEOREM With measures normalized as below,

$$\text{volume}(SL(2, F)\backslash SL(2, \mathbb{A})) = D^{3/2}\pi^{-m}\zeta_F(2),$$

where D is the discriminant of $\mathfrak{o}$ over $\mathbb{Z}$.

Proof Throughout the proof, we will use conventions and notation regarding "standard" measures and characters as in Appendix A.2; see that same appendix concerning the adelic form of the Poisson summation formula, which we also use.

Using the notational conventions above, for a prime $\mathfrak{p}$ of F let

$$G_{\mathfrak{p}} = SL(2, F_{\mathfrak{p}}), \qquad U_{\mathfrak{p}} = U(F_{\mathfrak{p}}) \cap G_{\mathfrak{p}}, \qquad A_{\mathfrak{p}} = Z(F_{\mathfrak{p}})T(F_{\mathfrak{p}}) \cap G_{\mathfrak{p}}$$

and write $P_{\mathfrak{p}} = U_{\mathfrak{p}}A_{\mathfrak{p}} = A_{\mathfrak{p}}U_{\mathfrak{p}}$. Give $F_{\mathfrak{p}}$ and $F_{\mathfrak{p}}^{x}$ the standard Haar measures. Then for any finite or infinite prime $\mathfrak{p}$,

$$d(u(x)) = dx$$

gives a two-sided Haar measure on $U_{\mathfrak{p}}$, where dx refers to the Haar measure on $F_{\mathfrak{p}}$. The expression

$$d(t(y)) = \|y\|_{\mathfrak{p}}^{-1}\, dy$$

gives a two-sided Haar measure on $A_{\mathfrak{p}}$, where now dy is the Haar measure on $F_{\mathfrak{p}}^{x}$, and

$$t(y) = \begin{pmatrix} y & 0 \\ 0 & y^{-1} \end{pmatrix}.$$

Here $\| \ \|_{\mathfrak{p}}$ is the normalization of the norm at $\mathfrak{p}$ so that the product formula holds. Then, for any prime $\mathfrak{p}$,

$$d(u(x)t(y)) = \|y\|_{\mathfrak{p}}^{-2}\, d(u(x))\, d(t(y))$$

gives a left Haar measure on $P_{\mathfrak{p}}$. The *Haar measure on* $G = SL(2, \mathbb{A})$ is not quite the product measure of all these local measures, since there is considerable intersection of $K(\mathfrak{p})$ and $P_{\mathfrak{p}}$. Rather, we define the Haar measure implicitly by defining a Haar integral via

$$f \to [\mathfrak{o}_{\mathfrak{p}}^{*} : \mathfrak{o}_{\mathfrak{p}}]^{1/2} \int_{P_{\mathfrak{p}}} \int_{K(\mathfrak{p})} f(pk)\, dk\, dp$$

where the constant factor arises because the measure of $K(\mathfrak{p}) \cap P_{\mathfrak{p}}$ in $P_{\mathfrak{p}}$ is not 1 but $[\mathfrak{o}_{\mathfrak{p}}^{*} : \mathfrak{o}_{\mathfrak{p}}]^{-1/2}$. We let $G_{\mathfrak{p}}$ act on row vectors v in $F_{\mathfrak{p}}^{2}$ by right multiplication. Write

$$W = G \cap U(\mathbb{A}) \qquad A = G \cap Z(\mathbb{A})T(\mathbb{A}) \qquad H = \prod_{\mathfrak{p}} K(\mathfrak{p})$$

(restricted direct products)

$$\Gamma = SL(2, F).$$

The first observation is that we can identify $W \backslash G$ with $V = \mathbb{A}^2 - \{0\}$ via the map

$$Wg \to (0, 1)g.$$

Since G is unimodular and W is unimodular, there is a G-invariant (positive regular Borel) measure η on $W \backslash G \approx \mathbb{A}^2 - \{0\}$ so that for a continuous function f on G with compact support we have

$$\int_G f(g)\, dg = \int_V \int_W f(u(x)g)\, dx\, d\eta(Wg).$$

Since an additive Haar measure dv on $\mathbb{A}^2$ is another G-invariant measure, it must be that

$$d\eta(v) = c\, dv$$

for some positive constant c.

Let dv denote the standard additive (product) Haar measure on $\mathbb{A}^2$ (equivalently, on the quotient $\mathbb{A}^2/F^2$), giving $\mathbb{A}^2/F^2$ total measure 1.

Now let f be a Schwartz function on $\mathbb{A}^2$ (as in Appendix A.2), and consider

$$I(f) = \int_{\Gamma\backslash G} \sum_{v \in F^2} f(vg)\, dg$$

where dg denotes the Haar measure on G and also the invariant measure on $\Gamma\backslash G$. On one hand, by the adelic Poisson summation formula (from Appendix A.2), this is

$$I(f) = \int_{\Gamma\backslash G} \sum_{v \in F^2} \hat{f}(vg^{T-1})\, dg.$$

Since $g \to g^{T-1}$ is an automorphism of order 2 of G, it must preserve Haar measure, so we may replace g by g^{T-1} in the latter integral to find $I(f) = I(\hat{f})$. On the other hand, by integration theory on homogeneous spaces,

$$I(f) - f(0)\,\mathrm{vol}(\Gamma\backslash G) = I(f) - \int_{\Gamma\backslash G} f(0)\, dg = \int_{\Gamma\backslash G} \sum_{v \in F^2 - 0} f(vg)\, dg =$$

$$= \int_{W \cap \Gamma\backslash G} f((0, 1)g)\, dg =$$

$$= \int_{W\backslash G} f((0, 1)g)\, dg \times \int_{W \cap \Gamma\backslash W} 1\, du =$$

$$= \int_{W\backslash G} f((0, 1)g)\, dg \times \int_{\mathbb{A}/F} 1\, dx = c \int_{\mathbb{A}^2} f(v)\, dv =$$

$$= c \int_{\mathbb{A}^2} f(v)\, dv = c\hat{f}(0).$$

Combining these two results, we have

$$\mathrm{vol}(\Gamma\backslash G)\{f(0) - \hat{f}(0)\} = c\{f(0) - \hat{f}(0)\}.$$

Therefore, as we may certainly find f so that $f(0) - \hat{f}(0)$ is non-zero, we have

$$\mathrm{vol}(\Gamma\backslash G) = c.$$

To determine the constant c, take

$$f(v) = \prod_{\mathfrak{p}} f_{\mathfrak{p}}(v_{\mathfrak{p}})$$

where

$$f_{\mathfrak{p}}(v_{\mathfrak{p}}) = \exp(-\pi \|v_{\mathfrak{p}}\|_{\mathfrak{p}}^2) \qquad \text{for } \mathfrak{p} \text{ real}$$

$$f_{\mathfrak{p}}(v_{\mathfrak{p}}) = (\text{characteristic function of } \mathfrak{o}_{\mathfrak{p}}^2)(v_{\mathfrak{p}}) \qquad \text{for } \mathfrak{p} \text{ finite.}$$

Consider the relation

$$\int_{W\backslash G} f((0,1)g)\, dg = c \int_V f(v)\, dv.$$

The righthand side is cD^{-1}, where D is the discriminant of $\mathfrak{o}$ over $\mathbb{Z}$ (and D^{-1} is the volume of $\prod_{\mathfrak{p}<\infty} \mathfrak{o}_{\mathfrak{p}}^2$). Since $f((0,1)g)$ is right K-invariant as a function of $g \in G$, the definitions of the measures involved allow us to transform the lefthand side to an integral over A; it is

$$D^{1/2} \int_A f((0,1)t(y))\|y\|^{-2}\, d(t(y)) = D^{1/2} \int_{\mathbb{J}} f(0, y^{-1})\|y\|^{-2}\, d^{\times}y =$$

$$= D^{1/2} \int_{\mathbb{J}} f(0, y)\|y\|^{2}\, d^{\times}y$$

where $d^{\times}y$ denotes the product measure on the ideles $\mathbb{J}$ of F, and $\| \;\; \|$ is the idele norm. This adelic integral is the product of all the corresponding local integrals and is a *zeta integral* which can be computed. The choice of functions $f_{\mathfrak{p}}$ at infinite primes was made so that those local integrals are all π^{-1}; the finite-prime part of the integral is $\zeta_F(2)$. Therefore, we have

$$D^{-1}c = D^{1/2}\pi^{-m}\zeta_F(2)$$

from which the assertion follows. ■

2.3 The Spherical Hecke Algebra

Fix a finite prime $\mathfrak{p}$ of $\mathfrak{o}$.

Definition Define the (spherical) *Hecke algebra* at $\mathfrak{p}$ as the complex vector space

$$H = H(\mathfrak{p}) = \{\text{continuous } \mathbb{C}\text{-valued functions } \varphi \text{ on } GL(2, F_{\mathfrak{p}})\text{: } \varphi \text{ is left and right } GL(2, \mathfrak{o}_{\mathfrak{p}})\text{-invariant, } \varphi \text{ is compactly supported}\}$$

with multiplication given by (right) convolution

$$(\varphi * \psi)(g) = \int_{GL(2,F_\mathfrak{p})} \varphi(gh^{-1})\psi(h)\,dh$$

where dh is a Haar measure on $GL(2, F_\mathfrak{p})$ giving $GL(2, \mathfrak{o}_\mathfrak{p})$ measure 1. For a right $GL(2, \mathfrak{o}_\mathfrak{p})$-invariant measurable function f on $GL(2, F_\mathfrak{p})$, and for $\varphi \in H$, define an operator

$$(T_\varphi f)(g) = \int_{GL(2,F_\mathfrak{p})} f(gh^{-1})\varphi(h)\,dh.$$

Definition Let $V(\mathfrak{p}, \omega)$ be the space of $\mathbb{C}$-valued continuous functions f on $GL(2, \mathbb{A})$ so that f is left $GL(2, F)$-invariant, right $GL(2, \mathfrak{o}_\mathfrak{p})$-invariant, $f(\zeta g) = \omega(\zeta)f(g)$ for $\zeta \in Z(\mathbb{A})$ for some unitary character ω, and so that

$$\int |f(g)|^2\,dg$$

exists, where the integral is taken over $Z(\mathbb{A})GL(2, F)\backslash GL(2, \mathbb{A})$ with respect to some fixed Haar measure. (We identify two such functions if they are equal almost everywhere). This space has an inner product

$$\langle f_1, f_2 \rangle = \int f_1(g)\bar{f_2}(g)\,dg \qquad \text{(integral over } Z(\mathbb{A})GL(2, F)\backslash GL(2, \mathbb{A})).$$

THEOREM (H is indeed closed under convolution, and)

i. The algebra H, with convolution, is commutative.
ii. The operators T_φ form a commutative ring of bounded operators on $V(\mathfrak{p}, \omega)$, which is closed under adjoints with respect to $\langle\ ,\ \rangle$. In particular, the adjoint of T_φ is T_{φ^*}, where

$$\varphi^*(g) = \bar{\varphi}(g^{-1}).$$

COROLLARY Let V be a finite-dimensional subspace of $V(\mathfrak{p}, \omega)$ which is stable under the action of $H = H(\mathfrak{p})$. Then V has an orthogonal basis consisting of simultaneous eigenvectors for H. ∎

Remarks The proof of the theorem is very formal, depending only on the fact that $GL(2, F_\mathfrak{p})$ admits a Cartan decomposition; hence, the same proof works much more generally, applying to any unimodular topological group admitting a Cartan-like decomposition.

Proof of Theorem For brevity, write $G = GL(2, F_\mathfrak{p})$, $Z = Z(F_\mathfrak{p})$, $K = GL(2, \mathfrak{o}_\mathfrak{p})$, and let A be the group of diagonal matrices in $GL(2, F_\mathfrak{p})$. We will write all integrals as $\int \ldots dg$ without explicit reference to the invariant measures.

First, for $f \in V(\mathfrak{p}, \omega)$ and $\varphi \in H$ note that

$$f(\zeta g k h^{-1})\varphi(h) = \omega(\zeta) f(g h^{-1})\varphi(h)$$

for $g, h \in G$, $k = GL(2, \mathfrak{o}_\mathfrak{p})$, and $\zeta \in Z$. It is then clear that $f * \varphi$ is again in $V(\mathfrak{p}, \omega)$.

We check that H is closed under convolution. For f and $\varphi \in H$ the two-sided invariance of $f * \varphi$ is formal: For $g \in G$, $k_1, k_2 \in K$,

$$(f * \varphi)(k_1 g k_2) = \int_G f(k_1 g k_2 h^{-1})\varphi(h)\, dh = \int_G f(g k_2 h^{-1})\varphi(h)\, dh.$$

Replace h by hk_2 in the integral, to obtain

$$(f * \varphi)(g) = \int_G f(gh^{-1})\varphi(hk_2)\, dh = \int_G f(gh^{-1})\varphi(h)\, dh = (f * \varphi)(k_1 g k_2).$$

Next, let S be the support of f, and S' the support of φ. Then $(f * \varphi)(g)$ is surely 0 unless $h \in S'$ and $gh^{-1} \in S$, i.e., unless $g \in Sh \subset SS'$. Since both S and S' are compact, certainly SS' is compact. Thus, H is closed under convolution.

LEMMA Let $g \to g^\sigma$ be the transpose on G. Then, for $f \in H$ and $g \in G$, $f(g^\sigma) = f(g)$.

Proof This is formal: With $t \in A$, $k_1, k_2 \in K$,

$$f(k_1 t k_2) = f(t) = f(t^\sigma) = f(k_2^\sigma t^\sigma k_1^\sigma) = f(g^\sigma). \qquad \blacksquare$$

Now the commutativity of H is easy. By the lemma, since H is closed under convolution,

$$(f * \varphi)(g) = (f * \varphi)(g^\sigma).$$

Then

$$(f * \varphi)(g^\sigma) = \int_G f(g^\sigma h^{-1})\varphi(h)\, dh = \int_G f(g^\sigma h)\varphi(h^{-1})\, dh,$$

upon replacing h by h^{-1} in the integral; the Haar measure is invariant under the anti-automorphism $h \to h^{-1}$ of G. Then replace h by $g^{\sigma-1}h$ in the integral, to obtain

$$(f * \varphi)(g) = \int_G f(h)\varphi(h^{-1}g^\sigma)\, dh.$$

Replacing h by h^σ (noting preservation of Haar measure ...), and applying the lemma to f and φ, we have

$$(f * \varphi)(g) = \int_G f(h^\sigma)\varphi(h^{\sigma-1}g^\sigma)\, dh = \int_G f(h)\varphi(gh^{-1})\, dh = \int_G \varphi(gh^{-1})f(h)\, dh = (\varphi * f)(g).$$

It is easy to check that operators T_φ map $V(\omega, \mathfrak{p})$ to itself and are bounded operators. Next, we explicitly determine the adjoint of an operator T_φ. Let Φ and Ψ be in $V(\omega, \mathfrak{p})$. Then, letting $\langle\ ,\ \rangle$ denote the inner product in $V(\omega, \mathfrak{p})$, we have

$$\begin{aligned}\langle T_\varphi(\Phi), \Psi\rangle &= \int_{Z(\mathbb{A})GL(2,F)\backslash GL(2,\mathbb{A})} T_\varphi(\Phi)(g)\bar{\Psi}(g)\,dg = \\ &= \int_{Z(\mathbb{A})GL(2,F)\backslash GL(2,\mathbb{A})} \int_G \Phi(gh^{-1})\varphi(h)\bar{\Psi}(g)\,dh\,dg = \\ &= \int_G \int_{Z(\mathbb{A})GL(2,F)\backslash GL(2,\mathbb{A})} \Phi(gh^{-1})\varphi(h)\bar{\Psi}(g)\,dg\,dh = \\ &= \int_G \int_{Z(\mathbb{A})GL(2,F)\backslash GL(2,\mathbb{A})} \Phi(g)\varphi(h)\bar{\Psi}(gh)\,dg\,dh = \\ &= \int_{Z(\mathbb{A})GL(2,F)\backslash GL(2,\mathbb{A})} \int_G \Phi(g)\varphi(h)\bar{\Psi}(gh)\,dh\,dg\end{aligned}$$

by Fubini's theorem, and by replacing g by gh; replacing h by h^{-1} then yields

$$\begin{aligned}\langle T_\varphi(\Phi), \Psi\rangle &= \int_{Z(\mathbb{A})GL(2,F)\backslash GL(2,\mathbb{A})} \int_G \Phi(g)\varphi(h^{-1})\bar{\Psi}(gh^{-1})\,dh\,dg = \\ &= \int_{Z(\mathbb{A})GL(2,F)\backslash GL(2,\mathbb{A})} \Phi(g)\left(\int_G \bar{\varphi}(h^{-1})\Psi(gh^{-1})\,dh\right)^{-} dg = \\ &= \langle \Phi, T_{\varphi *}(\Psi)\rangle,\end{aligned}$$

where $(\ldots)^{-}$ denotes complex conjugation, and where $\varphi^*(g) = \bar{\varphi}(g^{-1})$ is certainly still in H. ■

2.4 Invariant Differential Operators

Consider a fixed infinite prime $\mathfrak{p}$. Then we have decompositions

$$GL(2, F_\mathfrak{p}) = SL(2, F_\mathfrak{p})Z(F_\mathfrak{p})\left\{1_2, \begin{pmatrix}1 & 0\\ 0 & -1\end{pmatrix}\right\} \qquad \text{for } \mathfrak{p} \approx \mathbb{R}$$

$$GL(2, F_\mathfrak{p}) = SL(2, F_\mathfrak{p})Z(F_\mathfrak{p}) \qquad \text{for } \mathfrak{p} \approx \mathbb{C}.$$

Definition Let $\mathscr{Z} = \mathscr{Z}_\mathfrak{p}$ be the algebra of left and right $SL(2, F_\mathfrak{p})$-invariant differential operators Y on $SL(2, F_\mathfrak{p})$ so that Y also commutes with conjugation by $\varepsilon = \begin{pmatrix}1 & 0\\ 0 & -1\end{pmatrix}$.

More precisely, define the left regular representation λ and the right regular representation ρ of $SL(2, F_\mathfrak{p})$ on functions f on $SL(2, F_\mathfrak{p})$ by

$$(\rho(h)f)(g) = f(gh) \qquad (\lambda(h)f)(g) = f(h^{-1}g).$$

Note that these two operations commute. Then a differential operator Y is in $\mathscr{Z}$ if and only if, for all g and h in G and for all smooth f on $SL(2, F_\mathfrak{p})$,

$$\rho(g)\lambda(h)Yf = \rho(g)Y\lambda(h)f = \lambda(h)Y\rho(g)f$$

$$\rho(\varepsilon)\lambda(\varepsilon)^{-1}Yf = Y\rho(\varepsilon)\lambda(\varepsilon)^{-1}f.$$

Definition Let f be a smooth function on $GL(2, F_\mathfrak{p})$. The function f is an eigenfunction for $\mathscr{Z}$ if there is an algebra homomorphism μ: $\mathscr{Z} \to \mathbb{C}$ so that for every $Y \in \mathscr{Z}$ we have $Yf = \mu(Y)f$.

Remarks As it happens, for $SL(2, F_\mathfrak{p})$ as opposed to more arbitrary groups, $\mathscr{Z}$ has a simple nature: It is the polynomial algebra in a single second-order differential operator C, th *Casimir operator*. In particular, it is then necessarily commutative. The explicit form of the Casimir operator for $\mathfrak{p} \approx \mathbb{R}$ is computed in Appendix A.4. That $\mathscr{Z}$ is a (commutative) polynomial ring (and, hence, commutative) is true much more generally, but is not as easy to prove as was the fact that the spherical Hecke algebras are commutative.

2.5 Adelic Fourier Expansions, Cuspforms, and Hecke Operators

First, we give the formal general definition of the Fourier expansion of a wide class of functions on $GL(2, \mathbb{A})$, and give the definition of (a weak notion of) *cuspform*. Then we consider the relationship between Fourier coefficients and eigenfunction properties with respect to Hecke operators.

Let $\tau_\mathfrak{p}$ be the standard characters on the additive groups $F_\mathfrak{p}$ (as in Appendix A.2). Let dx denote a Haar measure on $\mathbb{A}$ (or on the quotient $\mathbb{A}/F$), giving $\mathbb{A}/F$ total measure 1. For a continuous function φ on $\mathbb{A}$ which is F-invariant, we have an L^2-equality

$$\varphi(x) = \sum_{\xi \in F} \tau(\xi x) \int_{\mathbb{A}/F} \overline{\tau}(\xi u)\varphi(u)\, du.$$

Definition Let f be a left $GL(2, F)$-invariant continuous function on $GL(2, \mathbb{A})$. For $\xi \in F$, the ξ-th Fourier coefficient of f is the function $W_\xi(g)$ on $GL(2, \mathbb{A})$ defined by

$$W_\xi(g) = \int_{\mathbb{A}/F} \overline{\tau}(\xi x) f(u(x)g)\, dx.$$

PROPOSITION

i. For $u(x) \in U(\mathbb{A})$, we have $W_\xi(u(x)g) = \tau(\xi x) W_\xi(g)$.
ii. In an L^2-sense (in $x \in \mathbb{A}/F$), $f(g) = \sum_{\xi \in F} W_\xi(g)$.
iii. For $p = \left(\begin{smallmatrix} a & b \\ 0 & d \end{smallmatrix}\right)$ in $GL(2, F)$, $W_\xi(g) = W_{ad^{-1}\xi}(pg)$.

Proof The first assertion follows from the definition, and the second follows from the first assertion and from the preceding remark. The third follows from the uniqueness of the Fourier expansion, using the left $GL(2, F)$-invariance of f. ■

Definition A left $GL(2, F)$-invariant continuous function f on $GL(2, \mathbb{A})$ so that the integrals defining the Fourier coefficients exist almost everywhere, and so that $W_0(g) = 0$ almost everywhere, is a (weak) *cuspform*.

Now suppose that f is a cuspform (in the present weak sense), has central character ω which is unramified at a fixed finite prime $\mathfrak{p}$, is right invariant under $GL(2, \mathfrak{o}_\mathfrak{p})$, and is an eigenfunction for the (spherical) Hecke algebra H at $\mathfrak{p}$ with character ω. Let π be a local parameter at $\mathfrak{p}$. Let

$$f(g) = \sum_\xi W_\xi(g) = \sum_\xi W_1(\ell(\xi)g)$$

be the Fourier expansion of f, with previous notation. Write $W = W_1$, and let δ be a generator for the local inverse different at $\mathfrak{p}$. Consider the special element η of H given by

$$\eta = \text{characteristic function of } GL(2, \mathfrak{o}_\mathfrak{p})\ell(\pi^{-1})GL(2, \mathfrak{o}_\mathfrak{p}).$$

Since the function W is formed from f by *left* integration, W will also be an eigenfunction with respect to the *right* convolution algebra H. Let λ be the eigenvalue of W with respect to (right convolution by) η.

Remarks The function W'' in the following theorem is often referred to as a ($\mathfrak{p}$-adic) *Whittaker function*, because of the analogy with the archimedean case.

THEOREM Under these hypotheses, the function W factors as

$$W(g) = W'(g')W''(g'')$$

where g' is the non-$\mathfrak{p}$ part of $g \in GL(2, \mathbb{A})$, $g'' \in GL(2, F_\mathfrak{p})$, $g = g'g''$. Further, for any $\varepsilon \in \mathfrak{o}_\mathfrak{p}^\times$, we have

$$\begin{aligned} W''(\ell(\varepsilon\pi^m\delta)) &= \\ &= W''(\ell(\delta)) \times \{\alpha^m + \alpha^{m-1}\beta + \alpha^{m-2}\beta^2 + \cdots + \alpha\beta^{m-1} + \beta^m\} \quad \text{for } m \geq 0 \\ &= 0 \quad \text{for } m < 0 \end{aligned}$$

where α and β are the roots of $X^2 - N\mathfrak{p}^{-1}\lambda X + \omega(\pi)N\mathfrak{p}^{-1} = 0$.

Proof First, the left equivariance of W by $U(F_{\mathfrak{p}})$ with respect to τ and the right invariance by $K = K_{\mathfrak{p}}$ imply that W is completely determined by its values on elements of the form $g''\ell(y)$ with y a power of π, as we have the Iwasawa decomposition

$$GL(2, F_{\mathfrak{p}}) = U(F_{\mathfrak{p}})Z(F_{\mathfrak{p}})T(F_{\mathfrak{p}})K_{\mathfrak{p}}.$$

We will show that the values $W(g'\ell(\delta))$ together with the eigenvalue λ determine W completely.

First, for $u(x)$ in $GL(2, \mathfrak{o}_{\mathfrak{p}})$, we have

$$W(g'\ell(y)) = W(g'\ell(y)u(x)) = W(g'u(yx)\ell(y)) = W(u(yx)g'\ell(y)) =$$
$$= \tau_{\mathfrak{p}}(yx)W(g'\ell(y)).$$

Thus, if y is not in $\mathfrak{o}_{\mathfrak{p}}^*$, it must be that $W(g'\ell(y)) = 0$. Then, by general results on integration theory on homogeneous spaces, we have

$$\lambda W(g'\ell(y)) = W(g'\ell(y)) * \eta = \int_{GL(2,F_{\mathfrak{p}})} W(g'\ell(y)h^{-1})\eta(h)\,dh =$$
$$= W(g'\ell(y)\pi\ell(\pi^{-1})) + \sum_{b \bmod \pi} W(g'\ell(y)u(b)\ell(\pi))$$

by taking representatives of the form $\left(\begin{smallmatrix} a & b \\ 0 & d \end{smallmatrix}\right)$ with $a, d \in \{1, \pi\}$, $ad = \pi$, $b \in \mathfrak{o}_{\mathfrak{p}}/a\mathfrak{o}_{\mathfrak{p}}$ for h^{-1} with $h^{-1} \in K\ell(\pi)K/K$. This yields the fundamental relation

$$\lambda W(g'\ell(y)) = \omega(\pi)W(g'\ell(y/\pi)) + \left\{\sum_{b \bmod \pi} \tau_{\mathfrak{p}}(by)\right\} W(g'\ell(y\pi)).$$

Because of the restriction on the support of W, this gives

$$\lambda W(g'\ell(y)) = \omega(\pi)W(g'\ell(y/\pi)) + N\mathfrak{p}W(g'\ell(y\pi)).$$

This formula gives a recursion for the values of W, for $m \geq 0$ yielding (by two-dimensional elementary linear algebra)

$$W(g'\ell(\pi^m\delta)) = W(g'\ell(\delta)) \times \{\alpha^m + \alpha^{m-1}\beta + \alpha^{m-2}\beta^2 + \cdots + \alpha\beta^{m-1} + \beta^m\}$$

where α, β are the roots of $X^2 - N\mathfrak{p}^{-1}\lambda X + \omega(\pi)N\mathfrak{p}^{-1} = 0$. ∎

Now we consider an important special form of the "*bad prime* case," i.e., where f is not right $GL(2, \mathfrak{o}_{\mathfrak{p}})$-invariant. Suppose that χ is a character on

$$\mathbb{K} = \left\{g \in GL(2, \mathfrak{o}_{\mathfrak{p}}) : g = \begin{pmatrix} a & b \\ c & d \end{pmatrix} \text{ with } c \equiv 0 \bmod \mathfrak{p}^n\right\}$$

which is trivial on the subgroup

$$\left\{g \in GL(2, \mathfrak{o}_{\mathfrak{p}}): g = \begin{pmatrix} a & b \\ c & d \end{pmatrix} \text{ with } c \equiv 0 \bmod \mathfrak{p}^n, d \equiv 1 \bmod \mathfrak{p}^n\right\}.$$

Suppose that f is right equivariant by $\mathbb{K}$ with respect to χ:

$$f(gk) = \chi(k)f(g) \qquad (\text{for } g \in GL(2, \mathbb{A}), k \in \mathbb{K}).$$

Consider a special function

$$\eta = [K_{\mathfrak{p}} : \mathbb{K}] \times \text{characteristic function of } \mathbb{K}\ell(\pi^{-1})\mathbb{K}$$

where π is a fixed local parameter at $\mathfrak{p}$. Suppose that f is an eigenfunction with respect to right convolution by η, with eigenvalue λ. (Again, the function W will also be an eigenfunction with eigenvalue λ).

Remarks Consideration of this special function is necessary, for example, at primes $\mathfrak{p}$ where the central character ω is ramified. Unfortunately, maps of the form $f \to f * \eta$ cannot be relied upon to be self-adjoint or normal operators. Also, unlike the case that $\mathfrak{n} = \mathfrak{o} = \mathfrak{p}^0$, it is generally not the case that the convolution algebra of left and right $\mathbb{K}$-invariant compactly supported continuous functions on $GL(2, F_{\mathfrak{p}})$ is commutative.

THEOREM Take $g = g'\ell(y) \in GL(2, \mathbb{A})$ with g' the non-$\mathfrak{p}$ part of g, and with $y \in F_{\mathfrak{p}}^{x}$. Under the above hypotheses, the function W factors as

$$W(g) = W'(g')W''(\ell(y)).$$

Further,

$$\begin{aligned} W''(\ell(\pi^m\delta)) &= W''(\ell(\delta)) \times (\lambda/N\mathfrak{p})^m \quad \text{for } m \geq 0 \\ &= \qquad 0 \qquad\qquad\qquad\quad \text{for } m < 0. \end{aligned}$$

Proof First, for $u(x)$ in $GL(2, \mathfrak{o}_{\mathfrak{p}})$, necessarily $u(x) \in \mathbb{K}$. Then we have

$$W(g'\ell(y)) = W(g'\ell(y)u(x)) = W(g'u(yx)\ell(y)) = W(u(yx)g'\ell(y)) =$$

$$= \tau_{\mathfrak{p}}(yx)W(g'\ell(y)).$$

Thus, if y is not in $\mathfrak{o}_{\mathfrak{p}}^*$ it must be that $W(g'\ell(y)) = 0$.

Another easy observation, which follows from the right $\mathbb{K}$-equivariance hypothesis, is that $W(g'\ell(y))$ depends only on the coset $y\mathfrak{o}_{\mathfrak{p}}^{x}$.

It is easy to check that the coset space

$$\mathbb{K}\backslash\mathbb{K}\ell(\pi^{-1})\mathbb{K}$$

has representatives of the form $\ell(\pi^{-1})u(b)$ with $b \in \mathfrak{o}_\mathfrak{p}/\mathfrak{p}\mathfrak{o}_\mathfrak{p}$. Then, for y in $\mathfrak{o}_\mathfrak{p}^* \cap F_\mathfrak{p}^x$,

$$\lambda W(g'\ell(y)) = W(g'\ell(y)) * \eta = \sum_{b \bmod \pi} W(g'\ell(y)u(b)\ell(\pi)).$$

Note that the normalizing constant $[K_\mathfrak{p} : \mathbb{K}]$ is convenient to avoid ugly constants arising from consideration of measure. This yields the fundamental relation

$$\lambda W(g'\ell(y)) = \left\{ \sum_{b \bmod \pi} \tau_\mathfrak{p}(by) \right\} W(g'\ell(y\pi)).$$

Because of the restriction on the support of W, this gives

$$\lambda W(g'\ell(y)) = N\mathfrak{p} W(g'\ell(y\pi)).$$

This formula gives a simple recursion for the values of W. ■

Remarks To demonstrate that the hypotheses of the previous result are not as restrictive as might appear, we will verify the following assertion: Given an automorphic form f on $GL(2, \mathbb{A})$, there is $h \in GL(2, \hat{\mathfrak{o}})$ and a Hecke character α so that the automorphic form

$$f_2(g) = \alpha(\det g) f(gh)$$

is a finite linear combination of automorphic forms (which are also right translates of f) meeting the equivariance hypothesis of the previous theorem at every bad prime.

Proof Let $\mathfrak{n}$ be a non-zero ideal of $\mathfrak{o}$ so that f is right invariant by

$$\{g \in GL(2, \hat{\mathfrak{o}}): g \equiv 1 \text{ modulo } \mathfrak{n}\} \subset GL(2, \hat{\mathfrak{o}}).$$

Let t be a finite idele so that $t\hat{\mathfrak{o}} = \mathfrak{n}\hat{\mathfrak{o}}$. Then $f_1(g) = f(g\ell(t^{-1}))$ is right invariant under

$$\mathbb{K}_1(\mathfrak{n}^2) = \left\{ g = \begin{pmatrix} a & b \\ c & d \end{pmatrix} \in GL(2, \hat{\mathfrak{o}}): c \equiv 0, a \equiv d \equiv 1 \text{ modulo } \mathfrak{n}^2 \right\} \subset GL(2, \hat{\mathfrak{o}}).$$

The group

$$\mathbb{K}_0(\mathfrak{n}^2) = \left\{ g = \begin{pmatrix} a & b \\ c & d \end{pmatrix} \in GL(2, \hat{\mathfrak{o}}): c \equiv 0, \text{ modulo } \mathfrak{n}^2 \right\} \subset GL(2, \hat{\mathfrak{o}})$$

normalizes $\mathbb{K}_1(\mathfrak{n}^2)$ with finite abelian quotient H. Let X be the (finite) collection of characters on $\mathbb{K}_0(\mathfrak{n}^2)$ which are trivial on $\mathbb{K}_1(\mathfrak{n}^2)$. Then for $\beta \in X$ the automorphic form

$$f_\beta(g) = (^\#X)^{-1} \sum_{h \in H} \bar{\beta}(h) f_1(gh)$$

is right equivariant under $\mathbb{K}_0(\mathfrak{n}^2)$ by β. On the other hand, by standard arguments

$$f_1(g) = (^\#X)^{-1} \sum_{\beta \in X} f_\beta(g).$$

For fixed β, let

$$\beta\left(\begin{pmatrix} a & b \\ 0 & d \end{pmatrix}\right) = \beta_1(a)\beta_2(d)$$

on $\mathbb{K}_0(\mathfrak{n}^2)$. Define another automorphic form

$$\varphi(g) = f_\beta(g)\beta_1^{-1}(\det g).$$

Then φ is right invariant under

$$\left\{g = \begin{pmatrix} a & b \\ c & d \end{pmatrix} \in GL(2,\hat{\mathfrak{o}}) : c \equiv 0,\, d \equiv 1 \text{ modulo } \mathfrak{n}^2\right\} \subset GL(2,\hat{\mathfrak{o}}),$$

and is still right equivariant under $\mathbb{K}_0(\mathfrak{n}^2)$. This is what we had aimed to demonstrate. ■

2.6 General Definition of Adelic Automorphic Forms

For every prime $\mathfrak{p}$, let $\| \ \|_\mathfrak{p}$ be the usual normalization of the norm so that the product formula holds, and let $\| \ \|$ be the usual idele norm.

Definition Let f be a continuous $\mathbb{C}$-valued function on $GL(2,\mathbb{A})$ so that

$$f(\zeta\gamma g) = \omega(\zeta) f(g)$$

for all ζ in the center $Z(\mathbb{A})$ of $GL(2,\mathbb{A})$, and for all $\gamma \in GL(2,F)$, with some Hecke character ω: Say that f *has central character* ω. Further, suppose that the span over $\mathbb{C}$ of

$$\left\{g \to f(gk) : k \in \prod_{\mathfrak{p} \le \infty} K_\mathfrak{p}\right\}$$

is finite-dimensional, i.e., that f is *right K-finite*.

Definition If, in addition to the previous conditions, $|f|^2$ is integrable with respect to any right $GL(2,\mathbb{A})$-invariant measure on $Z(\mathbb{A})GL(2,F)\backslash GL(2,\mathbb{A})$, then f is a *square-integrable automorphic form.* If f is continuous and if, for any $y \in \mathbb{J}$ and for any compact subset C of $GL(2, \mathbb{A})$,

$$f(g_1 \ell(y) g_2) = O(\|y\|^\sigma + \|y\|^{-\sigma}) \qquad \text{(uniformly for } g_1, g_2 \in C)$$

for some $\sigma \in \mathbb{R}$, then f is an *automorphic form of moderate growth.* If f is square-integrable and

$$\int_{x \in \mathbb{A}/F} f(u(x)g)\, dx = 0$$

for almost all $g \in GL(2, \mathbb{A})$, then f is a *cuspform.*

Definition Now suppose that f is either a *cuspform*, a *square integrable automorphic form*, or an *automorphic form of moderate growth.* Let X be the set of finite primes $\mathfrak{p}$ so that the projection of $\mathbb{K}$ to $GL(2, F_\mathfrak{p})$ is exactly $GL(2, \mathfrak{o}_\mathfrak{p})$; this holds for all but finitely many $\mathfrak{p}$; primes $\mathfrak{p}$ in X are often called *good primes* (or *unramified primes*, or *primes not dividing the level*). For $\mathfrak{p} \in X$ let $H(\mathfrak{p})$ be the spherical Hecke algebra at $\mathfrak{p}$. If f is an eigenfunction for all Hecke operators T_φ arising from $H(\mathfrak{p})$ for all primes $\mathfrak{p} \in X$, then say that f is a *Hecke eigenfunction at good primes.* If f is an eigenfunction for the invariant differential operators at each of the infinite primes, then say that f is an *eigenfunction for the invariant differential operators.*

Definition If, for a finite prime $\mathfrak{p}$, f is *not* right $GL(2, \mathfrak{o}_\mathfrak{p})$-invariant, we say that $\mathfrak{p}$ is *bad*, or *ramified* (for f). Suppose that χ is a character on

$$\mathbb{K} = \left\{ g \in GL(2, \mathfrak{o}_\mathfrak{p}) \colon g = \begin{pmatrix} a & b \\ c & d \end{pmatrix} \text{ with } c \equiv 0 \bmod \mathfrak{p}^n \right\}$$

which is trivial on the subgroup

$$\left\{ g \in GL(2, \mathfrak{o}_\mathfrak{p}) \colon g = \begin{pmatrix} a & b \\ c & d \end{pmatrix} \text{ with } c \equiv 0 \bmod \mathfrak{p}^n,\ d \equiv 1 \bmod \mathfrak{p}^n \right\}.$$

Suppose that f is right equivariant by $\mathbb{K}$ with respect to χ:

$$f(gk) = \chi(k) f(g) \qquad (\text{for } g \in GL(2, \mathbb{A}),\ k \in \mathbb{K}).$$

Suppose that f is an eigenfunction for the *special Hecke operator* considered in the previous section. Then we will say that f is a *Hecke eigenfunction at* $\mathfrak{p}$. Now fix an ideal $\mathfrak{n}$ in $\mathfrak{o}$, and let χ be a character on

$$\mathbb{K} = \left\{ g \in GL(2, \hat{\mathfrak{o}}) \colon g = \begin{pmatrix} a & b \\ c & d \end{pmatrix} \text{ with } c \equiv 0 \bmod \mathfrak{n} \right\}$$

which is trivial on the subgroup

$$\left\{g \in GL(2, \mathfrak{o}): g = \begin{pmatrix} a & b \\ c & d \end{pmatrix} \text{ with } c \equiv 0 \bmod \mathfrak{n},\, d \equiv 1 \bmod \mathfrak{n}\right\}.$$

Suppose that f is right equivariant by $\mathbb{K}$ with respect to χ:

$$f(gk) = \chi(k)f(g) \qquad (\text{for } g \in GL(2, \mathbb{A}),\, k \in \mathbb{K}).$$

Suppose that f is an eigenfunction for the *special Hecke operator* at each $\mathfrak{p}$ dividing $\mathfrak{n}$. Then say simply that f is a *Hecke eigenfunction.*

Remarks At an infinite prime $\mathfrak{p} \approx \mathbb{R}$, the right K-finiteness implies that f is a finite linear combination of functions *of weight* κ *at* $\mathfrak{p}$ in the following sense: Say that f is of weight κ (at $\mathfrak{p}$) if

$$f(gk(\theta)) = f(g)e^{i\kappa\theta}$$

where

$$k(\theta) = \begin{pmatrix} \cos\theta & \sin\theta \\ -\sin\theta & \cos\theta \end{pmatrix} \in SO(2) = K_{\mathfrak{p}}.$$

(At an infinite prime $\mathfrak{p} \approx \mathbb{C}$ the notion of "weight" is somewhat more complicated, because $SU(2)$ is not abelian).

It is easy to check that, if F is totally real and if f is of weight $(\kappa_1, \ldots, \kappa_m)$ with central character ω, then f must be identically zero unless

$$(-1)^{\kappa_i} = \omega(-1_i)$$

where -1_i denotes the idele which is 1 at all primes except the i-th infinite prime, where it is -1. The analogous requirement for more general number fields is comparable.

2.7 Formalism of L-Functions Associated to Cuspforms

Now suppose that f is a cuspform on $GL(2, \mathbb{A})$ with central character ω, and is right $GL(2, F_{\mathfrak{p}})$-invariant for finite primes $\mathfrak{p}$ outside a finite set S containing the infinite primes. Suppose that f is a Hecke eigenfunction at primes $\mathfrak{p}$ not in S. Let W be *the* Fourier coefficient of f:

$$W(g) = \int_{\mathbb{A}/F} \bar{\tau}(x)f(u(x)g)\, dx$$

where τ is the standard additive character on $\mathbb{A}/F$, etc. As we have seen, the Hecke eigenfunction hypothesis assures that W factors over primes as

$$W(g) = W_S(g_S) \prod_{\mathfrak{p} \notin S} W_{\mathfrak{p}}(g_{\mathfrak{p}})$$

where $g = (\ldots, g_{\mathfrak{p}}, \ldots) \in GL(2, \mathbb{A})$, where $g_S = \prod_{\mathfrak{p} \in S} g_{\mathfrak{p}}$. For $\mathfrak{p} \notin S$, let $\alpha_{\mathfrak{p}}$ and $\beta_{\mathfrak{p}}$ be the complex numbers (with product $\omega(\pi) N\mathfrak{p}^{-1}$) which determine $W_{\mathfrak{p}}$, as discussed above.

Let t be a finite idele so that the associated ideal of $\mathfrak{o}$ is the inverse different $\mathfrak{o}^*$ of $\mathfrak{o}$ over $\mathbb{Z}$. We normalize $W_{\mathfrak{p}}$ so that $W_{\mathfrak{p}}(\left(\begin{smallmatrix} t & 0 \\ 0 & 1 \end{smallmatrix}\right)) = 1$.

Remarks The following computations are formal, as *convergence questions are neglected*; the point is to show how Euler product factorizations of Dirichlet series associated to f arise. We note that the present normalization is slightly different from that considered in the classical case, and that other normalizations are also commonly used.

Definition Let χ be a (not necessarily unitary) Hecke character which is unramified at primes not in S, and define the *standard L-function associated to f* (evaluated at χ) by

$$L(f, \chi) = \int_{\mathbb{J}/F^\times} \chi(y) f(\ell(y))\, d^\times y$$

where $d^\times y$ denotes the standard Haar measure on $\mathbb{J}$.

PROPOSITION In this situation, assuming absolute convergence, we have

$$L(f, \chi) = \int_{\mathbb{J}} \chi(y) W(\ell(y))\, d^\times y =$$

$$= \prod_{\mathfrak{p} \notin S} \left\{ \int_{F_{\mathfrak{p}}^\times} \chi(y) W_{\mathfrak{p}}(\ell(y))\, d^\times y \right\} \times \int_{\prod_{\mathfrak{p} \in S} F_{\mathfrak{p}}^\times} \chi(y) W_S(\ell(y))\, d^\times y.$$

Further, for $\mathfrak{p} \neq S$, let $t_{\mathfrak{p}}$ be the $\mathfrak{p}$-component of t and let π be a local parameter at $\mathfrak{p}$; then

$$\int_{F_{\mathfrak{p}}^\times} \chi(y) W_{\mathfrak{p}}(\ell(y))\, d^\times y = \chi(t_{\mathfrak{p}})(1 - (\alpha_{\mathfrak{p}} + \beta_{\mathfrak{p}})\chi(\pi)^{-1} + \omega(\pi) N\mathfrak{p}^{-1} \chi(\pi)^{-2})^{-1}.$$

Proof Formally, we have

$$L(f, \chi) = \int_{\mathbb{J}/F^\times} \chi(y) f\left(\begin{pmatrix} y & 0 \\ 0 & 1 \end{pmatrix}\right) d^\times y$$

$$= \int_{\mathbb{J}/F^\times} \chi(y) \sum_{\xi \in F^\times} W(\ell(\xi)\ell(y))\, d^\times y = \int_{\mathbb{J}} \chi(y) W(\ell(y))\, d^\times y$$

simply by changing variables in the integral. Since W factors over primes not in S, we obtain the factorization into local integrals.

To compute the local integral at a prime $\mathfrak{p}$ not in S, abbreviate $\alpha = \alpha_\mathfrak{p}$ and $\beta = \beta_\mathfrak{p}$. Without loss of generality we may assume that $\alpha \neq \beta$. Then the local integral is

$$\int_{F_\mathfrak{p}^\times \cap \mathfrak{o}_\mathfrak{p}} \chi(t_\mathfrak{p} y)(\alpha^{m+1} - \beta^{m+1})/(\alpha - \beta)\, d^\times y$$

where we write $m = \mathrm{ord}_\mathfrak{p}\, y$, from our determination of the structure of $W_\mathfrak{p}$. Since χ is unramified at $\mathfrak{p}$, by homogeneous space integration theory this integral is just a sum

$$\chi(t_\mathfrak{p}) \sum_{m \geq 0} \chi(\pi^m)(\alpha^{m+1} - \beta^{m+1})/(\alpha - \beta) = \chi(t_\mathfrak{p})((1 - \alpha\chi(\pi)^{-1})(1 - \beta\chi(\pi)^{-1}))^{-1}$$

by summing the two geometric series and simplifying. ■

Now, under some more specific normalizing hypotheses, we can obtain a complete factorization of the L-function attached to f and χ.

PROPOSITION Fix an ideal $\mathfrak{n}$ in $\mathfrak{o}$, and let χ be a character on

$$\mathbb{K} = \left\{ g \in GL(2, \mathfrak{o}) \colon g = \begin{pmatrix} a & b \\ c & d \end{pmatrix} \text{ with } c \equiv 0 \bmod \mathfrak{n} \right\}$$

which is trivial on the subgroup

$$\left\{ g \in GL(2, \mathfrak{o}) \colon g = \begin{pmatrix} a & b \\ c & d \end{pmatrix} \text{ with } c \equiv 0 \bmod \mathfrak{n},\ d \equiv 1 \bmod \mathfrak{n} \right\}.$$

Suppose that f is right equivariant by $\mathbb{K}$ with respect to χ:

$$f(gk) = \chi(k) f(g) \qquad (\text{for } g \in GL(2, \mathbb{A}),\ k \in \mathbb{K}).$$

Suppose that f is a Hecke eigenfunction at good primes, and is an eigenfunction for the special Hecke operators at bad primes (i.e., those dividing $\mathfrak{n}$). *Assuming* that $W_\mathfrak{p}(\left(\begin{smallmatrix} t & 0 \\ 0 & 1 \end{smallmatrix}\right)) \neq 0$ at finite $\mathfrak{p}$ in S, we normalize $W_\mathfrak{p}$ so that $W_\mathfrak{p}(\left(\begin{smallmatrix} t & 0 \\ 0 & 1 \end{smallmatrix}\right)) = 1$. Then the L-function attached to f factors over all finite primes of F. The $\mathfrak{p}$-factor at a prime $\mathfrak{p}$ dividing $\mathfrak{n}$ is of the form

$$\chi(t_\mathfrak{p})/(1 - a_\mathfrak{p} N\mathfrak{p}^{-s}).$$

Proof The hypotheses assure that the function W now factors over all primes, by the result of the previous section on the special Hecke operators. Then homogeneous space integration theory shows that the $\mathfrak{p}$-factor in the integral defining the L-

function is merely a geometric series

$$\chi(t_{\mathfrak{p}}) \sum_{n \geq 0} \lambda_{\mathfrak{p}}^n N\mathfrak{p}^{-s} = \chi(t_{\mathfrak{p}})/(1 - \lambda_{\mathfrak{p}} N\mathfrak{p}^{-s}). \qquad \blacksquare$$

Remarks As noted in the discussion of Fourier expansions and the special Hecke operators at bad primes, given an automorphic form f, there is $h \in GL(2, \mathbb{A}_0)$ and there is a Hecke character α so that $\alpha(\det g)f(gh)$ is a finite linear combination of automorphic forms meeting the equivariance hypothesis of this proposition.

Remarks To obtain explicit results on the infinite prime factors, one must use the invariant differential operators. Also, when one cannot easily achieve the situation of the last proposition, one may not have information about the function W_S regarding the bad primes. Therefore, one often considers only the L-function *at good primes*, given by

$$L_S(f, \chi) = \prod_{\mathfrak{p} \notin S} (1 - (\alpha_{\mathfrak{p}} + \beta_{\mathfrak{p}})\chi(\pi)^{-1} + \omega(\pi)N\mathfrak{p}^{-1}\chi(\pi)^{-2})^{-1}.$$

Now suppose that f is as above. For a good prime $\mathfrak{p}$, define a two-by-two complex matrix

$$\Phi_{\mathfrak{p}} = \begin{pmatrix} \alpha_{\mathfrak{p}} & 0 \\ 0 & \beta_{\mathfrak{p}} \end{pmatrix}.$$

The good-prime part of the standard L-function associated to f and χ may be written as

$$L_S(f, \chi) = \prod_{\mathfrak{p} \notin S} \det(1 - \chi(\pi)^{-1}\Phi_{\mathfrak{p}})^{-1},$$

and this expression depends only on the *conjugacy classes* of the matrices $\Phi_{\mathfrak{p}}$. We can easily generalize the form of this expression:

Definition Let ρ be a finite-dimensional representation of $GL(2, \mathbb{C})$. For f as above, define the higher L-function associated to f, ρ, and χ by

$$L_S(f, \rho, \chi) = \prod_{\mathfrak{p} \notin S} \det(1 - \chi(\pi)^{-1}\rho(\Phi_{\mathfrak{p}}))^{-1}.$$

PROPOSITION (Assuming convergence ...) Let $\rho = \rho_1 \oplus \rho_2$. Then

$$L_S(f, \rho, \chi) = L_S(f, \rho_1, \chi)L_S(f, \rho_2, \chi).$$

If $\rho(g) = \det(g)$, then

$$L_S(f, \rho, \chi) = L_S(\omega\chi/\| \ \|),$$

where the latter L-function is the abelian L-function associated to the Hecke character $\omega\chi/\| \ \|$, where $\| \ \|$ is the idele norm.

Proof The first assertion is formal; the second follows immediately from $\alpha_\mathfrak{p}\beta_\mathfrak{p} = \omega(\pi)/N\mathfrak{p}$. ■

Extrapolating this idea in a somewhat different direction, we have:

Definition Let f_i $(i = 1, \ldots, m)$ be a cuspform as above, with associated matrix $\Phi_{\mathfrak{p}i}$. Define the *tensor product L-function* by

$$L(f_1 \otimes \cdots \otimes f_m, \chi) = \prod_{\mathfrak{p} \notin S} \det(1 - \chi(\pi)^{-1}\Phi_{\mathfrak{p}1} \otimes \cdots \otimes \Phi_{\mathfrak{p}m})^{-1}.$$

(Clearly this and the previous definitions may be combined....)

Remarks Under the additional hypothesis that these cuspforms are eigenfunctions for the invariant differential operators and Hecke operators at good primes, it is conjectured that *all such L-functions should have analytic continuations to meromorphic functions with finitely many poles* (and have functional equations). Further, under various technical hypotheses, these L-functions should be *entire*.

It is known that all the *holomorphic* representations of $GL(2, \mathbb{C})$ are of the form

$$g \to \det(g)^m \times \mathrm{Sym}^n(g),$$

where Sym^n is the representation (contragredient to) the representation of $GL(2, \mathbb{C})$ on homogeneous polynomials P of degree n in two variables, described by

$$(gP)(x, y) = P((x, y)g).$$

Therefore, one might reasonably restrict one's attention to L-functions

$$L_S(f, \mathrm{Sym}^n, \chi).$$

At this time the best results concerning these symmetric power L-functions are due to F. Shahidi: (Assuming that f is an eigenfunction for the invariant differential operators) these "symmetric power" L-functions have meromorphic analytic continuations meromorphic for $n \leq 5$; see [S1] and [S2].

In [R], Rankin found the analytic continuation of product L-functions $L_S(f_1 \otimes \bar{f}_2, \chi)$. In [G2], this author found the analytic continuation of triple products $L_S(f_1 \otimes f_2 \otimes f_3, \chi)$. This was reformulated in [PR], and it is shown there that there are finitely many poles. The book [GPR] discusses general ideas in this direction.

3 Comparison of Classical and Adelic Viewpoints

This chapter is intended to make clear that the previous very general adelic definition of automorphic form on $GL(2, \mathbb{A})$ includes ("adelizations of") holomorphic Hilbert modular forms as a special case, and that the adelic notions of Hecke operators, Fourier expansions, etc., correspond perfectly to ("adelizations of") the classical ones.

More specifically, the first section uses the Strong Approximation Theorem for $SL(2)$ (from Appendix A.3) to demonstrate that there is an injection of spaces of automorphic forms on quotients of $\mathfrak{H}^m$ to spaces of automorphic forms on $GL(2, \mathbb{A})$; however, the whole space of automorphic forms on $GL(2, \mathbb{A})$ is a finite direct sum of spaces of automorphic forms on quotients of $\mathfrak{H}^m$.

Beyond this, one also sees that (essentially) the restrictions of the adelic Fourier coefficients are the classical ones.

Also, applying a Hecke operator and "adelizing" an automorphic form on $\mathfrak{H}^m$ is the same as "adelizing" and then applying the corresponding adelic Hecke operator. This offers a definite advantage, as there is no difficulty in describing the (spherical) Hecke operators on adelic automorphic forms, whereas class numbers and units questions intervene in trying to imitate the classical elliptic modular Hecke operators for classical Hilbert modular forms.

The "adelized" holomorphic Hilbert modular forms are easily distinguished: They are exactly those adelic automorphic forms annihilated by certain left-invariant differential operators. (There is a more intrinsic characterization involving the eigenvalues of a two-sided invariant operator, the Casimir operator, but such a characterization presumes in effect a "classification" of adelic cuspforms, entailing considerably more development of issues peripheral to our present goals.)

Section 3.5 makes explicit the form of the Fourier expansion of holomorphic automorphic forms on $GL(2, \mathbb{A})$.

The last section in this chapter briefly indicates what more can be said in comparison—for example, contrasting the formation of Dirichlet series as Mellin transforms of cuspforms which are eigenfunctions for Hecke operators. Even superficially, the adelic formulation has the distinct advantage of assuring an Euler product, while the classical version inadvertently omits (in general) some of the information in the adelic version.

3.1 Comparison of Function Spaces via Strong Approximation

Here we make a series of adjustments of viewpoint in order to see that the general notion of automorphic form on $GL(2, \mathbb{A})$ includes the classical Hilbert modular forms as special cases:

i. In a broad sense, "Hilbert modular forms" of weight $\kappa \in \mathbb{Z}^m$ on $\mathfrak{H}^m$ with respect to subgroups Γ may be converted into *left* Γ*-invariant* functions on $SL(2, \mathbb{R})^m$ which are *right* $SO(2)^m$*-equivariant* with respect to a one-dimensional representation of $SO(2)^m$ determined by the weight κ, *and vice-versa*. That is, instead of looking at $\Gamma\backslash\mathfrak{H}^m$, we may consider $\Gamma\backslash SL(2, \mathbb{R})^m$.

ii. Because of the Strong Approximation Theorem, every quotient $\Gamma\backslash SL(2, \mathbb{R})^m$ by a *congruence* subgroup is homeomorphic to a quotient $SL(2, F)\backslash SL(2, \mathbb{A})/\mathbb{K}$ for a suitable compact open subgroup $\mathbb{K}$ of $SL(2, \mathbb{A}_0)$, and conversely. We view this as asserting that $SL(2, F)\backslash SL(2, \mathbb{A})$ is a *universal object* for these quotients.

iii. Because the matrices involved in the definitions of the Hecke operators do not often lie in $SL(2)$ anyway, we might prefer to be considering $GL(2, F)\backslash GL(2, \mathbb{A})$; to make a Hilbert modular form (etc.) into an automorphic form on this quotient is the last objective of this section.

Comparison of Hecke operators and the notion of "holomorphic automorphic form" will be discussed later.

Let Γ be a discrete subgroup of $SL(2, \mathbb{R})^m$, let $\kappa \in \mathbb{Z}^m$, and let f be a function on $\mathfrak{H}^m$ so that

$$f|_\kappa \gamma = f$$

for every $\gamma \in \Gamma$. Define a function $f^\#$ on $SL(2, \mathbb{R})^m$ by

$$f^\#(g) = f(g(i))\mu(g, i)^{-\kappa}.$$

For $\theta \in \mathbb{R}^m$, using a multi-index notation, let

$$k(\theta) = \begin{pmatrix} \cos\theta & \sin\theta \\ -\sin\theta & \cos\theta \end{pmatrix} \in SO(2)^m.$$

PROPOSITION Such a function $f^\#$ on $SL(2, \mathbb{R})^m$ has the property that

$$f^\#(\gamma g k(\theta)) = f^\#(g)e^{i\kappa\theta}$$

(in a multi-index notation) for $\gamma \in \Gamma$ and $k(\theta) \in SO(2)^m$. Conversely, let φ be a function on $SL(2, \mathbb{R})^m$ so that

$$\varphi(\gamma g k(\theta)) = \varphi(g)e^{i\kappa\theta}$$

for $\gamma \in \Gamma$ and $k(\theta) \in SO(2)^m$. Then, for $y \in (0, \infty)^m$ and $x \in \mathbb{R}^m$,

$$\varphi^{\natural}(x+iy) = y^{-\kappa/2}\varphi\left(\begin{pmatrix}1 & x\\ 0 & 1\end{pmatrix}\begin{pmatrix}y^{1/2} & 0\\ 0 & y^{-1/2}\end{pmatrix}\right)$$

is a function on $\mathfrak{H}^m$ so that

$$\varphi^{\natural}|_{\kappa}\gamma = \varphi^{\natural}$$

for all $\gamma \in \Gamma$. Further, $f = (f^{\#})^{\natural}$.

Proof From the definition,

$$f^{\#}(\gamma g k(\theta)) = f(\gamma g k(\theta)(i))\mu(\gamma g k(\theta), i)^{-\kappa} = f(\gamma g(i))\mu(\gamma g k(\theta), i)^{-\kappa}$$

since $SO(2)^m$ is the isotropy group of $i \in \mathfrak{H}^m$. This is

$$(f|_{\kappa}\gamma)(g(i)\mu(\gamma, g(i))^{\kappa}\mu(\gamma g k(\theta), i)^{-\kappa} = f(g(i))\mu(\gamma, g(i))^{\kappa}\mu(\gamma g k(\theta), i)^{-\kappa},$$

since $f|_{\kappa}\gamma = f$. Then we have

$$f^{\#}(\gamma g k(\theta)) = f(g(i))\mu(\gamma, g(i))^{\kappa}\mu(\gamma, gk(\theta)(i))^{-\kappa}\mu(g, k(\theta)(i))^{-\kappa}\mu(k(\theta), i)^{-\kappa} =$$

$$= f(g(i))\mu(\gamma, g(i))^{\kappa}\mu(\gamma, g(i))^{-\kappa}\mu(g, i)^{-\kappa}\mu(k(\theta), i)^{-\kappa} =$$

$$= f(g(i))\mu(g, i)^{-\kappa}\mu(k(\theta), i)^{-\kappa}$$

using again the fact that $SO(2)^m$ is the isotropy group of $i \in \mathfrak{H}^m$, and using the "cocycle property"

$$\mu(gh, z) = \mu(g, h(z))\mu(h, z).$$

That is, we have

$$f^{\#}(\gamma g k(\theta)) = f^{\#}(g)(-i\sin\theta + \cos\theta)^{-\kappa} = f^{\#}(g)e^{i\kappa\theta}.$$

The other assertions follow similarly. ∎

Remarks What is not so clear from this is how we are to distinguish the functions $f^{\#}$ obtained from *holomorphic* Hilbert modular forms, or holomorphic *cuspforms*, etc.

On the other hand, there are noteworthy advantages in this viewpoint. On the crudest level, it is certainly easier to deal with *left-invariant* (and *right-equivariant*) functions, rather than functions which transform according to some rule like $f|_{\kappa}\gamma = f$. Another point, whose ramifications are subtler, is that $SL(2, \mathbb{R})^m$ acts *on the right* on $\Gamma\backslash SL(2, \mathbb{R})^m$, whereas there was no such action on $\Gamma\backslash\mathfrak{H}^m$. This allows us to give an action of $SL(2, \mathbb{R})^m$ on spaces of functions on $\Gamma\backslash SL(2, \mathbb{R})^m$ (though certainly right translation of an $SO(2)^m$-equivariant function by a typical element of $SL(2,\mathbb{R})^m$ destroys the right $SO(2)^m$-equivariance).

Let $\hat{\mathfrak{o}} = \prod_{p<\infty} \mathfrak{o}_p$, and let $F_\infty = \prod_{p|\infty} F_p \approx \mathbb{R}^m$. (We are considering only totally real number fields F.) Let Γ be a congruence subgroup of $SL(2,F)$ (note: *not* of $GL^+(2,F)$). For present purposes, it is convenient to observe that this is equivalent to the existence of a compact open subgroup $\mathbb{K}$ of $SL(2,\hat{\mathfrak{o}})$ so that

$$\Gamma = SL(2,F) \cap SL(2,F_\infty)\mathbb{K} \qquad (\text{inside } SL(2,\mathbb{A})).$$

Conversely, any such $\mathbb{K}$ gives rise to a congruence subgroup by this formula.

PROPOSITION For any Γ and $\mathbb{K}$ as just above, the map $\iota\colon \Gamma g \to SL(2,F)g\mathbb{K}$ gives rise to a homeomorphism

$$\iota\colon \Gamma\backslash SL(2,F_\infty) \approx SL(2,F)\backslash SL(2,\mathbb{A})/\mathbb{K}.$$

Proof It is immediate that ι is well-defined, since $\Gamma \subset SL(2,F)$, etc. First, we show that ι is injective. Suppose that $\gamma g k = g'$ with $\gamma \in SL(2,F)$, g and g' in $SL(2,F_\infty)$, $k \in \mathbb{K}$. Then for any finite prime $\mathfrak{p}$ the $\mathfrak{p}$-component of γk must be 1_2. That is, certainly

$$\gamma \in SL(2,F) \cap SL(2,F_\infty)\mathbb{K} = \Gamma.$$

This yields the injectivity. For surjectivity, we use the Strong Approximation Theorem: One has

$$SL(2,\mathbb{A}) = SL(2,F)\,SL(2,F_\infty)(SL(2,F_\infty)\mathbb{K}) = SL(2,F)SL(2,F_\infty)\mathbb{K}$$

because $SL(2,F_\infty)\mathbb{K}$ is an open set in $SL(2,\mathbb{A})$ and $SL(2,F)SL(2,F_\infty)$ is dense in $SL(2,\mathbb{A})$. ■

COROLLARY In this notation, let φ be a function on $SL(2,\mathbb{R})^m$ so that

$$\varphi(\gamma g k(\theta)) = \varphi(g)e^{i\kappa\theta}$$

for $\gamma \in \Gamma$ and $k(\theta) \in SO(2)^m$. Define a function $\varphi^\#$ on $SL(2,\mathbb{A})$ by

$$\varphi^\#(\gamma g k) = \varphi(g)$$

for $\gamma \in SL(2,F)$, $g \in SL(2,F_\infty) \subset SL(2,\mathbb{A})$, $k \in \mathbb{K}$. Then $\varphi^\#$ is well-defined, is left $SL(2,F)$-invariant, right $\mathbb{K}$-invariant, and right $SO(2)$-equivariant by the representation $k(\theta) \to e^{i\kappa\theta}$. Conversely, given $\varphi^\#$ satisfying these invariance and equivariance properties, there exists a *unique* φ on $SL(2,F_\infty)$ which induces $\varphi^\#$; φ is given just by restriction.

Proof The previous proposition shows that $\varphi^\#$ is well-defined, and that φ is completely determined by φ. The other assertions are then easy and formal. ■

Definition We say that φ and $\varphi^\#$ as in the previous proposition are of weight κ.

Let Z be the center of $GL(2)$, and let $\mathbb{J}$ be the ideles of F. We need to introduce some special congruence subgroups; there is a variety of legitimate choices, and we have simply chosen one of them. Define

$$\mathbb{K}(\mathfrak{n}) = \{\gamma \in SL(2, \mathfrak{o}): \gamma \equiv 1 \text{ modulo } \mathfrak{n}\}$$

$$\mathbb{K}_{00}(\mathfrak{n}) = \left\{\gamma \in SL(2, \hat{\mathfrak{o}}): \gamma = \begin{pmatrix} a & b \\ c & d \end{pmatrix} \text{ with } c \equiv b \equiv 0 \text{ modulo } \mathfrak{n}\right\}$$

$$\tilde{\mathbb{K}}(\mathfrak{n}) = \{\gamma \in GL(2, \hat{\mathfrak{o}}): \gamma \equiv 1 \text{ modulo } \mathfrak{n}\}$$

$$\tilde{\mathbb{K}}_{00}(\mathfrak{n}) = \left\{\gamma \in GL(2, \hat{\mathfrak{o}}): \gamma = \begin{pmatrix} a & b \\ c & d \end{pmatrix} \text{ with } c \equiv b \equiv 0 \text{ modulo } \mathfrak{n}\right\}.$$

For the moment, for further developments we only consider the special where

$$\mathbb{J}/F_\infty^+ F^\times \hat{\mathfrak{o}}^\times = \{1\},$$

where F_∞^+ is the connected component of the identity in $F_\infty^\times$. That is, we require that the "narrow ideal class group" be trivial. For example, this applies to the case $F = \mathbb{Q}$.

PROPOSITION Under the hypothesis that

$$\mathbb{J}/F_\infty^+ F^\times \hat{\mathfrak{o}}^\times = \{1\},$$

for every proper ideal $\mathfrak{n}$ of $\mathfrak{o}$, we have a natural homeomorphism

$$Z(F_\infty)GL(2, F)\backslash GL(2, \mathbb{A})/\tilde{\mathbb{K}}_{00}(\mathfrak{n}) \approx SL(2, F)\backslash SL(2, \mathbb{A})/\mathbb{K}_{00}(\mathfrak{n})\{\pm 1\}^m$$

where $\{\pm 1\}^m = Z(F_\infty) \cap SO(2)^m$.

Proof Note that $\{\pm 1\}^m$ is the intersection of $SL(2, F_\infty)$ and $Z(F_\infty)$, and also of $SO(2)^m$ and $Z(F_\infty)$. First, the map

$$\iota: SL(2, F)\backslash SL(2, \mathbb{A})/\mathbb{K}_{00}(\mathfrak{n})\{\pm 1\}^m \to Z(F_\infty)GL(2, F)\backslash GL(2, \mathbb{A})/\tilde{\mathbb{K}}_{00}(\mathfrak{n})$$

defined by

$$\iota: SL(2, F)g\mathbb{K}_{00}(\mathfrak{n})\{\pm 1\}^m \to Z(F_\infty)GL(2, F)\, g\tilde{\mathbb{K}}_{00}(\mathfrak{n})$$

is certainly well-defined. To show that ι is injective, suppose that

$$\zeta\gamma g k = g'$$

with g and $g' \in SL(2, \mathbb{A})$, $\zeta \in Z(F_\infty)$, $\gamma \in GL(2, F)$, $k \in \tilde{\mathbb{K}}_{00}(\mathfrak{n})$. Then $\det(\zeta\gamma k) = 1$, from which it follows that $\det(\gamma)$ is a totally positive unit ε in $\mathfrak{o}$. Take $t \in \mathfrak{o}^x$ so that $t^2 = \varepsilon$, and replace

$$\gamma \text{ by } \gamma' = \gamma\begin{pmatrix} t & 0 \\ 0 & t \end{pmatrix}^{-1} \qquad \zeta \text{ by } \zeta' = \zeta\begin{pmatrix} t & 0 \\ 0 & t \end{pmatrix}_\infty \qquad k \text{ by } k' = k\begin{pmatrix} t & 0 \\ 0 & t \end{pmatrix}_0$$

where the subscripts ∞ and 0 denote the infinite and finite-prime parts, respectively. We still have

$$\zeta'\gamma' g k' = g'.$$

Then $\gamma' \in SL(2, F)$; since $\zeta' \in GL(2, F_\infty)$ and $k' \in GL(2, \mathbb{A}_0)$ and $\det(\zeta' k') = 1$, it must be that ζ' and k' also are in $SL(2, \mathbb{A})$. This proves the injectivity. For the surjectivity, it is necessary and sufficient to show that

$$Z(F_\infty)GL(2, F)SL(2, \mathbb{A})\tilde{\mathbb{K}}_{00}(\mathfrak{n}) = GL(2, \mathbb{A}).$$

To this end, it would suffice to show that the determinants of elements of the lefthand side of the latter expression are arbitrary elements of the ideles $\mathbb{J}$. The collection of determinants so obtainable is exactly

$$F_\infty^+ F^x \hat{\mathfrak{o}}^x.$$

This is just $\mathbb{J}$, by hypothesis. ■

Remarks It is important to note that *not* every compact open subgroup of $SL(2, \hat{\mathfrak{o}})$ contains a subgroup of the form $\mathbb{K}_{00}(\mathfrak{n})$, but by definition does contain some $\mathbb{K}(\mathfrak{n})$.

Now let

$$M(\mathfrak{n}, \kappa) = \{\mathbb{C}\text{-valued functions } f \text{ on } SL(2, \mathbb{A})\text{: } f \text{ is left } SL(2, F)\text{-invariant, of weight } \kappa, \text{ and right } \mathbb{K}(\mathfrak{n})\text{-invariant}\}.$$

The quotient $H = \mathbb{K}_{00}(\mathfrak{n})/\mathbb{K}(\mathfrak{n})$ is a *finite abelian group*, so $\mathbb{K}_{00}(\mathfrak{n})$ acts on $M(\mathfrak{n}, \kappa)$. Since H is finite, by elementary linear algebra (despite the fact that $M(\mathfrak{n}, \kappa)$ is infinite-dimensional), $M(\mathfrak{n}, \kappa)$ decomposes as a finite direct sum

$$M(\mathfrak{n}, \kappa) = \bigoplus_\chi M(\mathfrak{n}, \kappa, \chi)$$

where χ runs over the (finitely many) characters of H, and (identifying such χ with characters on $\mathbb{K}_{00}(\mathfrak{n})$) where

$$M(\mathfrak{n}, \kappa, \chi) = \{f \in M(\mathfrak{n}, \kappa) : f(gk) = \chi(k)f(g) \text{ for } k \in \mathbb{K}_{00}(\mathfrak{n})\}.$$

PROPOSITION Suppose that

$$\mathbb{J}/F_\infty^+ F^x \hat{\mathfrak{o}}^x = \{1\}.$$

Let $f \in M(\mathfrak{n}, \kappa, \chi)$. Extend χ to $\tilde{\mathbb{K}}_{00}(\mathfrak{n})$ by defining

$$\chi\left(\begin{pmatrix} \lambda & 0 \\ 0 & 1 \end{pmatrix} k\right) = \chi(k)$$

where $k \in \mathbb{K}_{00}(\mathfrak{n})$, $\lambda \in \hat{\mathfrak{o}}^{\times}$. Define a function $f^{\#}$ on $GL(2, \mathbb{A})$ by

$$f^{\#}(\zeta\gamma g k) = f(g)\chi(k)$$

for $\gamma \in GL(2, F)$, $g \in SL(2, \mathbb{A})$, $k \in \tilde{\mathbb{K}}_{00}(\mathfrak{n})$, ζ in the *connected component* Z^{+} of the identity in $Z(F_{\infty})$. Then $f^{\#}$ is well-defined, is of weight κ, and satisfies

$$f^{\#}\left(\begin{pmatrix} t & 0 \\ 0 & t \end{pmatrix} \gamma g k\right) = f(g)\chi(k)\tilde{\chi}(t)$$

for $t \in \mathbb{J}$, $\gamma \in GL(2, F)$, $g \in GL(2, \mathbb{A})$, $k \in \tilde{\mathbb{K}}_{00}(\mathfrak{n})$, where $\tilde{\chi}$ is a (finite-order) unitary Hecke character so that

$$\tilde{\chi}\left(\begin{pmatrix} t & 0 \\ 0 & t \end{pmatrix}\right) = \chi\left(\begin{pmatrix} t & 0 \\ 0 & t \end{pmatrix}\right) \qquad \text{for } t \in \hat{\mathfrak{o}}^{\times}$$

$$\tilde{\chi} \equiv 1 \qquad \text{on } Z^{+}$$

$$\tilde{\chi}(e^{i\theta_1}, \ldots, e^{i\theta_m}) = \textstyle\prod_{j=1,\ldots,m} e^{i\kappa_j\theta_j} \qquad \text{where each } \theta_j \in \mathbb{Z}.$$

Conversely, given a function $f^{\#}$ with these properties, there exists a *unique* $f \in M(\mathfrak{n}, \kappa, \chi)$ which gives rise to $f^{\#}$. (We say that $f^{\#}$ has *central character* $\tilde{\chi}$.)

Proof Note that

$$Z(\mathbb{A}) = Z^{+}Z(F)Z(\hat{\mathfrak{o}})$$

by hypothesis. The $Z(F)$ may be absorbed into $GL(2, F)$; the $Z(\hat{\mathfrak{o}})$ may be moved across to the right, giving the factor of $\chi(t)$ from the definition of the extension of χ to $\tilde{\mathbb{K}}_{00}(\mathfrak{n})$; $\tilde{\chi}$ is defined on $\{\pm 1\}^m \subset SO(2)^m$ by the weight κ; this, together with the triviality on Z^{+}, defines $\tilde{\chi}$ as a Hecke character. Then the fact that $f^{\#}$ is well-defined is a consequence of the previous proposition, as is the assertion that there exists a unique f. ■

Remarks The relative simplicity of the previous result is seriously dependent upon the hypothesis that

$$\mathbb{J}/F_{\infty}^{+}F^{\times}\hat{\mathfrak{o}}^{\times} = \{1\}.$$

In general, the map on double coset spaces discussed above is neither injective nor surjective. Nevertheless, it is useful to note the perfect correspondence between the various notions in this special case.

The general form of these results is as follows: First we prove two didactic lemmas which illustrate two of the complications; then we give a development subsuming the previous discussion with modifications necessitated by the complications.

LEMMA For every proper ideal $\mathfrak{n}$ of $\mathfrak{o}$, we have a disjoint union

$$GL(2,\mathbb{A}) \approx \coprod_{\xi} Z(F_\infty)GL(2,F)SL(2,\mathbb{A})\xi\tilde{\mathbb{K}}_{00}(\mathfrak{n})$$

where ξ runs over a set X of elements of $GL(2,\mathbb{A})$ so that the collection $\{\det\xi\colon \xi\in X\}$ is a set $\{\eta\}$ of ideles representing the strict ideal classes of $\mathfrak{o}$, i.e., representatives for the narrow ideal class group $\mathbb{J}/F_\infty^+F^\times\hat{\mathfrak{o}}^\times$.

Proof The determinants of elements of the righthand side of the above expression are arbitrary elements of the ideles $\mathbb{J}$: The set of determinants so obtainable is exactly

$$\bigcup_{\eta} F_\infty^+F^\times\hat{\mathfrak{o}}^\times\eta.$$

The hypotheses on the ξ's and on the η's were precisely made to assure that this union is $\mathbb{J}$, and that the union is disjoint. This yields the assertion of the lemma. ∎

LEMMA For g and g' in $SL(2,\mathbb{A})$,

$$g'\in Z(F_\infty)GL(2,F)g\tilde{\mathbb{K}}_{00}(\mathfrak{n})$$

if and only if

$$g'\in\delta SL(2,F)g\mathbb{K}_{00}(\mathfrak{n})\{\pm1\}^m$$

where

$$\delta=\begin{pmatrix} y^{1/2} & 0\\ 0 & y^{-1/2}\end{pmatrix}\in SL(2,F_\infty)$$

for some totally positive unit $y\in\mathfrak{o}^\times$.

Proof This comes from taking determinants, by a similar argument as used before. ∎

PROPOSITION We have a homeomorphism

$$Z(F_\infty)GL(2,F)\backslash GL(2,\mathbb{A})/\tilde{\mathbb{K}}_{00}(\mathfrak{n}) \approx \coprod_{\xi} Z(F_\infty)\Gamma_\xi\backslash GL^+(2,F_\infty)$$

where ξ runs over a set X of elements of $GL(2, \mathbb{A})$ so that the collection $\{\det \xi\colon \xi \in X\}$ is a set $\{\eta\}$ of ideles representing the strict ideal classes of $\mathfrak{o}$, i.e., representatives for the narrow ideal class group $\mathbb{J}/F_\infty^+ F^\times \hat{\mathfrak{o}}^\times$, and where Γ_ξ is a discrete subgroup of $GL^+(2, F_\infty)$ given by

$$\Gamma_\xi = GL^+(2, F) \cap GL^+(2, F_\infty)\xi\tilde{\mathbb{K}}_{00}(\mathfrak{n})\xi^{-1}.$$

Proof Let

$$X_\xi = Z(F_\infty)\Gamma_\xi \backslash GL^+(2, F_\infty).$$

Define

$$\iota\colon \coprod_\xi X_\xi \to Z(F_\infty)GL(2, F)\backslash GL(2, \mathbb{A})/\tilde{\mathbb{K}}_{00}(\mathfrak{n})$$

by

$$\iota(Z(F_\infty)\Gamma_\xi g) = Z(F_\infty)GL(2, F)g\xi\tilde{\mathbb{K}}_{00}(\mathfrak{n}).$$

The definition of Γ_ξ is designed exactly so that this map is well-defined. By considering determinants, it is clear that the union

$$\bigcup_\xi Z(F_\infty)GL(2, F)g\xi\tilde{\mathbb{K}}_{00}(\mathfrak{n})$$

is disjoint. Also by considering determinants (as in the two lemmas above), and using strong approximation for $SL(2)$,

$$\bigcup_\xi \iota(X_\xi) = \bigcup_\xi Z(F_\infty)GL(2, F)GL^+(2, F_\infty)\xi\tilde{\mathbb{K}}_{00}(\mathfrak{n}) =$$

$$= \bigcup_\xi Z(F_\infty)GL(2, F)GL^+(2, F_\infty)SL(2, \mathbb{A})\xi\tilde{\mathbb{K}}_{00}(\mathfrak{n}) = GL(2, \mathbb{A}).$$

Thus, ι is surjective and well-defined. We have to prove that

$$Z(F_\infty)\Gamma_\xi g \to Z(F_\infty)GL(2, F)g\xi\tilde{\mathbb{K}}_{00}(\mathfrak{n})$$

is injective. If

$$g'\xi = \zeta\gamma g\xi k$$

for g, g' in $GL^+(2, F_\infty)$, $\zeta \in Z(F_\infty)$, $\gamma \in GL(2, F)$, $k \in \tilde{\mathbb{K}}_{00}(\mathfrak{n})$, then

$$\gamma \in GL^+(2, F_\infty)\xi\tilde{\mathbb{K}}_{00}(\mathfrak{n})\xi^{-1} \cap GL(2, F) = \Gamma_\xi.$$

From this point the result follows easily. ■

COROLLARY The quotients $Z(F_\infty)GL(2, F)\backslash GL(2, \mathbb{A})/\tilde{\mathbb{K}}_{00}(\mathfrak{n})SO(2)^m$ are finite disjoint unions of quotients of $\mathfrak{H}^m$ by congruence subgroups. ■

Definition Fixing $\mathfrak{n}$ and $\{\xi\}$ as in the previous proposition, let

$$\Gamma_{1\xi} = GL^+(2, F) \cap GL^+(2, F_\infty)\xi\tilde{\mathbb{K}}(\mathfrak{n})\xi^{-1}$$

$$M(\Gamma_{1\xi}, \kappa) = \{\text{left } \Gamma_{1\xi}\text{-invariant functions on } GL^+(2, F_\infty) \text{ of weight } \kappa\}.$$

Now Γ_ξ normalizes $\Gamma_{1\xi}$, is of finite index over it, and the quotient $H = \Gamma_\xi/\Gamma_{1\xi}$ is an abelian group. Therefore,

$$M(\Gamma_{1\xi}, \kappa) = \bigoplus_\chi M(\Gamma_\xi, \kappa, \chi)$$

where χ runs over characters of H (identified with characters on Γ_ξ which are trivial on $\Gamma_{1\xi}$)

$$M(\Gamma_\xi, \kappa, \chi) = \{f \in M(\Gamma_{1\xi}, \kappa) \colon f(\gamma^{-1}g) = \chi(\gamma)f(g) \text{ for all } \gamma \in \Gamma_\xi, g \in GL^+(2, F_\infty)\}.$$

On the other hand, let χ be a character on $\tilde{\mathbb{K}}_{00}(\mathfrak{n})$, and let

$$\begin{aligned}\tilde{M}(\mathfrak{n}, \kappa, \chi) = \{&\text{functions } f \text{ on } GL(2, \mathbb{A}) \colon f \text{ is left } Z^+GL(2, F)\text{-invariant,}\\ &\text{weight } \kappa, \text{ and } f(gk) = \chi(k)f(g) \text{ for } k \text{ in } \tilde{\mathbb{K}}_{00}(\mathfrak{n})\}.\end{aligned}$$

COROLLARY Let $\{\xi\}$ be as above, and let χ be a character of $\tilde{\mathbb{K}}_{00}(\mathfrak{n})$ which is trivial on elements of the form

$$\begin{pmatrix} t & 0 \\ 0 & 1 \end{pmatrix} \in \tilde{\mathbb{K}}_{00}(\mathfrak{n}).$$

There is a bijection

$$\iota \colon \tilde{M}(\mathfrak{n}, \kappa, \chi) \to \bigotimes_\xi M(\Gamma_\xi, \kappa, \chi_\xi)$$

given by

$$\iota(f)_\xi(g) = f(g\xi)$$

for $g \in GL^+(2, F_\infty)$, where

$$\chi_\xi(\gamma) = \chi(\text{finite-prime part of } \xi^{-1}\gamma\xi).$$

Proof Apart from unraveling the definitions, this is an immediate consequence of the previous proposition. ■

COROLLARY The space $\tilde{M}(\mathfrak{n}, \kappa, \chi)$ decomposes further as

$$\tilde{M}(\mathfrak{n}, \kappa, \chi) = \bigotimes_{\omega} \tilde{M}(\mathfrak{n}, \kappa, \chi, \omega),$$

taking a finite direct sum over unitary Hecke characters ω so that

$$\omega(t) = \chi\left(\begin{pmatrix} t & 0 \\ 0 & t \end{pmatrix}\right)$$

for $t \in \hat{\mathfrak{o}}^{\times}$, where

$$\tilde{M}(\mathfrak{n}, \kappa, \chi, \omega) = \left\{f \in \tilde{M}(\mathfrak{n}, \kappa, \chi) : f\left(\begin{pmatrix} t & 0 \\ 0 & t \end{pmatrix} g\right) = \omega(t) f(g)\right\}.$$

Proof Let α be the restriction of χ to $Z(\mathbb{A}) \cap \tilde{\mathbb{K}}_{00}(\mathfrak{n})$. The restriction of an element f of $\tilde{M}(\mathfrak{n}, \chi)$ to $Z(\mathbb{A}) \approx \mathbb{J}$ is trivial on $F_{\infty}^{+} F^{\times}$ by definition, and is equivariant by α under translation by $\hat{\mathfrak{o}}^{\times}$. Since $\mathbb{J}/F_{\infty}^{+} F^{\times} \hat{\mathfrak{o}}^{\times}$ is finite and α is continuous, there are at most finitely many Hecke characters ω with the indicated property. Letting $\{h\}$ be a set of representatives for the finite set

$$Z^{+} Z(F) \backslash Z(\mathbb{A}) / Z(\mathbb{A}) \cap \tilde{\mathbb{K}}_{00}(\mathfrak{n}) \approx \mathbb{J}/F_{\infty}^{+} F^{\times} \hat{\mathfrak{o}}^{\times},$$

it is elementary to see that the projection from $\tilde{M}(\mathfrak{n}, \kappa, \chi)$ to $\tilde{M}(\mathfrak{n}, \kappa, \chi, \omega)$ is given by

$$f(g) \to {}^{\#}\{h\}^{-1} \sum_{h} f(hg) \omega(h^{-1})$$ ■

COROLLARY In the space $\tilde{M}(\mathfrak{n}, \kappa, \chi, \omega)$, the central character ω determines the character χ on $\tilde{\mathbb{K}}_{00}(\mathfrak{n})/\tilde{\mathbb{K}}(\mathfrak{n})$ completely, and determines the parity of the components κ_j of the weight κ. Therefore, the space $\mathrm{Wfm}(\mathfrak{n}, \kappa, \omega)$ defined as

$$\mathrm{Wfm}(\mathfrak{n}, \kappa, \omega) = \tilde{M}(\mathfrak{n}, \kappa, \chi, \omega)$$

unambiguously determines χ. Also, the finite-prime part of the conductor of ω divides $\mathfrak{n}$ (since ω is trivial on $Z(\mathbb{A}) \cap \tilde{\mathbb{K}}(\mathfrak{n})$). Therefore, for a fixed weight κ and unitary Hecke character ω the space

$$\mathrm{Wfm}(\kappa, \omega) = \bigcup_{\mathfrak{n}} \mathrm{Wfm}(\mathfrak{n}, \kappa, \omega) \qquad \text{(union over } \mathfrak{n} \text{ divisible by the conductor of } \omega)$$

is unambiguously defined. ■

Remarks In the previous discussion, if we had extended the character on $\mathbb{K}_{00}(\mathfrak{n})$ to $\tilde{\mathbb{K}}_{00}(\mathfrak{n})$ in a different way, the central character would be changed by the square of a Hecke character trivial on F_∞^x. On the other hand, for $f \in \mathrm{Wfm}(\kappa, \omega)$ and for any Hecke character β trivial on F_∞^x, the function $f_\beta(g) = \beta(\det g)f(g)$ has central character $\beta^2\omega$. Thus, the choice of extension has only a trivial effect: No generality is lost in the previous discussion.

3.2 Hecke Operators

In this section we demonstrate the equivalence of the classical and adelic versions of Hecke operators in simple situations. In particular, we suppose that F is such that the *narrow class number is 1*; that is, that

$$\mathbb{J}/F_\infty^+ F^x \hat{\mathfrak{o}}^x = \{1\}.$$

(This was, in effect, the hypothesis we used to give a classical theory of Hecke operators for $SL(2, \mathfrak{o})$.) Then, from the discussion of the previous section, we have

$$SL(2,\mathfrak{o})\backslash\mathfrak{H}^m \approx SL(2,\mathfrak{o})\backslash SL(2,F_\infty)/SO(2)^m \approx$$

$$\approx SL(2,F)\backslash SL(2,\mathbb{A})/SO(2)^m SL(2,\hat{\mathfrak{o}}) \approx$$

$$\approx Z^+ GL^+(2,\mathfrak{o})\backslash GL^+(2,F_\infty)/SO(2)^m \approx$$

$$\approx Z(F_\infty)GL(2,F)\backslash GL(2,\mathbb{A})/SO(2)^m GL(2,\hat{\mathfrak{o}})$$

and

$$SL(2,\mathfrak{o})\backslash SL(2,F_\infty) \approx SL(2,F)\backslash SL(2,\mathbb{A})/SL(2,\hat{\mathfrak{o}}) \approx$$

$$\approx Z^+ GL^+(2,\mathfrak{o})\backslash GL^+(2,F_\infty) \approx Z(F_\infty)GL(2,F)\backslash GL(2,\mathbb{A})/GL(2,\hat{\mathfrak{o}})$$

From that same discussion, for a left $Z^+ GL^+(2,\mathfrak{o})$-invariant function f on $GL^+(2,F_\infty)$, we define a function $f^\#$ on $GL(2,\mathbb{A})$ by

$$f^\#(\zeta\gamma g k) = f(g)$$

for $\zeta \in Z(F_\infty)$, $\gamma \in GL(2,F)$, and $k \in GL(2,\hat{\mathfrak{o}})$; this function $f^\#$ is well-defined, is left $Z(F_\infty)GL(2,F)$-invariant, and right $GL(2,\hat{\mathfrak{o}})$-invariant.

Definition Let $\mathfrak{n}$ be an ideal of $\mathfrak{o}$ which is a power of a prime $\mathfrak{p}$. Let $\Delta(\mathfrak{n})$ be the collection of 2×2 matrices over $\mathfrak{o}$ whose determinant is totally positive and generates $\mathfrak{n}$. Define a (semi-classical) Hecke operator on left $GL^+(2,\mathfrak{o})$-invariant function f on $GL^+(2,F_\infty)$ by

$$(T_\mathfrak{n} f)(g) = \sum_\delta f(\delta g) \qquad (\text{for } g \in GL^+(2,F_\infty),\ \delta \text{ summed over } GL^+(2,\mathfrak{o})\backslash\Delta(\mathfrak{n})).$$

On the other hand, let $\Delta'(\mathfrak{n})$ be the set of 2×2 matrices over $\mathfrak{o}_{\mathfrak{p}}$ whose determinant generates $\mathfrak{n}\mathfrak{o}_{\mathfrak{p}}$, and let $\varphi_{\mathfrak{n}}$ be the characteristic function of $\Delta'(\mathfrak{n})$. Let μ be a Haar measure on $GL(2, F_{\mathfrak{p}})$ so that the compact open subgroup $GL(2, \mathfrak{o}_{\mathfrak{p}})$ has measure 1. Define an adelic Hecke operator on left $GL(2, F)$-invariant, and right $GL(2, \hat{\mathfrak{o}}_{\mathfrak{p}})$-invariant function Φ by the *convolution* operator

$$(T'_{\mathfrak{n}}\Phi)(g) = \int_{GL(2,F_{\mathfrak{p}})} \Phi(gh^{-1})\varphi(h)\,d\mu(h).$$

PROPOSITION Under the above hypotheses, for a left $GL^{+}(2, \mathfrak{o})$-invariant function f on $GL^{+}(2, F_{\infty})$, we have

$$(T_{\mathfrak{n}}f)^{\#} = T'_{\mathfrak{n}}(f^{\#}).$$

That is, on the adele group the (prime power) Hecke operators are given as convolution operators on the right.

Proof Since $f^{\#}$ is completely determined by its values on $GL^{+}(2, F_{\infty})$, it suffices to compute both $(T_{\mathfrak{n}}f)^{\#}(g)$ and $(T'_{\mathfrak{n}}(f^{\#}))(g)$ just for g in $GL^{+}(2, F_{\infty})$. Regarding the integral

$$(T'_{\mathfrak{n}}(f^{\#})(g) = \int_{GL(2,F_{\mathfrak{p}})} f^{\#}(gh^{-1})\varphi(h)\,d\mu(h)$$

we first make two observations:

i. Since $g \in GL^{+}(2, F_{\infty})$ and $h \in GL(2, F_{\mathfrak{p}})$, g and h^{-1} commute;
ii. Since $f^{\#}$ is right $GL(2, \mathfrak{o}_{\mathfrak{p}})$-invariant and φ is left- and right $GL(2, \mathfrak{o}_{\mathfrak{p}})$-invariant, we could integrate over h in

$$GL(2, \mathfrak{o}_{\mathfrak{p}})\backslash\Delta'(\mathfrak{n}).$$

Let π be a totally positive generator for $\rho \subset \mathfrak{o}$, and take $\mathfrak{n} = \pi^{n}\mathfrak{o}$; let $\pi_{\mathfrak{p}}$ be the image of π in $F_{\mathfrak{p}}$. Since $\mathfrak{o}_{\mathfrak{p}}$ is a principal ideal domain, this quotient has representatives

$$X = \left\{\begin{pmatrix} a & b \\ 0 & d \end{pmatrix} : a, d \text{ are powers of } \pi_{\mathfrak{p}},\ ad = \pi_{\mathfrak{p}}^{n},\ b \in \mathfrak{o}_{\mathfrak{p}} \bmod d\mathfrak{o}_{\mathfrak{p}}\right\}.$$

Since we have normalized the Haar measure by taking $\mu(GL(2, \mathfrak{o}_{\mathfrak{p}})) = 1$, the induced measure on this finite discrete quotient is counting measure. Also, note that $\mathfrak{o}_{\mathfrak{p}}/d\mathfrak{o}_{\mathfrak{p}}$ has representatives in $\mathfrak{o}$. Therefore,

$$T'_{\mathfrak{n}}(f^{\#})(g) = \int_{GL(2,F_{\mathfrak{p}})} f^{\#}(gh^{-1})\varphi(h)\,d\mu(h) = \sum_{\delta \in X} f^{\#}(g\delta^{-1}).$$

By the Chinese Remainder Theorem, for each $\delta \in X$ there is an upper triangular matrix $p \in \Delta(\mathfrak{n}) \subset GL^{+}(2, F)$ so that

$$p = \delta \qquad \text{(when projected to } GL(2, F_{\mathfrak{p}})\text{)}$$

$$p \in GL(2, \mathfrak{o}_v) \qquad \text{(when projected to } GL(2, F_v) \text{ for finite primes } v \neq \mathfrak{p}\text{)}.$$

Let p_∞ be the projection of p to $GL^+(2, F_\infty)$, and let

$$p' = \text{projection of } p \text{ to } \prod_{\text{finite primes } v \neq \mathfrak{p}} GL(2, \mathfrak{o}_{\mathfrak{p}}).$$

Then, for each $\delta \in X$,

$$f^\#(g\delta^{-1}) = f^\#(\delta^{-1}g) = f^\#(p\delta^{-1}g)$$

by the left $GL^+(2, F)$-invariance of $f^\#$. This is

$$f^\#(p_\infty p'g) = f^\#(p_\infty gp') = f^\#(p_\infty g) = f(p_\infty g)$$

since $p_\infty g \in GL^+(2, F_\infty)$. It is easy to see that the map $\delta \to p$ induces a bijection

$$X \to GL^+(2, \mathfrak{o})\backslash\Delta(\mathfrak{n}).$$

Therefore, we can conclude that

$$T'_{\mathfrak{n}}(f^\#) = \sum_{\delta \in X} f^\#(g\delta^{-1}) = \sum_{\delta \in X} f(p_\infty g) = (T_{\mathfrak{n}}f)(g) = (T_{\mathfrak{n}}f)^\#(g),$$

since $g \in GL^+(2, F_\infty)$. This gives the assertion. ∎

3.3 Holomorphic Automorphic Forms

Let f be a twice differentiable function on $\mathfrak{H}^m$ so that $f|_\kappa\gamma = f$ for all γ in some subgroup Γ of $SL(2, \mathbb{R})^m$, for some weight $\kappa \in \mathbb{Z}^m$. Define (as above) an associated function $f^\#$ on $SL(2, \mathbb{R})^m$ by

$$f^\#(g) = f(g(i))\mu(g, i)^{-\kappa}.$$

In this section we will use "Iwasawa coordinates" on $SL(2, \mathbb{R})$ or on $SL(2, \mathbb{R})^m$ given by

$$\begin{pmatrix} 1 & x \\ 0 & 1 \end{pmatrix}\begin{pmatrix} y^{1/2} & 0 \\ 0 & y^{-1/2} \end{pmatrix}\begin{pmatrix} \cos\theta & \sin\theta \\ -\sin\theta & \cos\theta \end{pmatrix}.$$

(In Appendix A.4 we determine the invariance properties of the differential operators given below in these coordinates.)

PROPOSITION Let L_j be the Maass lowering operator

$$L_j = e^{-2i\theta}\left(y_j\frac{\partial}{\partial x_j} + iy_j\frac{\partial}{\partial y_j} - \frac{1}{2}\frac{\partial}{\partial\theta_j}\right)$$

on the j-th factor $SL(2,\mathbb{R})$ of $SL(2,\mathbb{R})^m$. Then, f is holomorphic in z_j if and only if $L_j f^\# = 0$.

Proof As verified in Appendix A.4, each L_j is a left-invariant differential operator on smooth functions on $SL(2,\mathbb{R})^m$. In the Iwasawa coordinates, from the definition of $f^\#$,

$$f^\#\left(\begin{pmatrix}1 & x\\ 0 & 1\end{pmatrix}\begin{pmatrix}y^{1/2} & 0\\ 0 & y^{-1/2}\end{pmatrix}\begin{pmatrix}\cos\theta & \sin\theta\\ -\sin\theta & \cos\theta\end{pmatrix}\right) = y^{\kappa/2}f(x+iy)e^{i\kappa\theta}.$$

Let

$$\delta = (0,\ldots,0,1,0,\ldots,0)$$

with the 1 at the j-th entry. Then

$$L_j f^\# = e^{-2i\theta}(y_j y^{\kappa/2}\, e^{i\kappa\theta}\,\partial f/\partial x_j + iy_j e^{i\kappa\theta}\,\partial(y^{\kappa/2}f)/\partial y_j - \tfrac{1}{2}y^{\kappa/2}f\partial e^{i\kappa\theta}/\partial\theta_j) =$$

$$= e^{-2i\theta}(y_j y^{\kappa/2}\, e^{i\kappa\theta}\,\partial f/\partial x_j + iy_j y^{\kappa/2} e^{i\kappa\theta}\,\partial f/\partial y_j +$$

$$+ iy_j(\tfrac{1}{2}\kappa_j)y^{\kappa/2-\delta}fe^{i\kappa\theta} - \tfrac{1}{2}y^{\kappa/2}f(i\kappa_j)e^{i\kappa\theta}) =$$

$$= e^{-2i\theta}y_j y^{\kappa/2}\, e^{i\kappa\theta}\{\partial f/\partial x_j + i\partial f/\partial y_j\}.$$

The differential operator in curly brackets is (up to a constant) just $\partial/\partial\bar{z}$. ■

PROPOSITION Let Y_j be the Casimir operator

$$Y_j = y_j^2(\partial^2/\partial x_j^2 + \partial^2/\partial y_j^2) - \tfrac{1}{2}y_j\partial^2/\partial x_j\partial\theta_j$$

on the j-th factor $SL(2,\mathbb{R})$ of $SL(2,\mathbb{R})^m$. If f is holomorphic in z_j, then

$$Y_j f = \tfrac{1}{2}\kappa_j(\tfrac{1}{2}\kappa_j - 1)f.$$

Proof Direct calculation: Let δ be as in the previous proof. Write

$$f^\#\left(\begin{pmatrix}1 & x\\ 0 & 1\end{pmatrix}\begin{pmatrix}y^{1/2} & 0\\ 0 & y^{-1/2}\end{pmatrix}\begin{pmatrix}\cos\theta & \sin\theta\\ -\sin\theta & \cos\theta\end{pmatrix}\right) = y^{\kappa/2}f(x+iy)e^{i\kappa\theta}.$$

Then, using subscripts to denote partial derivatives,

$$Y_j f^{\#}\left(\begin{pmatrix}1 & x\\ 0 & 1\end{pmatrix}\begin{pmatrix}y^{1/2} & 0\\ 0 & y^{-1/2}\end{pmatrix}\begin{pmatrix}\cos\theta & \sin\theta\\ -\sin\theta & \cos\theta\end{pmatrix}\right) =$$

$$= y^{\delta}(y^{\delta}\{f_{x_jx_j} + (\tfrac{1}{2}\kappa_j)(\tfrac{1}{2}\kappa_j - 1)y^{-2\delta}f + 2(\tfrac{1}{2}\kappa_j)y^{-\delta}f_{y_j} + f_{y_jy_j}\} - i\kappa_j f_{x_j})y^{\kappa/2}e^{i\kappa\theta}.$$

If f is holomorphic in z_j, then $f_{y_j} = if_{x_j}$, and this becomes

$$= \{y^{2\delta}[f_{x_jx_j} + f_{y_jy_j}] + (\tfrac{1}{2}\kappa_j)(\tfrac{1}{2}\kappa_j - 1)f\}\, y^{\kappa/2}\, e^{i\kappa\theta} =$$

$$= y^{2\delta}[f_{x_jx_j} + f_{y_jy_j}]y^{\kappa/2}\, e^{i\kappa\theta} + (\tfrac{1}{2}\kappa_j)(\tfrac{1}{2}\kappa_j - 1)f^{\#}.$$

Also

$$f_{x_jx_j} + f_{y_jy_j} = 0,$$

and we have the asserted eigenvalue property. ■

3.4 Holomorphic Hecke Eigenfunctions

We make explicit some facts implicit in the accumulation of results we have so far:

THEOREM Fix a compact open subgroup $\mathbb{K}$ of $GL(2, \hat{\mathfrak{o}})$, and fix a weight κ. Let S be the finite set of finite primes from which the projection of $\mathbb{K}$ to $GL(2, F_{\mathfrak{p}})$ is not $GL(2, \mathfrak{o})$. Fix a central character ω which is unramified at finite primes not in S. Let

$$\tilde{H} = \bigotimes_{\mathfrak{p}\text{ finite not in } S} \text{(spherical Hecke algebra at } \mathfrak{p}).$$

Then

i. The space V of right $\mathbb{K}$-invariant weight κ holomorphic cuspforms with central character ω on $GL(2, \mathbb{A})$ is finite-dimensional.
ii. The space V admits an orthogonal decomposition $V = \oplus_\lambda V(\lambda)$ into eigenspaces

$$V(\lambda) = \{v \in V : v * \eta = \lambda(\eta)v \text{ for every } \eta \text{ in } \tilde{H}\}$$

over algebra representations $\lambda: \tilde{H} \to \mathbb{C}$.
iii. Within each $V(\lambda)$ which is not merely $\{0\}$, there is at least one non-zero vector which is a simultaneous eigenvector for the special Hecke operators at all primes in S.

Proof We have proven the finite dimensionality in a classical setting. Since the adelic quotient space

$$Z(\mathbb{A})GL(2,F)\backslash GL(2,\mathbb{A})/\mathbb{K}$$

is just a *finite* union of classical quotients $\Gamma\backslash\mathfrak{H}^m$, this yields the finite dimensionality. The algebra H is commutative and closed under adjoints with respect to the Petersson inner product. Further, as the operators from H commute with all the lowering operators, H stabilizes the finite-dimensional space V; by finite-dimensional linear algebra H must decompose V as claimed. The special Hecke operators at primes in S commute with each other, and commute with the Hecke operators in H, so must stabilize each $V(\lambda)$; again by the finite dimensionality of $V(\lambda)$ there must be at least one non-zero simultaneous eigenvector if $V(\lambda)$ is not trivial. ■

Remarks In this observation there is indeed use made of the property of holomorphy; in no other cases have we established the finite dimensionality of any space of cuspforms stable under Hecke operators.

3.5 Fourier Expansions of Holomorphic Automorphic Forms

Both as an explicit example and for subsequent application, it is of interest to consider the precise nature of Fourier expansions of holomorphic automorphic forms. This is the easiest example to consider, since the "infinite-prime part" of the Fourier coefficients is essentially exponential functions. For any other type of automorphic form these exponential functions would be replaced by more complicated transcendental functions.

We will consider only holomorphic automorphic forms f of weight $(\kappa,\ldots,\kappa)$ on $GL(2,\mathbb{A})$, with central character ω which is trivial on

$$Z^+ = \{\text{connected component of the identity in } Z(F_\infty)\}.$$

Suppose that f is right-invariant under

$$\mathbb{K} = \left\{g \in GL(2,\hat{\mathfrak{o}}): g \equiv \begin{pmatrix} * & 0 \\ 0 & 1 \end{pmatrix} \text{ modulo } \mathfrak{n}\right\}.$$

Let $\mathbb{J}^+$ be the subgroup of the ideles consisting of elements whose infinite-prime parts are all positive. Consider elements g of $GL(2,\mathbb{A})$ of the form

$$g = u(x)\ell(y) = \begin{pmatrix} 1 & x \\ 0 & 1 \end{pmatrix}\begin{pmatrix} y & 0 \\ 0 & 1 \end{pmatrix}$$

with $x \in \mathbb{A}$ and $y \in \mathbb{J}^+$. Note that it follows from Strong Approximation (Appendix A.3) that

$$GL(2,\mathbb{A}) = Z(\mathbb{A})GL(2,F)\{\text{ such elements } g\}\mathbb{K}SO(2)^m.$$

That is, the values of f on such elements g determines the automorphic form completely.

PROPOSITION With g as just above,

$$f(g) = \|y\|^{\kappa/2}\{c_0(y) + \sum_{\xi}^{+} \tau(\xi x)V(\xi y)\exp(-2\pi \operatorname{Tr} \xi_\infty y_\infty)\}$$

where the sum is over totally positive elements ξ of F, where the functions $V(y)$ and $c_0(y)$ of $y \in \mathbb{J}^+$ depend only upon the finite-prime part y_0 of y.

Proof For fixed x_0 and y_0, as a function of $g_\infty \in GL^+(2,\mathbb{R})^m$ the automorphic form f is a holomorphic automorphic form on $GL^+(2,\mathbb{R})$, so (by sections 3.1 and 3.3, essentially reducing to the classical case) is of the form

$$\|y_\infty\|^{\kappa/2}\{c_0(y_0) + \sum_{\xi}^{+} \tau_\infty(\xi_\infty x_\infty)c(\xi, x_0, y_0)\exp(-2\pi \operatorname{Tr} \xi_\infty y_\infty)\}.$$

As a function of $x \in \mathbb{A}$, f does have a Fourier expansion, so there is an expansion of the form

$$f(g) = \|y\|^{\kappa/2}\{c_0'(y_0) + \sum_{\xi \in F^\times} \tau(\xi x)c'(\xi, y_0)\exp(-2\pi \operatorname{Tr} \xi_\infty y_\infty)\}$$

where we have also divided $c(\xi, x_0, y_0)$ by $\|y_0\|^{\kappa/2}$. Left multiplication of g by $\ell(t)$ with totally positive $t \in F^\times$ leaves f invariant and transforms the Fourier expansion to

$$f(g) = \|y\|^{\kappa/2}\{c_0(t_0 y_0) + \sum_{\xi \in F^\times} \tau(\xi t x)c'(\xi, (ty)_0)\exp(-2\pi \operatorname{Tr} t_\infty \xi_\infty y_\infty)\} =$$

$$= \|y\|^{\kappa/2}\{c_0(t_0 y_0) + \sum_{\xi \in F^\times} \tau(\xi x)c'(\xi t^{-1}, (ty)_0)\exp(-2\pi \operatorname{Tr} \xi_\infty y_\infty)\}$$

by replacing ξ by ξt^{-1}. Therefore, by the uniqueness of Fourier expansions,

$$c'(\xi, y_0) = c'(1, (\xi y)_0).$$

■

PROPOSITION In the notation of the previous proposition, if $\kappa > 0$, and if V is identically zero on $\mathbb{J}^+$, then f itself is identically zero.

Proof If V is identically zero on $\mathbb{J}^+$, then for any fixed finite-part g_0 of g the classical automorphic form $f^*(z)$ on $\mathfrak{H}^m$ given by

$$f^*(x_\infty + iy_\infty) = \|y_\infty\|^{-\kappa/2} f(u(x_\infty)\ell(y_\infty)g_0)$$

(with $x_\infty \in \mathbb{R}^m$ and $y_\infty \in (0,\infty)^m$) is a constant. (This "unadelization" is a reversing of the "adelization" discussed in section 3.1). Yet f^* is also of weight $\kappa > 0$. Let

$$\gamma = \begin{pmatrix} 1 & 0 \\ v & 1 \end{pmatrix}$$

be an element of $SL(2, \mathfrak{o})$ so that $f^*|_\kappa \gamma = f^*$; there is a non-zero ideal $\mathfrak{a}$ so that this holds for every $v \in \mathfrak{a}$. Then for $z \in \mathfrak{H}^m$ we have

$$c_0 = f^*(z) = (f^*|_\kappa \gamma)(z) = f^*(\gamma z)(vz + 1)^{-\kappa} = c_0(vz + 1)^{-\kappa}.$$

Fixing z and letting v become large (in a Euclidean sense) shows that $c_0 = 0$ also. That is, if V is identically zero, then $c_0 = 0$.

If V and c_0 are both zero, then

$$f(u(x)\ell(y)) = 0$$

for all $x \in \mathbb{A}$ and $y \in \mathbb{J}^+$. Since f has central character ω and weight κ, this implies that

$$f\left(\begin{pmatrix} a & b \\ 0 & d \end{pmatrix}\begin{pmatrix} \cos\theta & \sin\theta \\ -\sin\theta & \cos\theta \end{pmatrix}\right) = 0$$

where $ad^{-1} \in \mathbb{J}^+$ and $\theta \in \mathbb{R}^m$. Since f is left $GL(2, F)$-invariant and right $\mathbb{K}$-invariant, f is certainly identically zero on

$$SL(2, F)SL(2, \mathbb{R})^m(\mathbb{K} \cap SL(2, \mathbb{A})) = SL(2, \mathbb{A}),$$

invoking Strong Approximation. Then, further, f is identically zero on

$$GL(2, F)Z^+ SL(2, \mathbb{A})\{\ell(y): y \in \mathbb{J}^+\}\mathbb{K}.$$

To verify that the latter set is all of $GL(2, \mathbb{A})$, it suffices to check that any idele may be obtained as a determinant of an element of this set. The collection of determinants obtained from this set is obviously

$$F^\times F_\infty^+ \mathbb{J}^+ \hat{\mathfrak{o}}^\times = F^\times F_\infty^+ \mathbb{J}_0.$$

The quotient

$$\mathbb{J}/(F^\times F_\infty^+ \hat{\mathfrak{o}}^\times)$$

is the narrow ideal class group of $\mathfrak{o}$. Dirichlet's theorem on primes in a generalized arithmetic progression asserts that this (finite) quotient has representatives in $\mathbb{J}_0$. Therefore,

$$GL(2, F)Z^+ SL(2, \mathbb{A})\{\ell(y): y \in \mathbb{J}^+\}\mathbb{K} = \mathbb{J}$$

as claimed. Therefore, f is identically zero on $GL(2, \mathbb{A})$. ■

Definition Suppose that f is a Hecke eigenfunction (i.e., for the spherical Hecke algebras at primes not dividing $\mathfrak{n}$, and for the special Hecke operators at primes dividing $\mathfrak{n}$). From section 2.5, this implies that the function V in the previous proposition factors over (finite) primes as

$$V(y) = V(y_0) = \prod_{\mathfrak{p}<\infty} V_{\mathfrak{p}}(y_{\mathfrak{p}}).$$

Let δ be a finite idele so that $\delta\hat{\mathfrak{o}} = \mathfrak{d}^*\hat{\mathfrak{o}}$. Then say that f is a *normalized Hecke eigenform* if, for all finite primes $\mathfrak{p}$,

$$V_{\mathfrak{p}}(\delta_{\mathfrak{p}}) = 1.$$

Remarks In section 2.5, we took a different normalization, which was simpler for the purposes there; in section 2.5 we chose a normalization which would give

$$\|\delta\|^{\kappa/2} V_{\mathfrak{p}}(\delta_{\mathfrak{p}}) = 1.$$

From the general discussion of Hecke operators and Fourier expansions in section 2.5, if f is a normalized Hecke eigenform *in the present sense*, then:

i. For finite $\mathfrak{p}$ not dividing $\mathfrak{n}$ there are $\alpha, \beta \in \mathbb{C}$ (depending on $\mathfrak{p}$) so that

$$\|\delta\|^{-\kappa/2}\|y\|^{\kappa/2} V_{\mathfrak{p}}(\delta_{\mathfrak{p}}\pi^n) = \begin{cases} (\alpha^{n+1} - \beta^{n+1})/(\alpha - \beta) & \text{for } n \geq 0 \\ 0 & \text{for } n < 0. \end{cases}$$

ii. For $\mathfrak{p}$ dividing $\mathfrak{n}$ there is a complex number α so that

$$\|\delta\|^{-\kappa/2}\|y\|^{\kappa/2} V_{\mathfrak{p}}(\delta_{\mathfrak{p}}\pi^n) = \begin{cases} \alpha^n & \text{for } n \geq 0 \\ 0 & \text{for } n < 0. \end{cases}$$

This much did not depend on f being holomorphic, nor did the finite-prime part of the (standard) L-functions obtained from cuspforms by a (Mellin) integral transform in section 2.7.

Remarks It may seem that by restricting our attention to $y \in \mathbb{J}^+$ we have neglected something. For *holomorphic* automorphic forms f this is not so, as can be seen through the following. Let F^+ be the totally positive elements of F. Then there is an *injection*

$$F^+\backslash\mathbb{J}^+ \to F^\times\backslash\mathbb{J}.$$

Further, it is easy to see that $F^\times\mathbb{J}^+ = \mathbb{J}$, so this map is actually a *bijection*. Let

$$f(g) = W_0(g) + \sum_{\xi \in F^\times} \tau(x\xi)W(\ell(\xi)g)$$

be the Fourier expansion of f, where W_0 is the 0-th Fourier coefficient. In the notation above (with a Hecke character χ), the integral giving the standard L-function attached to f is

$$\int_{\mathbb{J}/F^\times} \left\{ \sum_{\xi \in F^\times} W(\ell(\xi y)) \right\} \chi(y)\, d^\times y = \int_{\mathbb{J}^+/F^+} \left\{ \sum_{\xi \in F^\times} W(\ell(\xi y)) \right\} \chi(y)\, d^\times y =$$

$$= \int_{\mathbb{J}^+} \left\{ \sum_{\xi \in F^\times/F^+} W(\ell(\xi y)) \right\} \chi(y)\, d^\times y.$$

The salient feature of holomorphic forms is that for $y \in \mathbb{J}^+$, $W(\ell(\xi y)) = 0$ unless ξ is totally positive (or 0); therefore, the nature of the quotient $F^\times/F^+$ is irrelevant, and the integral is just

$$\int_{\mathbb{J}^+} W(\ell(y))\chi(y)\, d^\times y.$$

When f is a *normalized Hecke eigenform*, this integral factors over primes, and the finite-prime parts are evaluated as in section 2.7. For each infinite-prime, $\mathfrak{p}$, the $\mathfrak{p}$-factor of the integral considered in section 2.7 is of the form

$$\int_{(0,\infty)} y^s y^{\kappa/2} \exp(-2\pi y)\, d^\times y = (2\pi)^{-(s+\kappa/2)}\Gamma(s + \tfrac{1}{2}\kappa).$$

This is the same as was obtained in the classical style. Further, even if the weight is not of the form $(\kappa, \ldots, \kappa)$, the same computation arises.

3.6 Further Remarks

We have already seen that, at level one with narrow class number one, the classical and adelic Hecke operators are identical. However, the adelic operators *always* may be defined as at level one and have the same properties as at level one (at good primes). To define Hecke operators (even at good primes) for higher-level modular forms in a classical manner is not a trivial matter, and it is not easy to verify the expected properties. Certainly the adelic view is completely efficient in separating matters regarding distinct primes.

Also, if we are at level one with narrow class number one, then the L-functions obtained as Mellin transforms classically or adelically are identical (up to some normalizations). However, if, for example, the narrow class number h of $\mathfrak{o}$ is greater than one, the classical Mellin transform is only one of h parts of the adelic integral, and therefore the classical integral typically has no Euler product expansion. In any case, the adelic approach makes the existence of an Euler product *formal*, resembling the treatment of L-functions of number fields given in Tate's thesis.

For holomorphic cuspforms, the classical estimates suffice to assure convergence, analytic continuation, etc. of the Mellin transform L-functions. Adelically or classically, the *infinite prime* Euler factors are gamma functions with exponential functions.

A full answer to the question of the nature of Hecke operators and their eigenfunctions at *bad primes* is more difficult (necessitating the idea of "newform"), and we will not pursue it here (see [A], [M], [C]). As necessitated by circumstances, this issue and others are illuminated to some degree by going further in adjustment of viewpoint and technique, to associate L-functions to "irreducible cuspidal representations" ρ of $GL(2, \mathbb{A})$ (as in [JL]). In effect, one considers the right translation representation ρ of $GL(2, \mathbb{A})$ on the subspace of

$$L^2(Z(\mathbb{A})GL(2, F)\backslash GL(2, \mathbb{A}))$$

generated by right translations by $GL(2, \mathbb{A})$ of a cuspform f. The irreducibility of this representation space is equivalent to a suitable general version of a requirement that f be an eigenfunction for invariant differential operators and Hecke operators. Presuming that this Hilbert space representation is irreducible, (from [GGP]) one knows that it is the (completion of a restricted) tensor product of representations of the local factors $GL(2, F_{\mathfrak{p}})$; the $\mathfrak{p}$-th Euler factor of the associated L-function is prescribed from the nature of these "local" representations.

4 Eisenstein Series

In considering "higher-level" Eisenstein series (i.e., for congruence subgroups, with characters, etc.) it is *very* convenient to use the formalism of adeles, if only as a technical device. Indeed, this makes visible the fact that, under certain hypotheses which are desirable anyway, the Fourier coefficients of Eisenstein series have Euler products. That these Euler products are ratios of L-functions (over the base field) then comes out of adelic Mellin transform expressions for these. Also, this makes somewhat easier the consideration of "singular values" of Eisenstein series, i.e., using a "convergence factor" to sum what *would* be holomorphic Eisenstein series, but are of such low weight that the defining series does not converge. For example, weight $(2, \ldots, 2)$ automorphic forms are of interest since they can be made into invariant differential forms, but the Eisenstein series fail to converge just at this weight.

There is a certain genuinely unavoidable burden of notation regarding Eisenstein series, and there does not seem to be a universal normalization (nor perhaps is there any normalization which suits all purposes). We have maintained a semblance of reasonableness by having a notation which will always be roughly of the form

$$E(g;\ \text{weight; congruence conditions; } s\text{; central character; auxiliary character}).$$

Here $g \in GL(2, \mathbb{A})$, $s \in \mathbb{C}$. At first, the "congruence conditions" are only described via a finite-adele Schwartz function, but later are more specific. The auxiliary character datum is useful and necessary because of the distinction between $SL(2)$ and $GL(2)$ (see section 4.9). When the parameter s is specialized to $s = \frac{1}{2}(\text{weight}) > 1$, this Eisenstein series is a holomorphic automorphic form.

Most of the facts portrayed here have been known for a very long time in some form. The most recent developments here are in sections 4.7 and 4.10. In section 4.7, we give a formulation of a result essentially due to Hecke (though considered also by Eisenstein himself). The result of section 4.10 is the most striking idea here; it is due to Rankin, who used it (together with a beautiful old result [L] of Landau) to give the best estimate known at that time (1939) on Fourier coefficients of cuspforms.

4.1 Definitions and Integral Representations

First, we give a very general and formal definition of an Eisenstein series, and then a very useful integral representation for certain of these.

Definition Let $\varepsilon(g)$ be a $\mathbb{C}$-valued function on $GL(2, \mathbb{A})$ which is right $\mathbb{K}$-finite, left $P(F)$-invariant, left $U(\mathbb{A})$-invariant, and for $p \in P(\mathbb{A})$ we have

$$\varepsilon(pg) = \chi(p)\varepsilon(g)$$

for all $g \in GL(2, \mathbb{A})$ and for some character on (the abelian group) $P(\mathbb{A})/P(F)U(\mathbb{A})$. Then the Eisenstein series attached to ε is

$$E(g; \varepsilon) = \sum_{\gamma \in P(F)\backslash GL(2,F)} \varepsilon(\gamma g).$$

Remarks From this very general definition it is not so easy to understand the convergence, etc., of the defining series.

Let $S = S(\mathbb{A}^2)$ be the space of "Schwartz functions", i.e., that function space spanned by (finite) linear combinations of functions

$$\varphi(v) = \varphi_\infty(v_\infty)\varphi_0(v_0),$$

where

$$\varphi_\infty \in \text{the usual Schwarz space on } (\mathbb{R}^2)^n \approx \mathbb{R}^{2n},$$

$$\varphi_0 \text{ is continuous, uniformly locally constant, and compactly supported on } \mathbb{A}_0^2,$$

where $\mathbb{A}_0$ is the finite adeles, v_0 is the finite-adele component of v, and v_∞ is the archimedean component. For $g \in GL(2, \mathbb{A})$, define the *theta series attached to* φ by

$$\Theta(g, \varphi) = \sum_{v \in F^2} \varphi(vg),$$

where v is construed as being a "row vector." Consider the positive reals $\mathbb{R}^+$ as inbedded in $\mathbb{J}$ by

$$r \to (r^{1/n}, \ldots, r^{1/n}, 1, 1, \ldots),$$

where the 1's occur at the non-archimedean places. As $\mathbb{J}/F^\times = \mathbb{R}^+ \times \mathbb{J}_1/F^\times$ and the latter quotient is compact, every character on $\mathbb{J}/F^\times$ decomposes as $t \to \|t\|^{2s}\omega(t)$ for a *unitary* Hecke character ω trivial on the copy of $\mathbb{R}^+$ and for $s \in \mathbb{C}$. Let χ be another *unitary* Hecke character.

With such ω and for $s \in \mathbb{C}$, define *the normalized Eisenstein series attached to* φ, s, and ω by

$$E^*(g; \varphi; s, \omega, \chi) = \chi(\det g)\|\det g\|^s \int_{\mathbb{J}/F^\times} \|t\|^{2s}\bar{\omega}\chi^2(t)[\Theta(tg, \varphi) - \varphi(0)]\, d^\times t,$$

where $d^\times t$ is the usual normalization of the Haar measure on $\mathbb{J}$, and we identify $\mathbb{J}$ with the center of $GL(2, \mathbb{A})$. For the moment we disregard questions of convergence, and note some formal properties. For α in the center of $GL(2, \mathbb{A})$ (identifying the center with $\mathbb{J}$),

$$E^*(\alpha g; \varphi; s, \omega, \chi) = \omega(\alpha) E^*(g; \varphi; s, \omega, \chi).$$

That is, $E^*(g; \varphi; s, \omega, \chi)$ has central character ω. Further, by construction,

$$E^*(\gamma g k; \varphi; s, \omega, \chi) = E^*(g; \varphi; s, \omega, \chi)$$

for $\gamma \in GL(2, F)$, and for k in a sufficiently small open subgroup of $GL(2, \hat{\mathfrak{o}})$. (The second assertion follows from the continuity, uniform local constancy, and compact support on $\mathbb{A}_0^2$.) Note that we do not refer to the infinite component in the second of these assertions.

Now fix $\kappa = (\kappa_1, \ldots, \kappa_m) \in \mathbb{Z}^m$, and suppose that $\varphi = \varphi_\infty \varphi_0$, where φ_0 is a Schwartz function on $\mathbb{A}_0^2$, and

$$\varphi_\infty = \prod_{\mathfrak{p}\ \text{infinite}} \varphi_\mathfrak{p},$$

where for each infinite prime $\mathfrak{p} = \mathfrak{p}_i$ we have

$$\varphi_\mathfrak{p}(v_1, v_2) = \exp(-\pi(v_{1\mathfrak{p}}^2 + v_{2\mathfrak{p}}^2))(-iv_{1\mathfrak{p}} + v_{2\mathfrak{p}})^{\kappa_i} \qquad (\kappa_i \geq 0),$$

$$\varphi_\mathfrak{p}(v_1, v_2) = \exp(-\pi(v_{1\mathfrak{p}}^2 + v_{2\mathfrak{p}}^2))(iv_{1\mathfrak{p}} + v_{2\mathfrak{p}})^{-\kappa_i} \qquad (\kappa_i \leq 0).$$

Then we also will have a right-equivariance assertion for the standard copies of $SO(2)$ inside the infinite-prime components of $GL(2, \mathbb{A})$. *With any choice of φ with these infinite-prime components, we will say that $E^*(g; \varphi, s, \omega)$ is of weight $\kappa = (\kappa_1, \ldots, \kappa_m)$*, and write

$$E^*(g; \kappa; \varphi_0; s, \omega, \chi) = E^*(g; \varphi; s, \omega, \chi).$$

PROPOSITION Granting convergence, the Eisenstein series just defined are Eisenstein series in the sense of the first definition. In particular, in the notation above, we have

$$\varepsilon(g) = \chi(\det g) \|\det g\|^s \int_\mathbb{J} \|t\|^{2s} \overline{\omega} \chi^2(t) \varphi(t(0, 1)g)\, d^\times t.$$

Proof The key point here is the bijection

$$P(F) \backslash GL(2, F) \approx F^2 - \{0\}$$

given via

$$g \to (0, 1)g.$$

Clearly the integral

$$\varepsilon(g) = \chi(\det g)\|\det g\|^s \int_{\mathbb{J}} \|t\|^{2s}\overline{\omega\chi}^2(t)\varphi(t(0,1)g)\,d^\times t$$

has the requisite properties. The sum defining $E^*(g;\varphi;s,\omega,\chi)$ is visibly of the desired form. ∎

Remarks As the simplest example, consider $F = \mathbb{Q}$, $4 \le \kappa \in \mathbb{Z}$, for p finite

$$\varphi_p = \text{characteristic function of } \mathbb{Z}_p^2$$

and φ_∞ chosen as above for weight κ. Take $\omega = \chi = 1$. Then one can compute that, with $x \in \mathbb{R}$, $y > 0$,

$$E^*\left(\begin{pmatrix}1 & x\\ 0 & 1\end{pmatrix}\begin{pmatrix}y^{1/2} & 0\\ 0 & y^{-1/2}\end{pmatrix};\varphi;s,\omega,\chi\right) = \pi^{-s}\Gamma(s)\zeta(2s)\sum_{c,d}{}'' y^s/|cz+d|^{2s+\kappa}(c\bar{z}+d)^{-\kappa} =$$

$$= \pi^{-s}\Gamma(s)\sum_{c,d}{}' y^s/|cz+d|^{2s+\kappa}(c\bar{z}+d)^{-\kappa} =$$

$$= \pi^{-s}\Gamma(s)\zeta(2s)\sum_{c,d}{}'' y^s/(cz+d)^{s+\kappa/2}(c\bar{z}+d)^{s-\kappa/2}$$

where $\sum'$ denotes the sum over $\mathbb{Z}^2 - \{0\}$, $\sum''$ denotes the sum over relatively prime integers, and in the last expression one makes the sensible choices of branches of the square roots. In particular, for $s = \frac{1}{2}\kappa$, the last expression makes visible the fact that we obtain $y^{\kappa/2}$ times the holomorphic Eisenstein series of level one and weight κ for $SL(2,\mathbb{Z})$.

PROPOSITION For $\mathrm{Re}(s) > 1$ the Eisenstein series $E^*(g;\varphi;s,\omega,\chi)$ is absolutely convergent, and uniformly so for g in compact subsets of $GL(2,\mathbb{A})$. It is a left $GL(2,F)$-invariant function of g, and is equivariant by the central character ω. Let $\mathbb{K}$ be a compact open subgroup of $\mathbb{A}_0^2$ so that $\varphi(vk) = \varphi(v)$ for $k \in \mathbb{K}$ and $v \in \mathbb{A}^2$, and $\chi(\det \mathbb{K}) = 1$. Then the Eisenstein series is right $\mathbb{K}$-invariant.

Proof All the other assertions follow easily from the definitions and previous discussion, granting convergence. We take $1 < s \in \mathbb{R}$ in what follows.

First, in the usual way

$$\mathbb{J}/F^\times \approx (0,\infty) \times (\text{compact group}).$$

It is easy to check that if we integrate $\varphi(tv)$ over t in this compact group the resulting function $\tilde{\varphi}$ of v is still a Schwartz function. Therefore, it suffices to consider the function

$$\int_{(0,\infty)} \|t\|^{2ms} \sum_{0 \ne v \in F^2} \tilde{\varphi}(tvg)\,d^\times t.$$

Since $\tilde{\varphi}$ is a Schwartz function, for each $N > 0$ there is a constant c_N so that for $v \in \mathbb{A}^2$

$$|\tilde{\varphi}(v)| \leq c_N \,\mathrm{ch}(v_0; X)(1 + \|v_\infty\|)^{-N}$$

where v_0 is the finite-adele part of v, v_∞ is the infinite-prime part, $v_\infty \to \|v_\infty\|$ is the usual Euclidean norm on $\mathbb{R}^{2m}$, X is a compact subset of $\mathbb{A}_0^2$, and $\mathrm{ch}(*; X)$ is the characteristic function of X. Thus, we need only consider the convergence of

$$\int_{(0,\infty)} \|t\|^{2ms} \sum_{0 \neq v \in F^2} \mathrm{ch}(v_0 g_0; X)(1 + \|t v_\infty g_\infty\|)^{-N}\, d^\times t =$$

$$= \int_{(0,\infty)} \|t\|^{2ms}(1 + 2mt)^{-N}\, d^\times t \times \sum_{0 \neq v \in F^2} \mathrm{ch}(v_0; X g_0^{-1}) \|v_\infty g_\infty\|^{-2ms}$$

(replacing t by $t/\|v_\infty g_\infty\|$). For $s > 1/2m$ and for N large enough the integral is nicely convergent. The sum is of an elementary sort, being a finite sum of sums over lattices in $\mathbb{R}^{2m}$; therefore, it is convergent for $2ms > 2m$, i.e., for $s > 1$. It is easy to check that the sum is a continuous function of g_0 and g_∞, etc. ∎

4.2 Analytic Continuation and Functional Equation

THEOREM An adelic Eisenstein series $E^*(g; \varphi; s, \omega, \chi)$ has an analytic continuation in $s \in \mathbb{C}$ to a meromorphic function, and has functional equation

$$E^*(g; \varphi; s, \omega, \chi) = E^*(g^{T-1}; \hat{\varphi}; 1 - s, \bar{\omega}, \bar{\chi}).$$

The poles, if any, are simple, and are at $s = 0, 1$ with respective residues

$$\chi(\det g)\,\|\det g\|\,\varphi(0)\delta(\omega) \qquad (\text{at } s = 0)$$

$$\hat{\varphi}(0)\delta(\omega) \qquad (\text{at } s = 1),$$

where

$$\delta(\omega) \int_{\mathbb{J}_1/F^\times} \omega(t)\, d^\times t = \delta(\bar{\omega})$$

is zero unless ω is trivial on $\mathbb{J}_1 = \{t \in \mathbb{J}\colon \|t\| = 1\}$. Away from the poles, the analytically continued Eisenstein series is a real-analytic function on $GL(2, \mathbb{A}_\infty)$; likewise, if $\varphi(vk) = \varphi(v)$ for all $v \in \mathbb{A}^2$ and for all k in a compact open subgroup $\mathbb{K}$ of $GL(2, \hat{\mathfrak{o}})$, and if $\chi(\det \mathbb{K}) = 1$, then $E^*(g; \varphi; s, \omega, \chi)$ is right $\mathbb{K}$-invariant for s away from the poles.

Remarks The form of this proof is essentially identical to that of the proof of analytic continuation and functional equation of L-functions of number fields given by zeta integrals, as in Tate's thesis.

Some obvious hypotheses assure that $\varphi(0) = \hat{\varphi}(0) = 0$; this is the case if the Eisenstein series is of weight κ with $\kappa \neq 0$. The constant $\delta(\omega)$ with ω the trivial character is the residue of the zeta function of F at $s = 1$.

As in the previous section, from the proof it will be clear that the Eisenstein series easily may be invariant under a strictly larger subgroup of $GL(2, \hat{\mathfrak{o}})$ than just the $\mathbb{K}$ occurring in the theorem.

Proof The asserted equality is a corollary of the adelic Poisson summation formula, as follows. Temporarily, let

$$P = \{t \in \mathbb{J}: \|t\| \geq 1\} \qquad Q = \{t \in \mathbb{J}: \|t\| < 1\}.$$

Let

$$\beta(t) = \|t\|^{2s}\bar{\omega}\chi^2(t) \qquad \alpha(g) = \chi(\det g)\|\det g\|^s.$$

Then from the definition of the Eisenstein series as a Mellin transform of the associated theta series, we have

$$E^*(g; \varphi; s, \omega, s) = \alpha(g) \int_{P/F^\times} \beta(t)[\Theta(tg; \varphi) - \varphi(0)]\, d^\times t + $$

$$+ \alpha(g) \int_{Q/F^\times} \beta(t)[\Theta(tg; \varphi) - \varphi(0)]\, d^\times t.$$

The latter integral is

$$\alpha(g) \int_{Q/F^\times} \beta(t) \sum_{\xi \in F^2} \varphi(t\xi g)\, d^\times t - \alpha(g)\varphi(0) \int_{Q/F^\times} \beta(t)\, d^\times t =$$

$$= \alpha(g)\|\det g\|^{-1} \int_{Q/F^\times} \beta(t) \sum_{\xi \in F^2} \|t\|^{-2}\hat{\varphi}(t^{-1}\xi g^{T-1})\, d^\times t - s^{-1}\alpha(g)\varphi(0)\delta(\omega)$$

by adelic Poisson summation and by directly evaluating the last integral. By replacing t by t^{-1} in the integral, this is

$$\alpha(g)\|\det g\|^{-1} \int_{P/F^\times} \|t\|^2\beta^{-1}(t)[\Theta(tg^{T-1}; \hat{\varphi}) - \hat{\varphi}(0)]\, d^\times t +$$

$$+ \alpha(g)\|\det g\|^{-1}\hat{\varphi}(0) \int_{P/F^\times} \|t\|^2\beta^{-1}(t)\, d^\times t - s^{-1}\alpha(g)\varphi(0)\delta(\omega) =$$

$$= \alpha(g)\|\det g\|^{-1} \int_{P/F^\times} \|t\|^2\beta^{-1}(t)[\Theta(tg^{T-1}; \hat{\varphi}) - \hat{\varphi}(0)]\, d^\times t -$$

$$-(1 - s)^{-1}\alpha(g)\|\det g\|^{-1}\hat{\varphi}(0)\delta(\bar{\omega}) - s^{-1}\alpha(g)\varphi(0)\delta(\omega).$$

Rewriting everything, we have

$$E^*(g;\varphi;s,\omega,\chi) =$$

$$= \alpha(g)\int_{P/F^\times} \beta(t)[\Theta(tg;\varphi) - \varphi(0)]\,d^\times t +$$

$$+ \alpha(g)\|\det g\|^{-1}\int_{P/F^\times} \|t\|^2\beta^{-1}(t)[\Theta(tg^{T-1};\hat{\varphi}) - \hat{\varphi}(0)]\,d^\times t -$$

$$- (1-s)^{-1}\alpha(g)\|\det g\|^{-1}\delta(\bar{\omega}) - s^{-1}\alpha(g)\varphi(0)\delta(\omega) =$$

$$= \chi(\det g)\|\det g\|^s\int_{P/F^\times} \|t\|^{2s}\bar{\omega}\chi^2(t)[\Theta(tg;\varphi) - \varphi(0)]\,d^\times t +$$

$$+ \bar{\chi}(\det g^{T-1})\|\det g^{T-1}\|^{1-s}\int_{P/F^\times} \|t\|^{2-2s}\omega\chi^2(t)[\Theta(tg^{T-1};\hat{\varphi}) - \hat{\varphi}(0)]\,d^\times t -$$

$$- (1-s)^{-1}\bar{\chi}(\det g)\|\det g^{T-1}\|^{1-s}\delta(\bar{\omega}) - s^{-1}\chi(\det g)\|\det g\|^s\varphi(0)\delta(\omega).$$

Granting momentarily that the two integrals are entire functions of $s \in \mathbb{C}$, the symmetry giving the functional equation is apparent. The last two terms give the poles with explicit residues.

Regarding entireness of the two integrals just above, as in the proof of convergence of the integral representations of the Eisenstein series, it suffices to consider the convergence of functions of the form

$$\int_{(1,\infty)} \|t\|^{2ms} \sum_{0\neq v\in F^2} \mathrm{ch}(v_0 g_0; X)(1 + \|tv_\infty g_\infty\|)^{-N}\,d^\times t$$

with $s \in \mathbb{R}$, X a compact subset of $\mathbb{A}_0^2$, $\mathrm{ch}(*;X)$ the characteristic function of X, $\|v_\infty\|$ the Euclidean norm on $\mathbb{R}^{2m}$, etc. This expression is dominated by

$$\int_{(1,\infty)} \|t\|^{2ms-N}\,d^\times t \times \sum_{0\neq v\in F^2} \mathrm{ch}(v_0 g_0; X)\|v_\infty g_\infty\|^{-N}.$$

Given s, for N large enough the integral converges nicely, as does the sum.

The last assertions follow immediately from the integral expressions just above. ■

4.3 Moderate Growth of Eisenstein Series

For various reasons, we need some technical information about the asymptotic behavior of Eisenstein series as functions on $GL(2,\mathbb{A})$, not only for $s \in \mathbb{C}$ in the region of convergence, but also for arbitrary $s \in \mathbb{C}$ away from the poles. Recall the notation

$$\ell(y) = \begin{pmatrix} y & 0 \\ 0 & 1 \end{pmatrix}.$$

THEOREM Let C be a compact set in $GL(2, \mathbb{A})$. Suppose that $s \in \mathbb{C}$ is not a pole of $E^*(g; \varphi; s, \omega, \chi)$. Then there are constants Ω and M so that, for $y \in \mathbb{J}$, $k_1, k_2 \in C$,

$$|E^*(k_1 \ell(y) k_2; \kappa; \varphi_0; s, \omega, \chi)| \leq \Omega(\|y\|^M + \|y\|^{-M}).$$

That is, these Eisenstein series are *automorphic forms of moderate growth*. Further, the constants Ω and M depend continuously on $s \in \mathbb{C}$ (away from poles).

Remarks In fact, it is not so hard to obtain a sharp estimate for M in terms of $\mathrm{Re}(s)$. However, we have no need of such a result; therefore, we can afford to be somewhat lax in the estimates involved in the proof.

Proof Temporarily, let

$$P = \{t \in \mathbb{J}\colon \|t\| \geq 1\} \qquad Q = \{t \in \mathbb{J}\colon \|t\| < 1\}.$$

As in the proof of the analytic continuation and functional equation, using the notation there, we have

$$E^*(g; \varphi; s, \omega, \chi) =$$

$$= \chi(\det g) \|\det g\|^s \int_{P/F^\times} \|t\|^{2s} \bar{\omega}\bar{\chi}^2(t) [\Theta(tg; \varphi) - \varphi(0)] \, d^\times t +$$

$$+ \bar{\chi}(\det g^{T-1}) \|\det g^{T-1}\|^{1-s} \int_{P/F^\times} \|t\|^{2-2s} \omega(t) [\Theta(tg^{T-1}; \hat{\varphi}) - \hat{\varphi}(0)] \, d^\times t -$$

$$- (1-s)^{-1} \bar{\chi}(\det g^{T-1}) \|\det g^{T-1}\|^{1-s} \delta(\bar{\omega}) - s^{-1} \chi(\det g) \|\det g\|^s \varphi(0) \delta(\omega)$$

where

$$\Theta(g; \varphi) = \sum_{v \in F^2} \varphi(vg).$$

The last two terms certainly have the asserted growth properties. Clearly it suffices to show that

$$\int_{P/F^\times} \|t\|^{2s} \sum_{0 \neq v \in F^2} |\varphi(vg)| \, d^\times t$$

has the asserted growth properties for $s \in \mathbb{R}$. As in the proof of convergence, and as in the proof of the analytic continuation, it suffices to consider functions of the form

$$\int_{(1,\infty)} \|t\|^{2ms} \sum_{0 \neq v \in F^2} \mathrm{ch}(v_0 g_0; X)(1 + \|tv_\infty g_\infty\|)^{-N} \, d^\times t$$

with $s \in \mathbb{R}$, X a compact subset of $\mathbb{A}_0^2$, $\mathrm{ch}(*; X)$ the characteristic function of X, $\|v_\infty\|$ the Euclidean norm on $\mathbb{R}^{2m}$, etc. This expression is dominated by

$$\int_{(1,\infty)} \|t\|^{2ms-N} \, d^\times t \times \sum_{0 \neq v \in F^2} \mathrm{ch}(v_0 g_0; X) \|v_\infty g_\infty\|^{-N}.$$

For given $s \in \mathbb{R}$, the integral is nicely convergent for N large enough. For $N > 2m$, it is elementary that the sum is nicely convergent, depending continuously on g.

For the assertion of moderate growth we need a slightly sharper assertion about the dependence of the sum of $g = k_1 \ell(y) k_2$ with $k_1, k_2 \in C$. It certainly suffices to take X of the form

$$X = \mathfrak{n}^{-1}(\hat{\mathfrak{o}} \oplus \hat{\mathfrak{o}})$$

where $\hat{\mathfrak{o}} = \prod_{\mathfrak{p} < \infty} \mathfrak{o}_{\mathfrak{p}}$. Then the sum is over a lattice in F^2 (deleting 0) depending on g_0 of a norm depending on g_∞. First, we have the obvious estimate

$$\|v_\infty g_\infty\|^{-N} \leq (\text{minimal absolute value of an eigenvalue of } g_\infty)^{-N} \times \|v_\infty\|^{-N}.$$

It is easy to check that this depends on $k_{1\infty} \ell(y_\infty) k_{2\infty} = g_\infty$ in the desired manner, i.e., is uniformly (for k_1, k_2 in the compact set C) dominated by an expression

$$\text{constant} \times (\|y_\infty\|_\infty^N + \|y_\infty\|_\infty^{-N}).$$

Similarly, the sum

$$\sum_{0 \neq v \in F^2} \mathrm{ch}(v_0 g_0; X) \|v_\infty\|^{-N}$$

is readily bounded by a rational function of the $\mathfrak{p}$-norms of the elementary divisors of g_0; and, the $\mathfrak{p}$-norms of the elementary divisors of expressions $k_{10} \ell(y_0) k_{20}$ have uniform (for k_1, k_2 in the compact set C) estimates in terms of $\|y_0\|$ and $\|y_0\|^{-1}$. This gives the moderate growth property of the Eisenstein series.

That the estimates depend continuously on $s \in \mathbb{C}$ is clear from the fact that the number N used in the estimates must satisfy only a linear inequality relating it to $\mathrm{Re}(s)$, etc. ∎

4.4 Fourier Expansion

The Fourier coefficients of Eisenstein series have a very explicit form: Frequently these Fourier coefficients have Euler products; also, the "constant term" (i.e., the 0-th Fourier coefficient) is essentially a sum of two L-functions of the underlying field F. We observe these properties through a sequence of results which gradually become more specific.

THEOREM Let $\rho(t) = \|t\|^{2s}\bar{\omega}\bar{\chi}^2(t)$. The adelic Eisenstein series attached to an adelic Schwartz function φ, to an idele character ω, and to $s \in \mathbb{C}$ has the Fourier expansion

$$\bar{\chi}(\det g)\|\det g\|^{-s}E^*\left(\begin{pmatrix}1 & x\\ 0 & 1\end{pmatrix}g; \varphi; s, \omega, \chi\right) =$$

$$= \Psi_{0,+}(g; \varphi; s, \omega, \chi) + \Psi_{0,-}(g; \varphi; s, \omega, \chi) + \sum_{\xi \in F^\times} \Psi\left(\begin{pmatrix}\xi & 0\\ 0 & 1\end{pmatrix}g; \varphi; s, \omega, \chi\right)\tau(\xi x),$$

where

$$\Psi(g; \varphi; s, \omega, \chi) = \int_{\mathbb{A}\times\mathbb{J}} \bar{\tau}(ut^{-1})\rho(t)\|t\|^{-1}\varphi((t,u)g)\,d^\times t\,d^+u$$

$$\Psi_{0,+}(g; \varphi; s, \omega, \chi) = \int_{\mathbb{J}/F^\times} \rho(t)\varphi((0,t)g)\,d^\times t\,d^+x$$

$$\Psi_{0,-}(g; \varphi; s, \omega, \omega) = \int_{\mathbb{A}\times\mathbb{J}} \beta\rho(t)\varphi(t(1,u)g)\,d^\times t\,d^+u$$

and d^+x gives the usual normalization of the additive measure on the adeles. The integrals for $\xi \neq 0$ are entire. Those for $\xi = 0$ are meromorphic, with at most simple poles with residues

$$\varphi(0)\delta(\omega) \qquad \text{(at } s = 0)$$

$$\hat{\varphi}(0)\delta(\omega)\bar{\chi}(\det g)\|\det g\|^{-1} \qquad \text{(at } s = 1).$$

Remarks This Fourier expansion may be used to give another (somewhat more complicated) proof of the analytic continuation, etc., of these Eisenstein series. The only complication occurs at infinite primes, where confluent hypergeometric functions appear. The proof in section 4.2 of analytic continuation and functional equation of Eisenstein series completely avoids overt discussion of these. (In fact, the result of section 4.2 may be combined with the present calculations to infer some properties of the confluent hypergeometric functions.)

Proof First, consider the Fourier expansion of the subsum of $\bar{\chi}(\det g)\|\det g\|^{-s}E^*$ where $v_1 = 0$ and $v_2 \in F^\times$. It is

$$\int_{\mathbb{A}/F\times\mathbb{J}/F^\times} \rho(t)\bar{\tau}(\xi x) \sum_{v_2 \in F^\times} \varphi(t(0, v_2)g)\,d^\times t\,d^+x.$$

If $\xi \neq 0$, this is 0. If $\xi = 0$, this gives the first term on the righthand side of the theorem, since the measure of $\mathbb{A}/F$ is normalized to be 1. Now consider the Fourier transform of the subsum where $v_1 \neq 0$, $v_2 \in F$. It is

$$\int_{\mathbb{A}/F \times \mathbb{J}/F^\times} \rho(t)\bar{\tau}(\xi x) \sum_{v_1 \neq 0, v_2 \in F} \varphi(t(v_1, v_2 + v_1 x)g)\, d^\times t\, d^+ x =$$

$$= \int_{\mathbb{A} \times \mathbb{J}/F^\times} \rho(t)\bar{\tau}(\xi x) \sum_{v_1 \neq 0} \varphi(t(v_1, v_1 x)g)\, d^\times t\, d^+ x =$$

$$= \int_{\mathbb{A} \times \mathbb{J}} \rho(t)\bar{\tau}(\xi x)\varphi(t(1, x)g)\, d^\times t\, d^+ x.$$

If $\xi = 0$, then we have the second term on the righthand side. If $\xi \neq 0$, then replace x by $x/\xi t$. This gives the sum on the righthand side.

The integral expressions for the Fourier coefficients with $\xi \neq 0$ indicate an adelic Fourier transform in the second argument of $\varphi \circ g$; let $\tilde{\varphi}$ denote this transform (depending on g). Then the ξ-th Fourier coefficient of $\|\det g\|^{-s}\bar{\chi}(\det g)E^*$ is

$$\int_{\mathbb{J}} \rho(t)\, \|t\|^{-1}\, \tilde{\varphi}(t, t^{-1}\xi)\, d^\times t.$$

Thus, (for $\xi \neq 0$) these integrals are integrals of a character in $t \in \mathbb{J}$ against a sum of products of the form

(Schwartz function of t) $\times$ (Schwartz function of t^{-1}).

A sum as indicated of such integrals converges for all $s \in \mathbb{C}$, by some elementary estimates.

The first part of the 0-th Fourier coefficient is an integral of the type which occurs in an adelic Mellin transform representation of an L-function of F, whence the assertion of its meromorphy and properties of its poles. The second integral in the constant term indicates a Fourier transform in the second argument of $\varphi \circ g$, and then this integral is also of this type. ∎

COROLLARY If the Schwartz function φ factors over primes, then the Fourier coefficients of the Eisenstein series also factor over primes. That is, the Fourier coefficients $\Psi_{0,+}$, $\Psi_{0,-}$, and Ψ have "Euler product expansions"; they are functions ψ on $GL(2, \mathbb{A})$ which factor as

$$\psi(\ldots, g_{\mathfrak{p}}, \ldots) = \prod_{\mathfrak{p}} \psi_{\mathfrak{p}}(g_{\mathfrak{p}})$$

for some functions $\psi_{\mathfrak{p}}$ on $GL(2, F_{\mathfrak{p}})$.

Proof This is clear from the integral expressions in the previous theorem. ∎

COROLLARY In the special case where

$$g = \begin{pmatrix} 1 & x \\ 0 & 1 \end{pmatrix}\begin{pmatrix} y & 0 \\ 0 & 1 \end{pmatrix} \in GL(2, \mathbb{A}),$$

the Fourier expansion becomes

$$E^*(g; \varphi; s, \omega, \chi) =$$

$$= \chi(y)\|y\|^s \int_{\mathbb{J}} \|t\|^{2s}\overline{\omega\chi}^2(t)\varphi(0,t)\,d^\times t +$$

$$+ \chi^2\omega^2(y)\|y\|^{1-s} \int_{\mathbb{A}\times\mathbb{J}} \|t\|^{2s-1}\overline{\omega\chi}^2(t)\varphi(t,u)\,d^\times t\,d^+u +$$

$$+ \chi(y)\|y\|^s \sum_{\xi\in F^\times} \tau(\xi x) \int_{\mathbb{A}\times\mathbb{J}} \overline{\tau}(ut^{-1})\|t\|^{2s-1}\overline{\omega\chi}^2(t)\varphi(t\xi y, u)\,d^\times t\,d^+u. \qquad \blacksquare$$

COROLLARY Suppose that

$$\varphi(v_1, v_2) = \varphi_\infty(v_1, v_2)\alpha(v_1)\beta(v_2)$$

with Schwartz functions α, β on $\mathbb{A}_0$. Put $\rho(t) = \|t\|^{2s}\overline{\omega\chi}^2(t)$. Then we have the Fourier expansion

$$E^*\left(\begin{pmatrix} 1 & x \\ 0 & 1 \end{pmatrix}\begin{pmatrix} y & 0 \\ 0 & 1 \end{pmatrix}; \varphi; s, \omega, \chi\right) =$$

$$= \chi(y)\|y\|^s\alpha(0)\left(\int_{\mathbb{J}_\infty} \rho(t)\varphi_\infty(0,t)\,d^\times t\right)\left(\int_{\mathbb{J}_0} \rho(t)\beta(t)\,d^\times t\right) +$$

$$+ \hat{\beta}(0)\|y\|^{1-s}\chi^2\omega^2(y)\left(\int_{\mathbb{A}_\infty\times\mathbb{J}_\infty} \rho(t)\|t\|^{-1}\varphi_\infty(t,u)\,d^\times t\,d^+u\right) \times$$

$$\times \left(\int_{\mathbb{J}_0} \rho(t)\|t\|^{-1}\alpha(t)\,d^\times t\right) +$$

$$+ \chi(y)\|y\|^s \sum_{\xi\in F^\times} \tau(\xi x)\left(\left(\int_{\mathbb{A}_\infty\times\mathbb{J}_\infty} \overline{\tau}(ut^{-1})\rho(t)\|t\|^{-1}\varphi_\infty(t\xi y, u)\,d^\times t\,d^+u\right) \times\right.$$

$$\left.\times \left(\int_{\mathbb{J}_0} \rho(t)\alpha(t\xi y)\hat{\beta}(t^{-1})\,d^\times t\,d^+u\right)\right).$$

In particular, the finite-prime integrals in the first two ("constant") terms are just L-functions over F attached to Schwartz functions and characters. $\blacksquare$

Remarks Again, the integrals appearing in the first two terms of the last expression are recognizable as "zeta integrals"; i.e., these are integrals which give L-functions of the number field F along with their gamma factors.

The finite-prime integrals in the higher Fourier coefficients are finite sums. The infinite-prime integrals in the higher Fourier coefficients are (products over infinite primes of) Whittaker functions; we proved the functional equation and analytic continuation without explicit use of the properties of these functions, and rather may derive their properties from the fact that they occur in such a Fourier expansion, etc.

4.5 Hecke-Theoretic Aspects

The proof of the following proposition is surprisingly easy. We note that the hypotheses do apply (for almost all primes $\mathfrak{p}$) to the Eisenstein series we have expressed as integrals.

PROPOSITION Suppose that the function ε used in defining an Eisenstein series $E(g;\varepsilon)$ is right $GL(2,\mathfrak{o}_\mathfrak{p})$-invariant for a finite prime $\mathfrak{p}$, and that

$$\varepsilon(\ell(yy')) = \varepsilon(\ell(y))\varepsilon(\ell(y'))$$

for $y \in F_\mathfrak{p}^x$, where

$$\ell(y) = \begin{pmatrix} y & 0 \\ 0 & 1 \end{pmatrix}$$

as previously. Then this Eisenstein series is an eigenfunction for the spherical Hecke algebra at the prime $\mathfrak{p}$. If η is in the spherical Hecke algebra at $\mathfrak{p}$, the eigenvalue of $E(g;\varepsilon)$ with respect to η is

$$\int_{GL(2,F_\mathfrak{p})} \varepsilon(h^{-1})\eta(h)\,dh.$$

Proof By the definition,

$$E(g;\varepsilon) = \sum_{\gamma \in P(F)\backslash GL(2,F)} \varepsilon(\gamma g).$$

Since $E(g;\varepsilon)$ is a sum of *left* translates of ε, and the Hecke operators act by *right* translation, it would suffice to show that ε itself is an eigenfunction for the Hecke operators.

Let η be in the spherical Hecke algebra at $\mathfrak{p}$. Then, by definition,

$$(\varepsilon * \eta)(g) = \int_{GL(2,F_\mathfrak{p})} \varepsilon(gh^{-1})\eta(h)\,dh.$$

Certainly it suffices to consider the case that $g \in GL(2, F_\mathfrak{p})$. Let $g = utk$, with t diagonal, $u \in U(F_\mathfrak{p})$, and $k \in GL(2, \mathfrak{o}_\mathfrak{p})$. Replace h by $k^{-1}h$; using the left and right $K_\mathfrak{p} = GL(2, \mathfrak{o}_\mathfrak{p})$-invariance of η, and the right $K_\mathfrak{p}$-invariance of ε, and the left $U(F_\mathfrak{p})$-invariance of ε, we have

$$\int_{GL(2,F_\mathfrak{p})} \varepsilon(gh^{-1})\eta(h)\,dh = \int_{GL(2,F_\mathfrak{p})} \varepsilon(th^{-1})\eta(h)\,dh.$$

Now let $h^{-1} = u't'k'$, with $u' \in U(F_\mathfrak{p})$, t' diagonal, and $k' \in K_\mathfrak{p}$. Then

$$\varepsilon(th^{-1}) = \varepsilon(tu't'k') = \varepsilon((tu't^{-1})tt') = \varepsilon(tt') = \varepsilon(t)\varepsilon(t') = \varepsilon(t)\varepsilon(h^{-1}).$$

Therefore,

$$\int_{GL(2,F_\mathfrak{p})} \varepsilon(gh^{-1})\eta(h)\,dh = \varepsilon(t) \int_{GL(2,F_\mathfrak{p})} \varepsilon(h^{-1})\eta(h)\,dh =$$

$$= \varepsilon(g) \int_{GL(2,F_\mathfrak{p})} \varepsilon(h^{-1})\eta(h)\,dh. \quad \blacksquare$$

4.6 Application of Differential Operators

In this section, let $\varepsilon_j = (0, \ldots, 0, 1, 0, \ldots, 0) \in \mathbb{Z}^m$ with the 1 at the j-th place.

Definition For $j = 1, \ldots, m$, let L_j (resp., R_j, Y_j) be the lowering (resp., raising, Casimir) operator on the j-th copy of $GL(2, \mathbb{R})$ occurring in $GL(2, \mathbb{A}_\infty)$ (as discussed earlier, defined and computed in coordinates in Appendix A.4). As in Chapter 3, if f is an automorphic form on $GL(2, \mathbb{A})$, then say that f is *holomorphic* if $L_j f = 0$ for $j = 1, \ldots, m$.

Remarks We have seen in section 3.3 that this definition is completely compatible with "adelizing" classical holomorphic modular forms: If f is obtained by "adelizing" a holomorphic modular form on $\mathfrak{H}^m$, then the previous condition is met, etc.

PROPOSITION Let $E^*(g; \kappa; \varphi_0; s, \omega, \chi)$ be an Eisenstein series "of weight κ" as defined above, and define another normalization $E^\natural(g; \kappa; \varphi_0; s, \omega, \chi)$ of it by

$$\left(\prod_{j=1,\ldots,m} \pi^{-(s+|\kappa_j|/2)}\Gamma(s + \tfrac{1}{2}|\kappa_j|)\right) E^\natural(g; \kappa; \varphi_0; s, \omega, \chi) = E^*(g; \kappa; \varphi_0; s, \omega, \chi).$$

Then we have the relations

$$L_j E^\natural(g; \kappa; \varphi_0; s, \omega, \chi) = i(s - \tfrac{1}{2}\kappa_j) E^\natural(g; \kappa - 2\varepsilon_j; \varphi_0; s, \omega, \chi)$$

$$R_j E^\natural(g; \kappa; \varphi_0; s, \omega, \chi) = -i(s + \tfrac{1}{2}\kappa_j) E^\natural(g; \kappa + 2\varepsilon_j; \varphi_0; s, \omega, \chi)$$

$$Y_j E^\natural(g; \kappa; \varphi_0; s, \omega, \chi) = s(s - 1) E^\natural(g; \kappa; \varphi_0; s, \omega, \chi).$$

Remarks We note in the course of the proof that these Eisenstein series are identically zero unless ω is totally even for κ even and totally odd for κ odd.

Proof By uniqueness of analytic continuation, it suffices to prove these relations for Re(s) sufficiently positive.

Since these operators are left-invariant, and since we have seen that the Eisenstein series is a sum over

$$\gamma \in P\backslash GL(2, F) \approx (P \cap GL^+(2, F))\backslash GL^+(2, F)$$

of left translates by γ of

$$q(g; \kappa; \varphi_0; s, \omega, \chi) = \chi(\det g)\|\det g\|^s \int_{\mathbb{J}} \|t\|^{2s}\overline{\omega}\overline{\chi}^2(t)\varphi(t(0, 1)g)\, d^\times t,$$

it suffices to show that q has these properties. Indeed, since the effect of the differential operator depends only on the infinite-prime part and is left $GL(2, \mathbb{R})^m$-invariant, it suffices to show that the function on $SL(2, \mathbb{R})$ defined by

$$\pi^{-(s+\kappa/2)}\Gamma(s + \tfrac{1}{2}\kappa)q_1(g; \kappa, s, \omega) = \int_{\mathbb{R}^\times} |t|^{2s}\overline{\omega}\overline{\chi}^2(t)\varphi_\infty(t(0, 1)g)\, d^\times t$$

has the desired property, where $0 \leq \kappa \in \mathbb{Z}$, and where for $\kappa \geq 0$

$$\varphi_\infty(v_1, v_2) = (-iv_1 + v_2)^\kappa \exp(-\pi(v_1^2 + v_2^2)).$$

Since $\varphi_\infty(t(0, 1)g)$ is left-invariant by elements $\left(\begin{smallmatrix} 1 & x \\ 0 & 1 \end{smallmatrix}\right)$ with $x \in \mathbb{R}$, we need consider only g of the form

$$g = \begin{pmatrix} y^{1/2} & 0 \\ 0 & y^{-1/2} \end{pmatrix}\begin{pmatrix} \cos\theta & \sin\theta \\ -\sin\theta & \cos\theta \end{pmatrix}$$

where $y > 0$, by the Iwasawa decomposition. With g of this form, with $\kappa \geq 0$,

$$\pi^{-(s+\kappa/2)}\Gamma(s + \tfrac{1}{2}\kappa)q_1(g; \kappa, s, \omega) =$$

$$= |y^{1/2}|^s \int_{\mathbb{R}^\times} |t|^{2s}\omega^{-1}(t)\varphi_\infty(t(-y^{-1/2}\sin\theta, y^{-1/2}\cos\theta))\, d^\times t =$$

$$= y^s \int_{\mathbb{R}^\times} |t|^{2s}\omega^{-1}(t)\exp(-\pi t^2)t^\kappa(\cos\theta + i\sin\theta)^\kappa\, d^\times t =$$

$$= y^s e^{\kappa i\theta}\pi^{-(s+\kappa/2)}\Gamma(s + \tfrac{1}{2}\kappa) \qquad \text{if } \omega(-1) = (-1)^\kappa$$

$$= \qquad 0 \qquad \text{if } \omega(-1) \neq (-1)^\kappa.$$

A similar calculation for $\kappa \leq 0$ yields a uniform result for $\kappa \in \mathbb{Z}$:

$$\pi^{-(s+|\kappa|/2)}\Gamma(s+\tfrac{1}{2}|\kappa|)q_1(g;\kappa,s,\omega) =$$

$$= y^s e^{\kappa i\theta}\pi^{-(s+|\kappa|/2)}\Gamma(s+\tfrac{1}{2}|\kappa|) \qquad \text{if } \omega(-1)=(-1)^\kappa$$

$$= \qquad 0 \qquad \text{if } \omega(-1)\neq(-1)^\kappa.$$

In these Iwasawa coordinates, the lowering operator L and the raising operator R are given by

$$L = e^{-2i\theta}(y\partial/\partial x + iy\partial/\partial y - \tfrac{1}{2}\partial/\partial\theta)$$

$$R = e^{2i\theta}(y\partial/\partial x - iy\partial/\partial y - \tfrac{1}{2}\partial/\partial\theta).$$

Therefore, we have

$$Lq_1(g;\kappa,s,\omega) = Ly^s e^{\kappa i\theta} = e^{-2i\theta}(0 + isy^s e^{\kappa i\theta} - \tfrac{1}{2}y^s\kappa i e^{\kappa i\theta}) =$$

$$= i(s-\tfrac{1}{2}\kappa)y^s e^{(k-2)i\theta},$$

$$Rq_1(g;\kappa,s,\omega) = Ly^s e^{\kappa i\theta} = e^{2i\theta}(0 - isy^s e^{\kappa i\theta} - \tfrac{1}{2}y^s\kappa i e^{\kappa i\theta}) =$$

$$= -i(s+\tfrac{1}{2}\kappa)y^s e^{(k+2)i\theta}.$$

This proves the asserted relations for the raising and lowering operators.

In a similar manner the eigenvalues of such Eisenstein series for the Casimir operators are easily computed: Let Y be the Casimir operator on $SL(2,\mathbb{R})$ given in Iwasawa coordinates by

$$Y = y^2(\partial^2/\partial x^2 + \partial^2/\partial y^2) - y\partial^2/\partial x\,\partial\theta.$$

Then we merely need compute

$$Yy^s e^{i\kappa\theta} = y^2(s(s-1)y^{s-2}) - y0 = s(s-1)y^s e^{i\kappa\theta}. \qquad \blacksquare$$

4.7 Eisenstein Series of Low Weight

Now we are interested in case $\kappa_1 = \cdots = \kappa_m \leq 2$. These values are small enough so that the Eisenstein series are not absolutely convergent at $s = \frac{1}{2}\kappa_1$. For weights $\kappa_1 = \cdots = \kappa_m > 2$, the Eisenstein series are absolutely convergent at $s = \frac{1}{2}\kappa_1$ and so are "holomorphic," insofar as they are annihilated by all the lowering operators. We will see that the lower-weight Eisenstein series are also holomorphic under a very mild hypothesis.

Remarks Keeping in mind that there are no non-constant holomorphic automorphic forms of weight 0, and none of weight $(\kappa,\ldots,\kappa)$ with $\kappa < 0$, in the present context there is no reason to look at $E^\natural(g;\kappa;\varphi_0,\frac{1}{2}\kappa,\omega)$ with $\kappa < 1$.

As noted in the proof of the proposition of the previous section, an Eisenstein series of weight $(\kappa, \ldots, \kappa)$ with κ even vanishes identically unless the character ω is totally even; an Eisenstein series of weight $(\kappa, \ldots, \kappa)$ with κ odd vanishes identically unless the character ω is totally odd.

PROPOSITION Unless the character ω is trivial *and* $F = \mathbb{Q}$ *and* $\kappa = 2$, Eisenstein series $E^\natural(g; \kappa; \varphi_0; \frac{1}{2}\kappa, \omega, \chi)$ are holomorphic in g (i.e., are annihilated by all the lowering operators).

Proof What we must show is that for $j = 1, \ldots, m$

$$L_j E^\natural(g; \kappa; \varphi_0; s, \omega, \chi)|_{s=1/2\kappa} = 0.$$

From the proposition of the previous section we have

$$L_j E^\natural(g; \kappa; \varphi_0; s, \omega, \chi) = i(s - \tfrac{1}{2}\kappa_j) E^\natural(g; \kappa - 2\varepsilon_j; \varphi_0; s, \omega, \chi).$$

Put $\lambda = \kappa - 2\varepsilon_j$. The righthand side does indeed vanish unless

$$E^\natural(g; \lambda; \varphi_0; s, \omega, \chi) = \left(\prod_{j=1,\ldots,m} \pi^{-(s+\lambda_j/2)} \Gamma(s + \tfrac{1}{2}\lambda_j) \right)^{-1} E^*(g; \lambda; \varphi_0; s, \omega, \chi)$$

has a pole at $s = \frac{1}{2}\kappa_1$. The normalizing factor is finite for all $s \in \mathbb{C}$. From the theorem of section 4.2 on analytic continuation of Eisenstein series, the Eisenstein series E^* is entire unless $\kappa - 2\varepsilon_j = \lambda = (0, \ldots, 0)$ and $\omega \equiv 1$. This can happen only for $m = 1$ (i.e., $F = \mathbb{Q}$), $\kappa = 2$, $\omega \equiv 1$. ■

Remarks Nothing is asserted in the previous proposition about the possibility that $E^\natural(g; \kappa; \varphi_0; \frac{1}{2}\kappa, \omega, \chi)$ or $E^*(g; \kappa; \varphi_0; \frac{1}{2}\kappa, \omega, \chi)$ may be identically zero. For $\kappa = 2$ this issue can often be easily settled by showing that the "constant" Fourier coefficient is not zero. For $\kappa = 1$, such calculations are subtler: The archimedean part of higher Fourier coefficients is no longer absolutely convergent, and there is cancellation between the two parts of the 0-th Fourier coefficient. For $\kappa < 0$, often the trivial vanishing of an Eisenstein series is clear from the functional equation.

4.8 Fourier Expansion of Certain Holomorphic Eisenstein Series

Having noted the holomorphy of Eisenstein series of low weight, we return to calculation of Fourier expansions in some specific circumstances. In particular, we observe some arithmetic properties of the "higher" Fourier coefficients: Suitably normalized, they lie in cyclotomic fields, and transform in a simple manner under Galois automorphisms. The first theorem gives an explicit calculation "at level one," describing the higher Fourier coefficients as generalized "sums of powers of divisors," whereas the second theorem is less explicit, merely describing the behavior under Galois automorphisms. The third theorem complements the second, dealing with the case that (in particular normalization) the "constant term" is zero.

Remarks The arithmetic aspects of these results have numerous applications, both to special values of L-functions over totally real number fields, and to special values of L-functions attached to Hilbert modular forms. Some of these applications will be discussed later.

For an infinite idele y_∞, let $\mathrm{sgn}(y_\infty)$ denote the product of the signs of the components of y_∞. Fix a non-zero ideal $\mathfrak{n}$ in $\mathfrak{o}$. We will consider Eisenstein series

$$E^*(g; \kappa; \varphi_0; s, \omega, \chi)$$

of weight $\kappa \geq 2$ with specific choices of the finite-adele Schwartz function φ_0 factoring over primes as

$$\varphi_0 = \sum_{\mathfrak{p}<\infty} \varphi_\mathfrak{p}.$$

Let $\mathrm{ch}(*; X)$ denote the characteristic function of a set X. Let $c_0, d_0 \in \hat{\mathfrak{o}}$ so that for every $\mathfrak{p}$ dividing $\mathfrak{n}$ we have $c_0 \mathfrak{o}_\mathfrak{p} + d_0 \mathfrak{o}_\mathfrak{p} = \mathfrak{o}_\mathfrak{p}$. Then take

$$\varphi_\mathfrak{p}(v) = \mathrm{ch}(v; \mathfrak{o}_\mathfrak{p}) \otimes \mathrm{ch}(v; \mathfrak{o}_\mathfrak{p}) \qquad \text{for } \mathfrak{p} \text{ not dividing } \mathfrak{n}$$

$$= \mathrm{ch}(v; c_0 + \mathfrak{n}\mathfrak{o}_\mathfrak{p}) \otimes \mathrm{ch}(v'; d_0 + \mathfrak{n}\mathfrak{o}_\mathfrak{p}) \qquad \text{for } \mathfrak{p} \text{ dividing } \mathfrak{n}.$$

Let ω be a character on $\mathbb{J}/F^\times$ whose conductor divides $\mathfrak{n}$, and require that ω be totally even if κ is even, totally odd if κ is odd. Take χ ramified at most at infinite primes. Then, with this choice of φ_0,

Definition

$$E(g; \kappa; c_0, d_0, \mathfrak{n}; s, \omega, \chi) = E^*(g; \kappa; \varphi_0; s, \omega, \chi).$$

We have seen that, unless $F = \mathbb{Q}$ and $\kappa = 2$ and ω is trivial, $E(g; \kappa; c_0, d_0, \mathfrak{n}; \frac{1}{2}\kappa, \omega, \chi)$ is holomorphic (in the sense that it is annihilated by all the lowering operators).

PROPOSITION Take $h = \left(\begin{smallmatrix} a & b \\ c & d \end{smallmatrix}\right)$ in $GL(2, \hat{\mathfrak{o}})$. Then

$$E(gh; \kappa; c_0, d_0, \mathfrak{n}; s, \omega, \chi) = E(g; \kappa; c_0', d_0', \mathfrak{n}; s, \omega, \chi)$$

where

$$(c_0' d_0') \equiv (c_0 d_0) h^{-1} \text{ modulo } \mathfrak{n}.$$

Proof This is immediate from the definitions. ■

For the following two results, we exclude the case $\kappa = 2$, $F = \mathbb{Q}$, $\omega = 1$. Put

$$\rho = \bar{\omega}\bar{\chi}^2$$

and

$$L_{\mathfrak{n}}(s, \rho) = \prod 1/(1 - \rho(\mathfrak{p})N\mathfrak{p}^{-s}) \qquad \text{(product over finite } \mathfrak{p} \text{ not dividing } \mathfrak{n})$$

$$c = \text{index of } \{\alpha \in \hat{\mathfrak{o}}^{\times}: \alpha \equiv 1 \bmod \mathfrak{n}\} \text{ in } \hat{\mathfrak{o}}^{\times}.$$

Let $\mathbb{Q}_{ab}$ be the field generated over $\mathbb{Q}$ by all roots of unity. We recall that elements of the Galois group $\mathrm{Gal}(\mathbb{Q}_{ab}/\mathbb{Q})$ are all of the form

$$\sigma = \sigma_{\alpha}: \zeta \to \zeta^{\alpha}$$

where ζ is a primitive N-th root of unity and $\alpha \in \mathbb{Z}$ is prime to N. Thus, in the standard manner we may identify $\mathrm{Gal}(\mathbb{Q}_{ab}/\mathbb{Q})$ with $\hat{\mathbb{Z}}^{\times}$. Let D be the absolute discriminant of F. Let ch^{+} be the characteristic function of $\mathbb{J}^{+}$.

Finally, we introduce the general sums-of-powers-of-divisors functions. For a finite idele α and a Hecke character ρ put

$$\sigma_{\kappa-1}(\alpha; \rho) = \sum_{\mathfrak{a}} \bar{\rho}(\mathfrak{a}) N\mathfrak{a}^{\kappa-1}$$

the sum being over ideals $\mathfrak{a}$ so that $\mathfrak{a} \subset \mathfrak{o}^*$ and $\alpha\hat{\mathfrak{o}} \subset \mathfrak{a}\hat{\mathfrak{o}}$. It is clear that this function factors over primes; write

$$\sigma_{\kappa-1}(\alpha; \rho) = \prod_{\mathfrak{p} < \infty} \sigma_{\kappa-1}(\alpha_{\mathfrak{p}}; \rho)_{\mathfrak{p}}.$$

THEOREM Take $\mathfrak{n} = \mathfrak{o}$ (so both ω and χ are unramified at finite primes, and the choice of c_0 and d_0 is irrelevant). Let

$$g = u(x)\ell(y) = \begin{pmatrix} 1 & x \\ 0 & 1 \end{pmatrix}\begin{pmatrix} y & 0 \\ 0 & 1 \end{pmatrix} \in GL(2, \mathbb{A}).$$

Then the holomorphic Eisenstein series $E(g; \kappa; *, *, \mathfrak{o}; \tfrac{1}{2}\kappa, \omega, \chi)$ has Fourier expansion

$$\chi(y)\|y\|^{\kappa/2}\Bigg(\pi^{-m\kappa}\Gamma(\kappa)^m L_{\mathfrak{n}}(\kappa, \rho) + \\ + \mathrm{sgn}(y_{\infty})(-2i)^{\kappa m} D^{-1/2} \sum_{\xi} \sigma_{\kappa-1}(\xi_0 y_0; \rho)\tau(\xi x)\exp(-2\pi\xi_{\infty} y_{\infty})\Bigg),$$

where ξ is summed over elements of $F^{\times}$ so that $(\xi y)_{\infty}$ is totally positive, z_0 (resp., z_{∞}) is the finite (resp., infinite) part of an idele z. In particular, $\sigma_{\kappa-1}(\xi_0 y_0; \rho)$ lies in the field $\mathbb{Q}(\rho)$ generated over $\mathbb{Q}$ by the values of ρ, and for a Galois automorphism α of $\mathbb{Q}_{ab}$ over $\mathbb{Q}$ we have the *Galois equivariance property*

$$\sigma_{\kappa-1}(\xi_0 y_0; \rho)^{\alpha} = \sigma_{\kappa-1}(\xi_0 y_0; \rho^{\alpha}).$$

That is, for all $y \in \mathbb{J}_0$ the higher Fourier coefficients (without the factor of $\chi(y)$) lie in $\mathbb{Q}(\rho)$, and transform under Galois automorphisms exactly as does ρ.

THEOREM Let $\rho = \bar{\omega}\bar{\chi}^2$, and let

$$g = u(x)\ell(y) = \begin{pmatrix} 1 & x \\ 0 & 1 \end{pmatrix}\begin{pmatrix} y & 0 \\ 0 & 1 \end{pmatrix} \in GL(2, \mathbb{A}).$$

The holomorphic Eisenstein series $E(g; \kappa; 0, 1, \mathfrak{n}; \frac{1}{2}\kappa, \omega, \chi)$ has Fourier expansion

$$E(g; \kappa; 0, 1, \mathfrak{n}; \tfrac{1}{2}\kappa, \omega, \chi) =$$

$$= \chi(y)\|y\|^{\kappa/2}\Bigg(c\pi^{-m\kappa}\Gamma(\kappa)^m L_{\mathfrak{n}}(\kappa, \rho) +$$

$$+ (-2i)^{\kappa m} D^{-1/2} \sum_{\xi} \sigma_{\kappa-1,\mathfrak{n}}(\xi_0 y_0; \rho)\tau(\xi x)\exp(-2\pi \operatorname{Tr} \xi_\infty y_\infty)\Bigg),$$

where ξ is summed over elements of $F^\times$ so that $(\xi y_0)_\infty$ is totally positive, z_0 (resp., z_∞) is the finite (resp., infinite) part of an idele z, and where $\sigma_{\kappa-1,\mathfrak{n}}(*; \rho)$ is a function on finite ideles assuming values in $\mathbb{Q}_{ab}$. Further, for $\alpha \in \hat{\mathbb{Z}}^\times \subset \hat{\mathfrak{o}}^\times$ with associated $\sigma_\alpha \in \operatorname{Gal}(\mathbb{Q}_{ab}/\mathbb{Q})$ as above, for all $\beta \in \mathbb{J}_0$ we have

$$\sigma_{\kappa-1,\mathfrak{n}}(\beta; \rho)^\sigma = \rho^\sigma(\alpha)\sigma_{\kappa-1,\mathfrak{n}}(\beta; \rho^\sigma).$$

THEOREM Let $\rho = \bar{\omega}\bar{\chi}^2$ and let

$$g = u(x)\ell(y) = \begin{pmatrix} 1 & x \\ 0 & 1 \end{pmatrix}\begin{pmatrix} y & 0 \\ 0 & 1 \end{pmatrix} \in GL(2, \mathbb{A}).$$

For d_0 not a $\mathfrak{p}$-unit at some prime $\mathfrak{p}$ dividing $\mathfrak{n}$, the holomorphic Eisenstein series $E(g; \kappa; c_0, d_0, \mathfrak{n}; \frac{1}{2}\kappa, \omega, \chi)$ has a Fourier expansion of the form

$$E(g; \kappa; c_0, d_0, \mathfrak{n}; \tfrac{1}{2}\kappa, \omega, \chi) =$$

$$= \chi(y)\|y\|^{\kappa/2}(-2i)^{\kappa m} D^{-1/2} \sum_{\xi} C(\xi_0 y_0; \rho)\tau(\xi x)\exp(-2\pi \operatorname{Tr} \xi_\infty y_\infty),$$

where ξ is summed over elements of $F^\times$ so that $(\xi y)_\infty$ is totally positive, z_0 (resp., z_∞) is the finite (resp., infinite) part of an idele z, and where $C(*; \rho)$ is a function on finite ideles assuming values in $\mathbb{Q}_{ab}$. Further, for $\alpha \in \hat{\mathbb{Z}}^\times \subset \hat{\mathfrak{o}}^\times$ with associated $\sigma_\alpha \in \operatorname{Gal}(\mathbb{Q}_{ab}/\mathbb{Q})$ as above, for all $\beta \in \mathbb{J}_0$ we have

$$C(\beta; \rho)^\sigma = \rho^\sigma(\alpha)C(\beta; \rho^\sigma).$$

Proof We will give the proof of all three theorems simultaneously; the explicit form of the Fourier coefficients in the more general cases does occur in the proof, but the salient aspect is the behavior under Galois actions, and the form of the constant term, as indicated.

We are certainly in the situation of the last two corollaries of our prior calculation of Fourier coefficients of Eisenstein series: We have

$$\varphi(v_1, v_2) = \varphi_\infty(v_1, v_2)\alpha(v_1)\beta(v_2)$$

with Schwartz functions α, β on $\mathbb{A}_0$ which factor over primes. From the last of those two corollaries, we have the Fourier expansion

$$E^*\left(\begin{pmatrix}1 & x\\ 0 & 1\end{pmatrix}\begin{pmatrix}y & 0\\ 0 & 1\end{pmatrix}; \kappa; c_0, d_0, \mathfrak{n}; s, \omega, \chi\right) =$$

$$= \chi(y)\|y\|^s\alpha(0)\left(\int_{\mathbb{J}_\infty} \|t\|^{2s}\rho(t)\varphi_\infty(0,t)\,d^\times t\right)\left(\int_{\mathbb{J}_0} \|t\|^{2s}\rho(t)\beta(t)\,d^\times t\right) +$$

$$+ \hat{\beta}(0)\|y\|^{1-s}\omega^2\chi^2(y)\left(\int_{\mathbb{A}_\infty\times\mathbb{J}_\infty} \|t\|^{2s-1}\rho(t)\varphi_\infty(t,u)\,d^\times t\,d^+u\right)\times\left(\int_{\mathbb{J}_0} \|t\|^{2s-1}\rho(t)\alpha(t)\,d^\times t\right) +$$

$$+ \chi(y)\|y\|^s \sum_{\xi\in F^\times} \tau(\xi x)\left(\left(\int_{\mathbb{A}_\infty\times\mathbb{J}_\infty} \bar{\tau}(ut^{-1})\|t\|^{2s-1}\rho(t)\varphi_\infty(t\xi y, u)\,d^\times t\,d^+u\right)\times\right.$$

$$\left.\times\left(\int_{\mathbb{J}_0} \|t\|^{2s-1}\rho(t)\alpha(t\xi y)\hat{\beta}(t^{-1})\,d^\times t\,d^+u\right)\right)$$

and we know that all these integrals factor over primes, etc. An easy computation shows that the first summand here gives the first term in the asserted equality.

First, we consider the archimedean integrals in the second summand. (We suppress subscripts referring to the $\mathfrak{p}$-components and so on.) At each infinite prime, because of our assumption on the parities of ω and κ, the integral is

$$\int_{\mathbb{R}}\int_{\mathbb{R}^\times} |t|^{2s}(-ity + tx)^\kappa \exp(-\pi t^2(x^2+y^2))\,d^\times t\,d^+x =$$

$$= \pi^{-(s+\kappa/2)}\Gamma(s+\tfrac{1}{2}\kappa)\int_{\mathbb{R}} (x+iy)^{-(s+\kappa/2)}(x-iy)^{-(s-\kappa/2)}\,dx.$$

It is clear that this is a holomorphic function of s for $\mathrm{Re}(s) > 1$. Therefore, we may simply substitute $s = \frac{1}{2}\kappa$ to evaluate the integral: By residues, it is 0; therefore, the archimedean part of the second summand vanishes to order $m = [F:\mathbb{Q}]$ at $s = \frac{1}{2}\kappa$. More generally, for $\xi \in F^\times$, with $s = \frac{1}{2}\kappa$ the integral is

$$\pi^{-\kappa}\Gamma(\kappa)\int_{\mathbb{R}}(x+i\xi y)^{-\kappa}\exp(-2\pi ix)\,dx =$$

$$\begin{aligned} &= \pi^{-\kappa}(-2\pi i)^{\kappa}\Gamma(\kappa)^{-1}\exp(-2\pi\xi y) && \text{if } \xi y > 0 \\ &= \qquad 0 && \text{if } \xi y < 0, \end{aligned}$$

by *residues.* The finite-prime part of the second summand is the finite-prime part of a zeta integral, so is entire at $s = \frac{1}{2}\kappa$ if $\kappa > 2$ of $\omega \neq 1$, else has at most a simple pole. By hypothesis, the second summand vanishes at $s = \frac{1}{2}\kappa$.

Now we consider the finite-prime integrals in the higher-order terms. We may simply evaluate these at $s - \frac{1}{2}\kappa$. Suppressing subscripts $\mathfrak{p}$, etc., each of these is of the form

$$\int_{F_{\mathfrak{p}}^{x}}\|t\|^{\kappa-1}\rho(t)\alpha(t\xi y)\hat{\beta}(t^{-1})\,d^{x}t = \int_{F_{\mathfrak{p}}^{x}}\|t\|^{1-\kappa}\bar{\rho}(t)\alpha(t^{-1}\xi y)\hat{\beta}(t)\,d^{x}t$$

by replacing t by t^{-1}. Let π be a local parameter at $\mathfrak{p}$. First, if $\mathfrak{p}$ does not divide $\mathfrak{n}$, the Fourier transform of $\beta = \mathrm{ch}(*;\mathfrak{o}_{\mathfrak{p}})$ is $[\mathfrak{o}_{\mathfrak{p}}^{*}:\mathfrak{o}_{\mathfrak{p}}]^{-1/2}\,\mathrm{ch}(*;\mathfrak{o}_{\mathfrak{p}}^{*})$, and by homogeneous-space integration theory this integral is just

$$[\mathfrak{o}_{\mathfrak{p}}^{*}:\mathfrak{o}_{\mathfrak{p}}]^{-1/2}\sum_{n:\,y\xi/\pi^{n}\in\mathfrak{o}_{\mathfrak{p}}\text{ and }\pi^{n}\in\mathfrak{o}_{\mathfrak{p}}^{*}}\bar{\rho}(\pi^{n})N\mathfrak{p}^{n(\kappa-1)}.$$

(The constant factor arises by our normalization of additive Haar measures.) This *finite* sum certainly lies in the cyclotomic field generated by the values of ρ, and (for fixed $\xi \in F^{x}$ and $y \in F_{\mathfrak{p}}^{x}$) is either 0 or 1 for almost all primes $\mathfrak{p}$. Further, the sum transforms under Galois automorphisms of $\bar{\mathbb{Q}}$ over $\mathbb{Q}$ in precisely the same manner as does ρ.

Second, for $d_0 \equiv 1$ modulo $\mathfrak{n}$ and $c_0 \equiv 0$ modulo $\mathfrak{n}$, at a prime $\mathfrak{p}$ dividing $\mathfrak{n}$, one computes directly that

$$\hat{\beta}(v) = \mathrm{ch}(*;1+\mathfrak{n}\mathfrak{o}_{\mathfrak{p}})\hat{\ }(v) = [\mathfrak{o}_{\mathfrak{p}}^{*}:\mathfrak{n}^{2}\mathfrak{o}_{\mathfrak{p}}]^{-1/2}\tau(v)\,\mathrm{ch}(v;\mathfrak{n}^{-1}\mathfrak{o}_{\mathfrak{p}}^{*}).$$

Thus, the integral is

$$[\mathfrak{o}_{\mathfrak{p}}^{*}:\mathfrak{n}^{2}\mathfrak{o}_{\mathfrak{p}}]^{-1/2}\int_{F_{\mathfrak{p}}^{x}}\|t\|^{\kappa-1}\rho(t)\,\mathrm{ch}(t\xi y;\mathfrak{n}\mathfrak{o}_{\mathfrak{p}})\tau(t^{-1})\,\mathrm{ch}(t^{-1};\mathfrak{n}^{-1}\mathfrak{o}_{\mathfrak{p}}^{*})\,d^{x}t.$$

In this case, to reduce the integral to a sum we must note that the integrand is not constant on cosets of $\mathfrak{o}_{\mathfrak{p}}^{x}$, but only on cosets of a smaller compact open subgroup of $\mathfrak{o}_{\mathfrak{p}}^{x}$. Nevertheless, as $\xi \neq 0$, the integral is indeed a *finite* sum of numbers lying in $\mathbb{Q}_{ab}$. Take $\alpha \in \mathbb{Z}_{p}^{x} \subset \mathfrak{o}_{\mathfrak{p}}^{x}$, where p is the rational prime lying under $\mathfrak{p}$, and let $\sigma = \sigma_{\alpha}$ be the associated element of $\mathrm{Gal}(\mathbb{Q}_{ab}/\mathbb{Q})$. Since the integral is actually a finite sum, we have

$$\sigma\left(\int_{F_{\mathfrak{p}}^{\times}} \|t\|^{\kappa-1} \rho(t)\,\mathrm{ch}(t\xi y; \mathfrak{n}\mathfrak{o}_{\mathfrak{p}})\tau(t^{-1})\,\mathrm{ch}(t^{-1}; \mathfrak{n}^{-1}\mathfrak{o}_{\mathfrak{p}}^{*})\,d^{\times}t\right) =$$

$$= \int_{F_{\mathfrak{p}}^{\times}} \|t\|^{\kappa-1} \rho^{\sigma}(t)\,\mathrm{ch}(t\xi y; \mathfrak{n}\mathfrak{o}_{\mathfrak{p}})\tau(\alpha t^{-1})\,\mathrm{ch}(t^{-1}; \mathfrak{n}^{-1}\mathfrak{o}_{\mathfrak{p}}^{*})\,d^{\times}t.$$

Replacing t by $t\alpha$ shows that this is

$$\rho^{\sigma}(\alpha)\int_{F_{\mathfrak{p}}^{\times}} \|t\|^{\kappa-1} \rho^{\sigma}(t)\,\mathrm{ch}(t\xi y; \mathfrak{n}\mathfrak{o}_{\mathfrak{p}})\tau(t^{-1})\,\mathrm{ch}(t^{-1}; \mathfrak{n}^{-1}\mathfrak{o}_{\mathfrak{p}}^{*})\,d^{\times}t.$$

That is, the behavior of the integral under Galois automorphisms is independent of $\xi \in F^{\times}$. Since there are only finitely many primes dividing $\mathfrak{n}$, and since ω is trivial on $\mathfrak{p}$-units at primes $\mathfrak{p}$ not dividing $\mathfrak{n}$, this proves the Galois equivariance assertion of the second theorem.

In the simple case that $\mathfrak{n} = 1$, from the second form of the integral, the ξ-th Fourier coefficient is just

$$(-2i)^{\kappa m}[\mathfrak{o}^* : \mathfrak{o}]^{-1/2} \times$$
$$\times \prod_{\mathfrak{p}\text{ finite}} \left(\sum_{n:\, y_{\mathfrak{p}}\xi_{\mathfrak{p}}/\pi_{\mathfrak{p}} \in \mathfrak{o}_{\mathfrak{p}} \text{ and } \pi^n \in \mathfrak{o}_{\mathfrak{p}}^{*}} \bar{\rho}(\pi^n) N\mathfrak{p}^{n(\kappa-1)} \right)$$

which is readily verified to be as asserted in the first theorem. Therefore, this expression certainly lies in $\mathbb{Q}(\rho)$ and transforms in the same manner as does ρ.

Now consider the case that d_0 is not a $\mathfrak{p}$-unit at some $\mathfrak{p}$ dividing $\mathfrak{n}$. The same argument regarding higher Fourier coefficients applies to finite $\mathfrak{p}$ dividing $\mathfrak{n}$. Also, the "second" part of the constant term still must vanish. Since (in the notation above) $\alpha(0) = 0$, the whole constant term (i.e., with $\xi = 0$) is 0. (Note that the vanishing of the "first part" of the constant term is due to finite-prime contribution, while the vanishing of the second is due to infinite-prime contributions.) ■

Remarks From the proof it is clear that the higher Fourier coefficients factor over primes, and at finite primes $\mathfrak{p}$ not dividing $\mathfrak{n}$ the factor is (up to a normalizing constant) just the $\mathfrak{p}$-part $\sigma_{\kappa-1}(\alpha_{\mathfrak{p}}; \rho)_{\mathfrak{p}}$ of the sum-of-powers-of-divisors function.

4.9 Eisenstein Series and Cuspforms

The point of this section is that the holomorphic Eisenstein series considered in the previous section span the space of "all" holomorphic Eisenstein series. More precisely, we have:

THEOREM Take $\kappa \geq 2$. Suppose that either $\kappa > 2$ or $[F : \mathbb{Q}] > 1$ or $\omega \neq 1$. Let f be a holomorphic automorphic form of weight $(\kappa, \ldots, \kappa)$ on $SL(2, \mathbb{A})$, with central character ω, which is right-invariant under

$$\mathbb{K}(\mathfrak{n}) = \{g \in SL(2, \hat{\mathfrak{o}}) \colon g \equiv 1 \text{ modulo } \mathfrak{n}\}.$$

Then there is a linear combination E (over $\mathbb{C}$) of the holomorphic Eisenstein series

$E(g; \kappa; c_0, d_0, \mathfrak{n}; \tfrac{1}{2}\kappa, \omega, \chi)$ c_0 and d_0 in $\hat{\mathfrak{o}}/\mathfrak{n}\hat{\mathfrak{o}}$ so that $c_0\hat{\mathfrak{o}} + d_0\hat{\mathfrak{o}} = \hat{\mathfrak{o}}$, χ a finite order Hecke character so that the finite-prime part of the conductor of χ divides $\mathfrak{n}$

(as in the previous section) so that f-E is a holomorphic cuspform on $GL(2, \mathbb{A})$.

Remarks In a classical situation we have already shown that for weights $(\kappa_1, \ldots, \kappa_m)$ with not all κ_i the same, all holomorphic Hilbert modular forms on $\mathfrak{H}^m$ are cuspforms; by the discussion comparing classical and adelic version, the same result holds adelically. Also, except when κ is even and ω is totally even, or when κ is odd and ω is totally odd, a similar argument shows that the only holomorphic automorphic forms of that weight and central character are cuspforms.

Proof For a holomorphic Hilbert modular form as in the statement of the theorem, we define the 0-th Fourier coefficient of f by

$$W_0(g) = \int_{U(F)\backslash U(\mathbb{A})} f(ug)\, du.$$

From the results of the previous section, we will deduce that the 0-th Fourier coefficients of the present Eisenstein series with fixed χ and with (c_0, d_0) chosen among representatives for

$$X = \hat{\mathfrak{o}}^\times \backslash \{(c_0, d_0) \colon c_0, d_0 \text{ in } \hat{\mathfrak{o}}/\mathfrak{n}\hat{\mathfrak{o}} \text{ so that } c_0\hat{\mathfrak{o}} + d_0\hat{\mathfrak{o}} = \hat{\mathfrak{o}}\}$$

are *linearly independent*, as follows. Let

$$W_0(g; \kappa; c_0, d_0, \mathfrak{n}; \tfrac{1}{2}\kappa, \omega, \chi)$$

be the 0-th Fourier coefficient of $E(g; \mathrm{i}; c_0, d_0, \mathfrak{n}; \tfrac{1}{2}\kappa, \omega, \chi)$. Suppose that

$$\sum_{(c_0, d_0) \in X} A(c_0, d_0) W_0(g; \kappa; c_0, d_0, \mathfrak{n}; \tfrac{1}{2}\kappa, \omega, \chi) \equiv 0.$$

This would certainly hold for g of the form

$$g = \ell(y)h$$

with

$$h = \begin{pmatrix} a & b \\ c & d \end{pmatrix} \text{ in } GL(2, \hat{\mathfrak{o}}) \qquad \ell(y) = \begin{pmatrix} y & 0 \\ 0 & y^{-1} \end{pmatrix}.$$

The proposition of the previous section, together with the calculation of the constant terms in the second and third theorems, implies that (with such g)

$$W_0(g; \kappa; c_0, d_0, \mathfrak{n}; \tfrac{1}{2}\kappa, \omega, \chi) =$$

$$= \chi(y)\|y\|^{k/2} \times \text{non-zero constant} \qquad \text{for } (c_0 d_0)h^{-1} \equiv (0, 1) \text{ modulo } \mathfrak{n}$$

$$= \qquad 0 \qquad \text{otherwise.}$$

With our choice of X, this is impossible. Clearly the cardinality of X is the cardinality v of $(\mathfrak{o}/\mathfrak{n})^\times$.

The functions $y \to \chi(y)\|y\|^{\kappa/2}$ are obviously linearly independent. There are

$$hv$$

such χ, where h is the *narrow* ideal class number of $\mathfrak{o}$. *Altogether, we have*

$$N = hv^2$$

linearly independent 0-th Fourier coefficients of holomorphic Eisenstein series.

On the other hand, consider the 0-th Fourier *coefficient*

$$W_0(g) = \int_{U(F)\backslash U(\mathbb{A})} f(ug)\, du$$

of a holomorphic Hilbert modular form as in the statement of the theorem. The function W_0 must be left $U(\mathbb{A})$-invariant, left $P(F)$-invariant, right $\mathbb{K}(\mathfrak{n})$-invariant, equivarient under the center by the character ω. Further, there is an additional condition (explicated below) due to the fact that it arises from a holomorphic Hilbert modular form on $GL(2, \mathbb{A})$. We will give an upper bound for the dimension (over $\mathbb{C}$) of the vectorspace V of all such functions, and find that it is equal to N, the number of linearly independent 0-th Fourier coefficients of our Eisenstein series.

First, in

$$\mathbb{Z}(\mathbb{A})U(\mathbb{A})P(F)\backslash GL(2, \mathbb{A})/SO(2)^m \mathbb{K}(\mathfrak{n})$$

by the local Iwasawa decompositions obviously we can choose representatives of the form

$$g = \begin{pmatrix} y & 0 \\ 0 & 1 \end{pmatrix} k$$

where $k \in GL(2, \hat{\mathfrak{o}})/\mathbb{K}(\mathfrak{n})$, and $y \in F^\times\backslash\mathbb{J}/\hat{\mathfrak{o}}^\times$. Further, for any fixed $y \in \mathbb{J}$,

$$\ell(y)^{-1}U(\mathbb{A})\ell(y) \cap \mathbb{K} = \{u(x): x \in \hat{\mathfrak{o}}\}.$$

As we may also adjust k by $GL(2, \hat{\mathfrak{o}}) \cap Z(\mathbb{A})$, we need consider only representatives k in

$$\{u(x): x \in \hat{\mathfrak{o}}\}\backslash GL(2, \hat{\mathfrak{o}})/\mathbb{K}(\mathfrak{n})(GL(2, \mathfrak{o}) \cap Z(\mathbb{A})).$$

Let $\mathbb{J}^+$ be the ideles with all infinite parts positive; since $F^\times \mathbb{J}^+ = \mathbb{J}$, we need consider only representatives $y \in \mathbb{J}^+$. Now fix $k \in GL(2, \hat{\mathfrak{o}})/\mathbb{K}(\mathfrak{n})$, $y_0 \in \mathbb{J}_0$. From the comparison of adelic and classical notions of holomorphy in sections 3.3 and 3.5, for

$$g = u(x)\ell(y) \in GL^+(2, \mathbb{R})^m,$$

(with $x \in \mathbb{R}^m$ and $y \in F_\infty^+$) the function

$$\Psi(g) = y^{-\kappa/2} f(g\ell(y_0)k)$$

is a holomorphic modular form on $\mathfrak{H}^m$ in the classical sense, and its 0-th Fourier coefficient in a classical sense is a complex number $a(y_0 k)$. By the discussion of Fourier expansions in section 3.5, we must have

$$W_0(gy_0k) = y^{\kappa/2} a(\ell(y_0)k).$$

Therefore, the dependence on the archimedean part of $y \in \mathbb{J}^+$ is strongly restricted.

From this argument, we need determine only the cardinality of

$$(F^\times F_\infty^+ \backslash \mathbb{J}/\hat{\mathfrak{o}}^\times) \times \{u(x): x \in \hat{\mathfrak{o}}\}\backslash GL(2, \hat{\mathfrak{o}})/\mathbb{K}(\mathfrak{n})(GL(2, \mathfrak{o}) \cap Z(\mathbb{A}))$$

where F_∞^+ is the connected component of the identity in the infinite ideles $\mathbb{J}_\infty$. The cardinality of the first factor is exactly the narrow ideal class number h. An elementary counting argument shows that the cardinality of the second factor is v^2. Thus, we have shown that the dimension of V is less than or equal to hv^2.

The 0-th Fourier coefficients of our Eisenstein series lie in V, and there are dim V among them which are linearly independent. From this the assertion of the theorem is clear. ■

4.10 Rankin's Integral Representations of L-Functions

The proof of the theorem below contains Rankin's "unwinding trick," making use of the nature of Eisenstein series. This idea was one of the first significant new ideas in the subject to occur in this century (about 1936). The point (for the moment) is that the analytic continuation of the Eisenstein series may be made to yield the analytic continuation of some L-functions by expressing the L-functions as integrals against Eisenstein series in a "suitable" manner. The only previous method of proof of analytic continuation dealt with the standard L-functions, representing them as Mellin transform integrals. At the end of this section a variant of the first theorem is stated, which will be used later; its proof is identical to that of the first theorem. Though we have not given the most general formulation possible for this unwinding idea, the form given here should adequately illustrate it, and this suffices for our subsequent needs.

Fix a non-zero ideal $\mathfrak{n}$ in $\mathfrak{o}$. Let ω be a character on $\mathbb{J}/F^\times$ whose conductor divides $\mathfrak{n}$, is totally even if κ is even, and is totally odd if κ is odd. We will use an Eisenstein series like those just discussed in section 4.8: in the notation there, let

$$E(g; s) = E^*(g; \kappa; 0, 1, \mathfrak{n}; s, \omega, \mathbb{1})$$

be an Eisenstein series of weight $\kappa \geq 0$.

PROPOSITION Let f be a holomorphic cuspform of weight $\lambda \in \mathbb{Z}^m$, and let ψ be an automorphic form of moderate growth, of weight $\lambda - \kappa$ (i.e., $\lambda - (\kappa, \ldots, \kappa)$). Suppose that $\psi E(*; s)\bar{f}$ has trivial central character. Then the integral

$$\langle \psi E(*; s), f \rangle = \int_{Z(\mathbb{A})GL(2,F)\backslash GL(2,\mathbb{A})} \psi(g) E(g; s) \bar{f}(g)\, dg$$

is absolutely convergent and is a meromorphic function of $s \in \mathbb{C}$, with poles at worst at the poles of the Eisenstein series. (Proof below).

Keep the hypotheses of the previous proposition. Let δ be a finite idele so that $\delta\hat{\mathfrak{o}} = \mathfrak{o}^*\hat{\mathfrak{o}}$. Let $\mathfrak{n}_1$ and $\mathfrak{n}_2$ be two ideals whose least common multiple is $\mathfrak{n}$. Put

$$\mathbb{K}_i = \left\{ g \in GL(2, \hat{\mathfrak{o}}) \colon g = \begin{pmatrix} a & b \\ c & d \end{pmatrix} \text{with } c \equiv 0,\ d \equiv 1 \text{ modulo } \mathfrak{n}_i \right\}$$

$$\mathbb{K} = \left\{ g = \begin{pmatrix} a & b \\ c & d \end{pmatrix} \in GL(2, \hat{\mathfrak{o}}) \colon c \equiv 0,\ d \equiv \pm 1 \text{ modulo } \mathfrak{n} \right\}.$$

Suppose that f is right-invariant by $\mathbb{K}_2$ and that ψ is right-invariant by $\mathbb{K}_1$. Let ω_2 be the central character of f, and $\omega_1 = \omega_2/\omega$ the central character of ψ. Suppose that f and ψ are holomorphic *normalized Hecke eigenforms* (in the sense of section 3.5), of weights κ_2 and κ_1, respectively. Take κ_i so that $\kappa_1 + \kappa = \kappa_2$. First of all, this requires that in the Fourier expansions

$$f(g) = \sum_{\xi \in F^\times} W_2(\ell(\xi)g)$$

$$\psi(g) = (\text{0-th Fourier coefficient}) + \sum_{\xi \in F^\times} W_1(\ell(\xi)g)$$

with $y \in \mathbb{J}^+$ we have

$$W_i(\ell(y)) = \|y\|^{\kappa_i/2} \exp(-2\pi \operatorname{Tr} y_\infty) V_i(y_0)$$

and that V_i factors over finite primes as

$$V_i(y_0) = \prod_{\mathfrak{p}<\infty} V_{i\mathfrak{p}}(y_\mathfrak{p})$$

and

$$V_{i\mathfrak{p}}(1) = 1.$$

Further, the Hecke eigenfunction property implies (see section 2.5) that each factor

$$W_{i\mathfrak{p}}(\ell(y_\mathfrak{p})) = \|y_\mathfrak{p}\|^{\kappa_i/2} V_{i\mathfrak{p}}(y_\mathfrak{p})$$

of "the" (non-zero) Fourier coefficient has an explicit form. For $\mathfrak{p}$ not dividing $\mathfrak{n}_i$ it is

$$W_{i\mathfrak{p}}(\ell(\delta_\mathfrak{p} y_\mathfrak{p})) = W_{i\mathfrak{p}}(\ell(\delta_\mathfrak{p}))(\alpha^{n+1} - \beta^{n+1})/(\alpha - \beta) \qquad (\text{if } y_\mathfrak{p}\mathfrak{o}_\mathfrak{p} = \mathfrak{p}^n\mathfrak{o}_\mathfrak{p} \text{ and } n \geq 0)$$

$$0 \qquad (\text{if } y_\mathfrak{p}\mathfrak{o}_\mathfrak{p} = \mathfrak{p}^n\mathfrak{o}_\mathfrak{p} \text{ and } n < 0)$$

for some complex number α and β depending on i and $\mathfrak{p}$ satisfying

$$\alpha\beta = N\mathfrak{p}^{-1}\omega_i(\mathfrak{p}).$$

For $\mathfrak{p}$ not dividing $\mathfrak{n}_i$ it is

$$W_{i\mathfrak{p}}(\ell(\delta_\mathfrak{p} y_\mathfrak{p})) = W_{i\mathfrak{p}}(\ell(\delta_\mathfrak{p}))\alpha^{n+1} \qquad (\text{if } y_\mathfrak{p}\mathfrak{o}_\mathfrak{p} = \mathfrak{p}^n\mathfrak{o}_\mathfrak{p} \text{ and } n \geq 0)$$

$$0 \qquad (\text{if } y_\mathfrak{p}\mathfrak{o}_\mathfrak{p} = \mathfrak{p}^n\mathfrak{o}_\mathfrak{p} \text{ and } n < 0)$$

for some complex numbers α depending on i and $\mathfrak{p}$. In terms of the functions $V_{i\mathfrak{p}}$ *with the present normalization* $V_{i\mathfrak{p}}(1) = 1$ this is as follows. For $\mathfrak{p}$ not dividing $\mathfrak{n}_i$:

$$V_{i\mathfrak{p}}(\delta_\mathfrak{p} y_\mathfrak{p}) = (\alpha_{i\mathfrak{p}}^{n+1} - \beta_{i\mathfrak{p}}^{n+1})/(\alpha_{i\mathfrak{p}} - \beta_{i\mathfrak{p}}) \qquad (\text{if } y_\mathfrak{p}\mathfrak{o}_\mathfrak{p} = \mathfrak{p}^n\mathfrak{o}_\mathfrak{p} \text{ and } n \geq 0)$$

$$0 \qquad (\text{if } y_\mathfrak{p}\mathfrak{o}_\mathfrak{p} = \mathfrak{p}^n\mathfrak{o}_\mathfrak{p} \text{ and } n < 0)$$

where

$$\alpha_{i\mathfrak{p}} = N\mathfrak{p}^{\kappa_i/2}\alpha \qquad \beta_{i\mathfrak{p}} = N\mathfrak{p}^{\kappa_i/2}\beta$$

in terms of the parameters α and β in $W_{i\mathfrak{p}}$. Therefore,

$$\alpha_{i\mathfrak{p}}\beta_{i\mathfrak{p}} = \omega_i(\mathfrak{p})N\mathfrak{p}^{\kappa_i-1}.$$

For $\mathfrak{p}$ dividing $\mathfrak{n}_i$:

$$V_{i\mathfrak{p}}(\delta_\mathfrak{p} y_\mathfrak{p}) = \alpha_{i\mathfrak{p}}^n \qquad (\text{if } y_\mathfrak{p}\mathfrak{o}_\mathfrak{p} = \mathfrak{p}^n\mathfrak{o}_\mathfrak{p} \text{ and } n \geq 0)$$

$$0 \qquad (\text{if } y_\mathfrak{p}\mathfrak{o}_\mathfrak{p} = \mathfrak{p}^n\mathfrak{o}_\mathfrak{p} \text{ and } n < 0)$$

where $\alpha_{i\mathfrak{p}} = N\mathfrak{p}^{\kappa_i/2}\alpha$, in terms of the parameter α in $W_{i\mathfrak{p}}$. If a finite prime $\mathfrak{p}$ does not divide $\mathfrak{n}_1$ but does divide $\mathfrak{n}_2$, then (temporarily) say that $\mathfrak{p}$ is good$_1$; if $\mathfrak{p}$ does divide $\mathfrak{n}_1$ but does not divide $\mathfrak{n}_2$, then say that $\mathfrak{p}$ is good$_2$; if $\mathfrak{p}$ does not divide $\mathfrak{n}$, then say that $\mathfrak{p}$ is good; if $\mathfrak{p}$ divides both $\mathfrak{n}_1$ and $\mathfrak{n}_2$, then say that $\mathfrak{p}$ is bad.

Remarks For subsequent applications, a considerable burden is removed by the observation that in the following proof *we do not use the hypothesis that f and ψ are Hecke eigenfunctions at "bad" primes*, but only (the consequence of this hypothesis) that *the functions W_i and V_i have the indicated forms on the special elements* $\ell(y)$ with $y \in \mathbb{J}^+$.

THEOREM Let $s' = s - 1 + \frac{1}{2}(\kappa_1 + \kappa_2)$. With hypotheses as above, for $s \in \mathbb{C}$ away from poles of the Eisenstein series $E(*; s)$, we have

$$\langle \psi E(*;s), f \rangle =$$

$$= (\text{volume } \mathbb{K})\,(\text{index of } \{\alpha \in \hat{\mathfrak{o}}^\times : \alpha \equiv 1 \bmod \mathfrak{n}\} \text{ in } \hat{\mathfrak{o}}^\times)^{-1} \times$$

$$\times \|\delta\|^{s'} \times (\pi^{-(s+\kappa/2)}\Gamma(s + \tfrac{1}{2}\kappa)(4\pi)^{-s'}\Gamma(s'))^m \times$$

$$\times \prod_{\mathfrak{p}\text{ good}} 1/(1 - \alpha_{1\mathfrak{p}}\bar{\alpha}_{2\mathfrak{p}}N\mathfrak{p}^{-s'})(1 - \alpha_{1\mathfrak{p}}\bar{\beta}_{2\mathfrak{p}}N\mathfrak{p}^{-s'})(1 - \beta_{1\mathfrak{p}}\bar{\alpha}_{2\mathfrak{p}}N\mathfrak{p}^{-s'})(1 - \beta_{1\mathfrak{p}}\bar{\beta}_{2\mathfrak{p}}N\mathfrak{p}^{-s'}) \times$$

$$\times \prod_{\mathfrak{p}\text{ good}_1} 1/(1 - \alpha_{1\mathfrak{p}}\bar{\alpha}_{2\mathfrak{p}}N\mathfrak{p}^{-s'})(1 - \beta_{1\mathfrak{p}}\bar{\alpha}_{2\mathfrak{p}}N\mathfrak{p}^{-s'}) \times$$

$$\times \prod_{\mathfrak{p}\text{ good}_2} 1/(1 - \alpha_{1\mathfrak{p}}\bar{\alpha}_{2\mathfrak{p}}N\mathfrak{p}^{-s'})(1 - \alpha_{1\mathfrak{p}}\bar{\beta}_{2\mathfrak{p}}N\mathfrak{p}^{-s'}) \times$$

$$\times \prod_{\mathfrak{p}\text{ bad}} 1/(1 - \alpha_{1\mathfrak{p}}\bar{\alpha}_{2\mathfrak{p}}N\mathfrak{p}^{-s'}).$$

COROLLARY The "Rankin product" L-function $L(\psi \otimes \bar{f}, s)$ associated to ψ and $\bar{f}$ (given by the previous expression) has an analytic continuation to a meromorphic function of $s \in \mathbb{C}$. This L-function is entire unless $\omega_1 = \omega_2$ and the weight of f is the same as the weight of ψ. In the latter case there are at worst simple poles, at $s = 0, 1$, with residues

$$\langle \psi, f \rangle \times (\text{non-zero constant depending on } \mathfrak{n}).$$

Proof of Corollary From section 4.2, unless $\kappa = 0$ and $\omega = \mathbb{1}$, this Eisenstein series is entire, so the integral is entire. When $\kappa = 0$ and $\omega = \mathbb{1}$, from section 4.2 this Eisenstein series has poles at $s = 0, 1$ which are non-zero constants (as functions of $g \in GL(2, \mathbb{A})$). Therefore, the integral is a non-zero multiple of $\langle \psi, f \rangle$. ■

Remarks Since the Eisenstein series has a functional equation under $s \to 1 - s$, one can also obtain a functional equation relating $L(\psi \otimes \bar{f}, s)$ and (essentially)

$L(\psi \otimes \bar{f}, 1 - s)$ (in the present normalization). However, especially for $\mathfrak{n} \neq \mathfrak{o}$, under the functional equation the Eisenstein series is altered at bad primes; this creates technical though tractable complications.

Sometimes it is very useful to be able to express an inner product of automorphic forms as a residue of an L-function, or vice-versa. In particular, if any of the good-prime Hecke eigenvalues of ψ are different from those of f, then the Rankin product L-function is entire, since $\langle \psi, f \rangle = 0$.

Note The simplest classical version of this result has the following guise, which can be easily verified. Let f be a holomorphic cuspform on $\mathfrak{H}$ of weight κ with respect to $SL(2, \mathbb{Z})$, and let ψ be a holomorphic modular form of the same weight κ. Let the (numerical) Fourier coefficients of f be $\{b_n\}$ and those of ψ be $\{a_n\}$. Then the above "Rankin convolution" gives an integral representation of the Dirichlet series (with a gamma factor)

$$\sum_{n \geq 1} a_n \bar{b}_n / n^s.$$

This calculation does not presume that f and ψ are eigenfunctions for the Hecke operators. The original purpose of this calculation was to improve the then-known estimates on the Fourier coefficients of cuspforms, using the analytic properties of this L-function.

Proof of Proposition Recall that quotients $\Gamma \backslash \mathfrak{H}^m$ are covered by finitely many Siegel sets. From the comparison of quotients $\Gamma \backslash \mathfrak{H}^m$ and quotients $Z(\mathbb{A})GL(2, F) \backslash GL(2, \mathbb{A})$ via Strong Approximation for $SL(2)$ we can easily deduce the adelic analogue: There is a compact set C_1 in $GL(2, \mathbb{A})$, a compact set C_0 in $U(\mathbb{A})$, and a constant $c > 0$ so that

$$Z(\mathbb{A})GL(2, F)C_0\{\ell(y)\colon \|y\| \geq c\}C_1 = GL(2, \mathbb{A}).$$

The set $C_0\{\ell(y)\colon \|y\| \geq c\}C_1$ is an adelic version of a Siegel set. We gave a classical proof that *classical* holomorphic cuspforms are "of rapid decay" in Siegel sets in $\mathfrak{H}^m$; from this it follows that for an *adelic* holomorphic cuspform f, for $k_1, k_2 \in C$, $y \in \mathbb{J}$, $\zeta \in Z(\mathbb{A})$, $\gamma \in GL(2, F)$,

$$f(\zeta \gamma k_1 \ell(y) k_2)$$

is of rapid decay uniformly in $k_1 \in C_0$ and $k_2 \in C_1$ (and $\zeta \in Z(\mathbb{A})$, $\gamma \in GL(2, F)$) as $\|y\| + \|y\|^{-1}$ goes to infinity. We have seen that Eisenstein series are of "moderate growth." Therefore, as the integral over

$$Z(\mathbb{A})GL(2,F)\backslash GL(2,\mathbb{A})$$

is dominated by the integral over

$$C_0\{\ell(y)\colon \|y\| \geq c\}C_1$$

it is immediate that the integral is sufficiently nicely convergent, in particular so as to give a meromorphic function of $s \in \mathbb{C}$ away from the poles of the Eisenstein series. ■

Proof of Theorem Take $\mathrm{Re}(s) > 1$, so that the integral expression defining the Eisenstein series is nicely convergent. Put

$$\varepsilon(g) = \|\det g\|^s \int_{\mathbb{J}} \|t\|^{2s}\overline{\omega}(t)\varphi(t(0,1)g)\,d^\times t.$$

Recall that

$$E(g;s) = \sum_{\gamma \in P(F)\backslash GL(2,F)} \varepsilon(\gamma g).$$

From integration theory, we can "unwind" the integral:

$$\langle \psi E(*;s), f\rangle = \int_{Z(\mathbb{A})GL(2,F)\backslash GL(2,\mathbb{A})} \psi(g)E(g;s)\bar{f}(g)\,dg =$$

$$= \int_{Z(\mathbb{A})P(F)\backslash GL(2,\mathbb{A})} \psi(g)\varepsilon(g)\bar{f}(g)\,dg.$$

Now $\varepsilon(g)$ is supported on $P(\mathbb{A})K_\infty\mathbb{K}$, and the integrand is constant on right $K_\infty\mathbb{K}$ cosets, so this is equal to

$$(\text{measure of } \mathbb{K}) \times \int_{F^\times\backslash\mathbb{J}} \int_{U(F)\backslash U(\mathbb{A})} \psi(u(x)\ell(y))\varepsilon(\ell(y))\bar{f}(u(x)\ell(y))\,\|y\|^{-1}\,d^+x\,d^\times y,$$

where we have chosen the obvious representatives

$$u(x)\ell(y) = \begin{pmatrix} 1 & x \\ 0 & 1 \end{pmatrix}\begin{pmatrix} y & 0 \\ 0 & 1 \end{pmatrix}$$

for the quotient

$$Z(\mathbb{A})P(F)\backslash P(\mathbb{A}),$$

and the $\|y\|^{-1}$ appears so as to give a left Haar measure on $P(\mathbb{A})$ in these coordinates. (Here d^+x denotes the usual additive Haar measure on $\mathbb{A}$, and $d^\times y$ denotes the multiplicative Haar measure on $\mathbb{J}$.) Note that ε is left $U(\mathbb{A})$-invariant, so that

$$\varepsilon(u(x)\ell(y)) = \varepsilon(\ell(y)).$$

Now we expand f and ψ in their Fourier series. Integrating first in $x \in \mathbb{A}/F$, using the orthogonality of distinct additive characters (i.e., for distinct indices ξ), and using the normalization that the volume of $\mathbb{A}/F$ is 1, we have

$$(\text{measure of } \mathbb{K})^{-1}\langle \psi E(*; s), f\rangle =$$

$$= \int_{F^\times\backslash\mathbb{J}} \sum_{\xi\in F^\times} W_1(\ell(\xi y))\varepsilon(\ell(y))\overline{W_2}(\ell(\xi y))\|y\|^{-1}\, d^\times y.$$

As usual, let $\mathbb{J}^+$ be the ideles with positive infinite-prime components; as $F^\times\mathbb{J}^+ = \mathbb{J}$, the latter integral is the same as

$$\int_{F^+\backslash\mathbb{J}^+} \sum_{\xi\in F^\times} W_1(\ell(\xi y))\varepsilon(\ell(y))\overline{W_2}(\ell(\xi y))\|y\|^{-1}\, d^\times y$$

where F^+ is the totally positive element of F. This unwinds to

$$\int_{\mathbb{J}^+} \sum_{\xi\in F^\times/F^+} W_1(\ell(\xi y))\varepsilon(\ell(\xi y))\overline{W_2}(\ell(\xi y))\|y\|^{-1}\, d^\times y$$

in a further "unwinding," using the $\ell(F^\times)$-invariance of ε. Since $W_1(\ell(\xi y))$ and $\overline{W_2}(\ell(\xi y))$ are 0 for $y \in \mathbb{J}^+$ *unless* ξ is totally positive, this is actually

$$\int_{\mathbb{J}^+} W_1(\ell(y))\varepsilon(\ell(y))\overline{W_2}(\ell(y))\|y\|^{-1}\, d^\times y.$$

Further,

$$\varepsilon(\ell(y)) = \|\det \ell(y)\|^s \int_{\mathbb{J}^+} \|t\|^{2s}\overline{\omega}(t)\varphi(t(0,1)\ell(y))\, d^\times t =$$

$$= \|y\|^s \int_{\mathbb{J}^+} \|t\|^{2s}\overline{\omega}(t)\varphi(t(0,1))\, d^\times t.$$

In particular, from the explicit form of φ,

$$(\text{index of } \{\alpha \in \hat{\mathfrak{o}}^\times : \alpha \equiv 1 \bmod \mathfrak{n}\} \text{ in } \hat{\mathfrak{o}}^\times) \times \int_{\mathbb{J}} \|t\|^{2s}\bar{\omega}(t)\varphi(t(0,1))\,d^\times t =$$

$$= \left(\int_{\mathbb{R}^\times} |t|^{2s}\bar{\omega}(t)t^\kappa \exp(-\pi t^2)\,d^\times t\right)^m \times \prod_{\mathfrak{p}\text{ finite, good}} (1 - \bar{\omega}(\pi)N\mathfrak{p}^{-2s})^{-1} =$$

$$= (\pi^{-s+\kappa/2}\Gamma(s + \tfrac{1}{2}\kappa))^m \times \prod_{\mathfrak{p}\text{ finite, good}} (1 - \bar{\omega}(\pi)N\mathfrak{p}^{-2s})^{-1}.$$

We remove this factor for the following calculation.

By hypothesis, each W_i factors over primes very explicitly. We will treat only two of the five cases: infinite primes (where we use the assumption of holomorphy) and primes not dividing $\mathfrak{n}$ (where we use information from the spherical Hecke algebra). The other three cases are even easier: At primes which are good_1 we use information from the special Hecke operators for f and the spherical Hecke operators at ψ, etc.

At a prime $\mathfrak{p}$ not dividing $\mathfrak{n}$, as $\mathfrak{o}_\mathfrak{p}^\times$ has measure 1 in $F_\mathfrak{p}^\times$, by homogeneous space integration theory the $\mathfrak{p}$-integral is

$$\int_{F_\mathfrak{p}^\times} W_{1\mathfrak{p}}(\ell(y))\overline{W}_{2\mathfrak{p}}(\ell(y))\|y\|^{s-1}\,d^\times y =$$

$$= \int_{F_\mathfrak{p}^\times} V_{1\mathfrak{p}}(\ell(y))\overline{V}_{2\mathfrak{p}}(\ell(y))\|y\|^{s'}\,d^\times y =$$

$$= \sum_{n\geq 0} \|\pi^n\delta\|_\mathfrak{p}^{s'}(A^{n+1} - B^{n+1})(\bar{\alpha}^{n+1} - \bar{\beta}^{n+1})/(A - B)(\bar{\alpha} - \bar{\beta}) =$$

$$= \|\delta\|_\mathfrak{p}^{s'}(A - B)^{-1}(\bar{\alpha} - \bar{\beta})^{-1} \times$$

$$\times \{A\bar{\alpha}/(1 - A\bar{\alpha}X) - A\bar{\beta}/(1 - A\bar{\beta}X) - B\bar{\alpha}/(1 - B\bar{\alpha}X) + B\bar{\beta}/(1 - B\bar{\beta}X)\}$$

where

$$X = \|\pi\|_\mathfrak{p}^{s'}.$$

Using $AB = \omega_1(\pi)N\mathfrak{p}^{\kappa_1 - 1}$ and $\alpha\beta = \omega_2(\pi)N\mathfrak{p}^{\kappa_2 - 1}$, it is a matter of elementary algebra to see that this is just

$$\|\delta\|_\mathfrak{p}^{s'}(1 - AB\bar{\alpha}\bar{\beta}X^2)/(1 - A\bar{\alpha}X)(1 - A\bar{\beta}X)(1 - B\bar{\alpha}X)(1 - B\bar{\beta}X) =$$

$$= \|\delta\|_\mathfrak{p}^{s'}(1 - \omega_1(\pi)\bar{\omega}_2(\pi)N\mathfrak{p}^{\kappa_1+\kappa_2-2}X^2)/(1 - A\bar{\alpha}X)(1 - A\bar{\beta}X)(1 - B\bar{\alpha}X)(1 - B\bar{\beta}X) =$$

$$= \|\delta\|_\mathfrak{p}^{s'}(1 - \omega_1(\pi)\bar{\omega}_2(\pi)N\mathfrak{p}^{-2s})/(1 - A\bar{\alpha}X)(1 - A\bar{\beta}X)(1 - B\bar{\alpha}X)(1 - B\bar{\beta}X).$$

Since $\omega\omega_1 = \omega_2$, the factor

$$(1 - \omega_1(\pi)\bar{\omega}_2(\pi)N\mathfrak{p}^{-2s}) = (1 - \bar{\omega}(\pi)N\mathfrak{p}^{-2s})$$

in the numerator exactly cancels the $\mathfrak{p}$-factor arising from the integral in $t \in \mathbb{J}$ (which we had removed).

At an infinite-prime $\mathfrak{p}$, the $\mathfrak{p}$-factor is

$$\int_{(0,\infty)} |y|^{s'} \exp(-4\pi y)\, d^{\times} y = (4\pi)^{-s'}\Gamma(s').$$

This finishes the computation. ∎

5 Theta Series

Theta series are an important, explicitly constructed family of automorphic forms. Historically, one important use of theta series has been as a vehicle for the study of definite quadratic forms. In this volume, the main use will be in section 6.3 in the discussion of grossencharacter L-functions.

The first section of this chapter proves functional equations for the simplest theta series attached to a totally real field, and simultaneously proves an explicit quadratic reciprocity law, via a special case of Hecke's reciprocity law for quadratic exponential sums. This reciprocity law is used in the proof of the functional equations for more general theta series.

The second section is a proof of Hecke's identity, a Fourier transform calculation necessary to consider the most general pluriharmonic theta series.

The third section contains one main result of this chapter: Here it is shown that the most general theta series attached to a positive definite quadratic form, a homogeneous pluriharmonic polynomial, and a finite-adele Schwartz function is a Hilbert modular form of some explicitly estimable level. This is literally the case if the dimension of the quadratic form is even, as then the theta series are of integral weight; if the dimension is odd, then the theta series is of half-integral weight, and the automorphy factor involves a quadratic reciprocity symbol which complicates matters. In particular, half-integral weight automorphic forms cannot be turned into automorphic forms on $GL(2, \mathbb{A})$ in the manner discussed earlier for integral weight Hilbert modular forms.

The fourth section gives some explicit examples of theta series of level one over certain totally real number fields. The fifth section is a proof that spaces of integral weight theta series are stabilized by the natural action of $GL^+(2, F)$ and by the action of Hecke operators. The sixth section uses the stability of integral weight theta series under $GL^+(2, F)$ to show that a pluriharmonic theta series is always a cuspform if the pluriharmonic polynomial involved is not a constant. The last section gives a more refined treatment of examples of such cuspforms, whose associated L-functions are L-functions with grossencharacters.

A noteworthy feature of theta series is that it is relatively easy to understand the rationality properties of their Fourier coefficients. On the other hand, the fact that theta series are automorphic forms implies otherwise unexpected properties of these Fourier coefficients.

5.1 A Simple Theta Series, Reciprocity Laws

Here we simultaneously obtain a quadratic reciprocity law, and the fact that a certain simple theta series is a Hilbert modular form of weight $\frac{1}{2}$ (with a character).

Again, F is a totally real number field of degree m over $\mathbb{Q}$, with integers $\mathfrak{o}$, inverse different $\mathfrak{o}^*$, absolute discriminant D, and absolute trace Tr (extended $\mathbb{C}$-linearly to $\mathbb{C}^m \to \mathbb{C}$). We use a multi-index notation. For $z \in \mathfrak{H}^m$, write $\mathrm{e}(z) = \exp(2\pi i \operatorname{Tr} z)$. All square roots of numbers in $\mathbb{C} - \{x < 0\}$ are to be taken in the right half-plane.

Definition

$$\theta(z) = \sum_{v \in \mathfrak{o}} \mathrm{e}(\tfrac{1}{2} z v^2).$$

Definition For $d \in \mathfrak{o} - \{0\}$, define

$$\varepsilon(d) = (i \operatorname{sgn} d)^{1/2} 2^{-m/2} D^{-1/2} \sum_{v \in \mathfrak{o}^*/2\mathfrak{o}} \mathrm{e}(-v^2 d/4)$$

where $\operatorname{sgn}(d)$ is $\{\pm 1\}^m$-valued, and $(i \operatorname{sgn} d)^{1/2}$ is meant in a multi-index sense. For $d \in (\mathbb{R}^\times)^m$, let s be the number of negative components, and define $\tilde{\varepsilon}(d) = i^s$. Let Γ be the congruence subgroup of $SL(2, \mathfrak{o})$ defined by

$$\Gamma = \left\{ \begin{pmatrix} a & b \\ c & d \end{pmatrix} \in SL(2, \mathfrak{o})\colon b, c \in 2\mathfrak{o}^{*-1}, a\mathfrak{o}, d\mathfrak{o} \text{ prime to } 2\mathfrak{o}^{*-1} \right\}.$$

For a prime ideal $\mathfrak{p}$ and $\beta \in \mathfrak{o}$ define a *quadratic symbol*

$$\begin{aligned} (\beta/\mathfrak{p})_2 &= +1 \quad \text{if } \beta \bmod \mathfrak{p} \text{ is a square in } (\mathfrak{o}/\mathfrak{p})^\times \\ &= -1 \quad \text{if } \beta \bmod \mathfrak{p} \text{ is a non-square in } (\mathfrak{o}/\mathfrak{p})^\times \\ &= 0 \quad \text{if } \beta \in \mathfrak{p}. \end{aligned}$$

Extend the function $\mathfrak{p} \to (\beta/\mathfrak{p})_2$ to all non-zero ideals in $\mathfrak{o}$ by imposing multiplicativity. For $0 \neq \alpha \in \mathfrak{o}$, write $(\beta/\alpha)_2 = (\beta/\alpha\mathfrak{o})_2$.

THEOREM A For $\left(\begin{smallmatrix} a & b \\ c & d \end{smallmatrix}\right) \in \Gamma$ there are functional equations

$$\theta((az + b)(cz + d)^{-1}) = \tilde{\varepsilon}(d)\varepsilon(d)(-\tfrac{1}{2}c/d)_2 (cz + d)^{1/2} \theta(z)$$

$$\theta^2((az + b)(cz + d)^{-1}) = \chi(d)(cz + d)\theta^2(z)$$

where χ is a character defined modulo $4\mathfrak{o}^{*-1}$, and on $d \in \mathfrak{o}$ prime to $4\mathfrak{o}^{*-1}$ is given by

$$\chi(d) = \tilde{\varepsilon}(d)^2 \varepsilon(d)^2 = (\operatorname{sgn} d)^1(-1/d)_2.$$

THEOREM B Let α, β prime to $4\mathfrak{o}^{*-1}$ such that $\alpha\mathfrak{o} + \beta\mathfrak{o} = \mathfrak{o}$. Then

$$(\alpha/\beta)_2(\beta/\alpha)_2 = \varepsilon(\alpha\beta)/\varepsilon(\alpha)\varepsilon(\beta)$$

$$\varepsilon(\beta)^2 = (-1/\beta)_2.$$

Proof The proofs of the two theorems are given together. In the proof we need

i. a special case of Hecke's reciprocity law for quadratic exponential sums,
ii. elementary functional equations of theta series,
iii. quadratic symbols as ratios of exponential sums,
iv. containment of congruence subgroups in a certain special set of matrices in $SL(2, F)$. ∎

PROPOSITION (HECKE'S RECIPROCITY LAW) Let $0 \neq \alpha \in \mathfrak{o}$. Then

$$\varepsilon(\alpha) = |\alpha|^{-1/2} \sum_{v \in \mathfrak{o}/\alpha\mathfrak{o}} e(v^2/\alpha).$$

Proof For a fractional ideal $\mathfrak{a}$ of $\mathfrak{o}$ and for $z \in \mathfrak{H}^m$, consider

$$\theta(z; \mathfrak{a}) = \sum_{v \in \mathfrak{a}} e(\tfrac{1}{2}zv^2).$$

For $0 \neq \alpha \in \mathfrak{o}$ let $\mathfrak{a}'$ be any non-zero fractional ideal contained in $\alpha\mathfrak{o} \cap \mathfrak{a}^{-1}\mathfrak{o}^*$; then we have

$$\theta(2\alpha^{-1} + iy; \mathfrak{a}) = \sum_{v \in \mathfrak{a}} e(\tfrac{1}{2}(2\alpha^{-1} + iy)v^2) =$$

$$= \sum_{v \in \mathfrak{a}/\mathfrak{a}'} \sum_{w \in \mathfrak{a}'} e(\tfrac{1}{2}(2\alpha^{-1} + iy)(v + w)^2).$$

From elementary estimates it follows that

$$\lim_{y \to 0^+} y^{1/2}\theta(2\alpha^{-1} + iy) =$$

$$= \lim_{y \to 0^+} \sum_{v \in \mathfrak{a}/\mathfrak{a}'} y^{1/2} \sum_{w \in \mathfrak{a}'} e(\tfrac{1}{2}(2\alpha^{-1} + iy)(v + w)^2) =$$

$$= \sum_{v \in \mathfrak{a}/\mathfrak{a}'} e(v^2/\alpha) \times \lim_{y \to 0^+} y^{1/2} \sum_{w \in \mathfrak{a}'} \exp(-\pi \operatorname{Tr} yw^2) =$$

$$= D^{-1/2} N\mathfrak{a}'^{-1} \sum_{v \in \mathfrak{a}/\mathfrak{a}'} e(\operatorname{Tr} v^2/\alpha),$$

using the definition of the Riemann integral, where $N\mathfrak{a}'$ is the ideal norm of $\mathfrak{a}'$. On the other hand, from Poisson summation we have

$$\sum_{v\in\mathfrak{a}} \exp(-\pi \operatorname{Tr} yv^2) = y^{-1/2} D^{-1/2} N\mathfrak{a}^{-1} \sum_{v\in\mathfrak{a}^{-1}\mathfrak{d}^*} \exp(-\pi \operatorname{Tr} y^{-1} v^2)$$

and by analytic continuation in $z \in \mathfrak{H}^m$

$$\theta(z) = (-iz)^{-1/2} D^{-1/2} N\mathfrak{a}^{-1} \sum_{v\in\mathfrak{a}^{-1}\mathfrak{d}^*} \mathrm{e}(\tfrac{1}{2}(-z^{-1})v^2).$$

Taking $z = 2\alpha^{-1} + iy$ gives

$$\begin{aligned}\theta(2\alpha^{-1} + iy) &= (y - i2\alpha^{-1})^{-1/2} D^{-1/2} N\mathfrak{a}^{-1} \theta(-1/(2\alpha^{-1} + iy)) = \\ &= (y - 2i/\alpha)^{-1/2} D^{-1/2} N\mathfrak{a}^{-1} \theta([-2\alpha^{-1} + iy]/(4\alpha^{-2} + y^2)) = \\ &= (-2i/\alpha)^{-1/2} D^{-1/2} N\mathfrak{a}^{-1} \theta(-\tfrac{1}{2}\alpha + i\alpha^2 y/4 + O(y^2)) + O(y)\end{aligned}$$

as $y \to 0^+$. Now take $\mathfrak{a} = \mathfrak{o}$ and $\mathfrak{a}' = \alpha\mathfrak{o}$. Multiplying the previous equality by $y^{1/2}$ and letting $y \to 0^+$ gives (via some easy estimates)

$$\begin{aligned}&D^{-1/2} N\mathfrak{a}'^{-1} \sum_{v\in\mathfrak{o}/\alpha\mathfrak{o}} \mathrm{e}(v^2/\alpha) = \\ &\quad = (-2i/\alpha)^{-1/2} D^{-1/2} \times |2\alpha| D^{-1/2} N(2\mathfrak{o})^{-1} \sum_{v\in\mathfrak{o}^*/2\mathfrak{o}} \mathrm{e}(-\tfrac{1}{2}v^2\alpha).\end{aligned}$$

Upon simplifying, we have

$$\sum_{v\in\mathfrak{o}/\alpha\mathfrak{o}} \mathrm{e}(v^2/\alpha) = |\alpha|^{1/2} (i \operatorname{sgn}(\alpha))^{1/2} 2^{-m/2} D^{-1/2} \sum_{v\in\mathfrak{o}^*/2\mathfrak{o}} \mathrm{e}(-\tfrac{1}{2}v^2\alpha) = |\alpha|^{1/2} \varepsilon(\alpha)$$

by the definition of $\varepsilon(\alpha)$. ■

LEMMA Let $\beta \in \mathfrak{o}$ so that $\beta\mathfrak{o}$ is prime to $2\mathfrak{o}^{*-1}$. Let $\alpha = \mathfrak{o}$ be such that $\alpha\mathfrak{o}$ is prime to $\beta\mathfrak{o}$. Then

$$[\mathfrak{o} : \beta\mathfrak{o}]^{-1} \sum_{x\in\mathfrak{o}/\beta\mathfrak{o}} \mathrm{e}(\alpha x^2/\beta) = (\alpha/\beta)_2 [\mathfrak{o} : \beta\mathfrak{o}]^{-1} \sum_{x\in\mathfrak{o}/\beta\mathfrak{o}} \mathrm{e}(x^2/\beta).$$

Proof The proof is identical to the classical version for $F = \mathbb{Q}$, taking into account the fact that $\beta\mathfrak{o}$ is not divisible by any absolutely ramified primes, and is prime to $2\mathfrak{o}$. ■

COROLLARY Let $\alpha, \beta \in \mathfrak{o}$ be such that $\alpha\mathfrak{o} + \beta\mathfrak{o} = \mathfrak{o}$, and so that both $\alpha\mathfrak{o}$ and $\beta\mathfrak{o}$ are prime to $2\mathfrak{o}^{*-1}$. Then

$$\varepsilon(\alpha\beta) = (\alpha/\beta)_2 (\beta/\alpha)_2 \varepsilon(\alpha)\varepsilon(\beta).$$

Proof The proposition gives

$$\varepsilon(\alpha\beta) = [\mathfrak{o} : \alpha\beta\mathfrak{o}]^{-1} \sum_{x \in \mathfrak{o}/\alpha\beta\mathfrak{o}} e(x^2/\alpha\beta).$$

The approximation theorem yields

$$[\mathfrak{o} : \alpha\beta\mathfrak{o}]^{-1} \sum_{x \in \mathfrak{o}/\alpha\beta\mathfrak{o}} e(x^2/\alpha\beta) =$$

$$= [\mathfrak{o} : \alpha\mathfrak{o}]^{-1} \sum_{x \in \mathfrak{o}/\alpha\mathfrak{o}} e(x^2\beta/\alpha) \times [\mathfrak{o} : \beta\mathfrak{o}]^{-1} \sum_{x \in \mathfrak{o}/\beta\mathfrak{o}} e(x^2\alpha/\beta)$$

which by the lemma and the proposition is just

$$(\beta/\alpha)_2[\mathfrak{o} : \alpha\mathfrak{o}]^{-1} \sum_{x \in \mathfrak{o}/\alpha\mathfrak{o}} e(x^2/\alpha) \times (\alpha/\beta)_2[\mathfrak{o} : \beta\mathfrak{o}]^{-1} \sum_{x \in \mathfrak{o}/\beta\mathfrak{o}} e(x^2/\beta) =$$

$$= (\alpha/\beta)_2(\beta/\alpha)_2\varepsilon(\alpha)\varepsilon(\beta).$$ ∎

COROLLARY The function ε^2 is multiplicative on $\{\beta \in \mathfrak{o}: \beta \text{ is prime to } 2\mathfrak{o}^{*-1}\}$. For such β

$$\varepsilon(\beta)^2 = (-1/\beta)_2.$$

In particular, $\varepsilon(\beta) \in \{\pm 1, \pm i\}$, and $\varepsilon(\mathfrak{o}^{\times}) = 1$.

Proof From the previous corollary, $\varepsilon^2(\alpha\beta) = \varepsilon^2(\alpha)\varepsilon^2(\beta)$ if $\alpha\mathfrak{o} + \beta\mathfrak{o} = \mathfrak{o}$. On the other hand, from the definition, $\varepsilon(x)$ depends only on $\alpha + 4\mathfrak{o}^{*-1}$ and $\operatorname{sgn}\alpha$; we can easily replace α by $\alpha' \equiv \alpha$ modulo $4\mathfrak{o}^{*-1}$ with $\operatorname{sgn}\alpha' = \operatorname{sgn}\alpha$ so that $\alpha'\mathfrak{o} + \beta\mathfrak{o} = \mathfrak{o}$. This gives the multiplicativity of the function ε^2.

From the proposition, $\varepsilon(\mathfrak{o}^{\times}) = 1$, so $\varepsilon(1) = 1$. There is some power β^{2n} of β so that β^{2n} is totally positive and so that $\beta^{2n} \equiv 1 \bmod 4\mathfrak{o}^{*-1}$. Since $\varepsilon(\beta^{2n})$ depends only on $\operatorname{sgn}\beta^{2n}$ and upon $\beta^{2n} \bmod 4\mathfrak{o}^{*-1}$, by the multiplicativity of ε^2 we have

$$1 = \varepsilon(\beta^{2n}) = \varepsilon^2(\beta^n) = (\varepsilon^2(\beta))^n.$$

Therefore, $|\varepsilon(\beta)| = 1$. From the definition of ε, $\varepsilon(-\beta) = \overline{\varepsilon}(\beta)$; as $|\varepsilon| = 1$, this implies that $\varepsilon(-\beta) = \varepsilon(\beta)^{-1}$. On the other hand, by the proposition $\varepsilon(-\beta) = (-1/\beta)_2\varepsilon(\beta)$. Therefore, $\varepsilon(\beta)^2 = (-1/\beta)_2$, as claimed. ∎

Definition Let $\mathbb{A}$ be the adeles of F, and $\mathbb{A}_0$ the finite adeles of F. Let φ be a Schwartz function on $\mathbb{A}_0$ (as in Appendix A.2). For $z \in \mathfrak{H}^m$, *the theta series* attached to φ (and to the quadratic form $x \to x^2$) is defined to be

$$\Theta(z; \varphi) = \sum_{v \in F} \varphi(v)e(\tfrac{1}{2}zv^2).$$

The simplest case is θ itself, which is obtained by taking φ to be the characteristic function of $\prod_{\mathfrak{p}<\infty} \mathfrak{o}_{\mathfrak{p}}$.

Definition Take $y \in F^x$, $x \in F$. Put

$$t(y) = \begin{pmatrix} y & 0 \\ 0 & y^{-1} \end{pmatrix} \qquad u(x) = \begin{pmatrix} 1 & x \\ 0 & 1 \end{pmatrix} \qquad J = \begin{pmatrix} 0 & -1 \\ 1 & 0 \end{pmatrix}.$$

PROPOSITION For a Schwartz function φ on $\mathbb{A}_0$, let $\Theta(z;\varphi)$ be the associated theta series. For all $\alpha \in F^x$, and $w \in F$, we have elementary functional equations (in multi-index notation):

1. $\Theta(t(\alpha)z;\varphi) = \Theta(z;\psi)$, where $\psi(v) = \varphi(\alpha^{-1}v)$.
2. $\Theta(u(w)z;\varphi) = \Theta(z;\psi)$, where $\psi(v) = \varphi(v)\mathrm{e}(\frac{1}{2}wv^2)$.
3. $\Theta(Jz;\varphi) = (-iz)^{1/2}\Theta(z;\hat{\varphi})$.

Proof The first assertion follows from the definition of the theta series by replacing v by $\alpha^{-1}v$ in the definition of $\Theta(t(\alpha)z;\varphi)$. The second assertion follows from

$$\varphi(v)\mathrm{e}(\tfrac{1}{2}(z+w)v^2) = \varphi(v)\mathrm{e}(\tfrac{1}{2}zv^2) \times \mathrm{e}(\tfrac{1}{2}wv^2).$$

The third assertion follows from the adelic Poisson summation formula: For $v \in \mathbb{A}$, let

$$\Phi(v) = \varphi(v_0)\mathrm{e}(\tfrac{1}{2}zv_\infty^2),$$

where v_0 and v_∞ are the finite-adele and infinite-adele parts of v, respectively. The adelic Poisson summation formula asserts that

$$\Theta(z;\varphi) = \sum_{v \in F} \Phi(v) = \sum_{v \in F} \hat{\Phi}(v).$$

From the definitions of adelic Fourier transforms,

$$\hat{\Phi}(v) = \hat{\varphi}(v_0) \times \{\mathrm{e}(\tfrac{1}{2}zv_\infty^2)\}^\wedge$$

(i.e., taking Fourier transforms on $\mathbb{A}_0$ and $\mathbb{A}_\infty \approx \mathbb{R}^m$, respectively), since Φ factors into finite-adele and infinite-adele parts. We only need an explicit calculation of the infinite-adele part.

The function $\exp(-\pi \operatorname{Tr} v^2)$ is its own Fourier transform. By standard formal properties of the Fourier transform on $\mathbb{R}$, we find that the Fourier transform of the function $\exp(-\pi \operatorname{Tr} yv^2)$ is $y^{-1/2}\exp(-\pi \operatorname{Tr} y^{-1}v^2)$. Thus, we have computed the Fourier transform in the case $z = iy$. By analytic continuation in $z \in \mathfrak{H}^m$, we have

$$\{\mathrm{e}(\tfrac{1}{2}zv^2)\}^\wedge = (-iz)^{-1/2}\mathrm{e}(\tfrac{1}{2}(-z^{-1})v^2).$$

This gives the result. ■

Let P be the subgroup consisting of upper triangular matrices in $SL(2, F)$, U the subgroup consisting of unipotent upper triangular matrices, and T the subgroup consisting of diagonal matrices.

PROPOSITION The congruence subgroup Γ is contained in the set $PJUJ^{-1} \subset SL(2, \mathbb{Q})$.

Proof It is readily seen that $PJUJ^{-1}$ is exactly the collection of elements $\left(\begin{smallmatrix} a & b \\ c & d \end{smallmatrix}\right)$ of $SL(2, F)$ which have $d \neq 0$. The congruence condition $d \equiv 1$ modulo 2 certainly assures that $d \neq 0$. ∎

Now take $\alpha \in F^{\times}$, and $w, w' \in F$, and consider an element

$$\gamma = t(\alpha)u(w)(Ju(w')J^{-1}) = \begin{pmatrix} \alpha(1 - ww') & \alpha w \\ -w'/\alpha & \alpha^{-1} \end{pmatrix} = \begin{pmatrix} a & b \\ c & d \end{pmatrix} \in \Gamma.$$

We will compute $\theta(\gamma(z))$, using the functional equations in the proposition above. For each finite-prime $\mathfrak{p}$, let $\varphi_{\mathfrak{p}}$ be the characteristic function of $\mathfrak{o}_{\mathfrak{p}}$, and $\varphi = \Pi_{\mathfrak{p}}\varphi_{\mathfrak{p}}$. We have

$$\begin{aligned}
\theta(\gamma(z)) &= \Theta(\gamma(z); \varphi) = \Theta(t(\alpha)u(w)Ju(w')J^{-1}z; \varphi) = \\
&= \Theta(u(w)Ju(w')J^{-1}z; \varphi \circ \alpha^{-1}) = \\
&= \Theta(Ju(w')J^{-1}z; \varphi \circ \alpha^{-1} \times \mathrm{e}(\tfrac{1}{2}wv^2)) = \\
&= (-iu(w')J^{-1}z)^{1/2} \times \Theta(u(w')J^{-1}z; [\varphi \circ \alpha^{-1} \times \mathrm{e}(\tfrac{1}{2}wv^2)]\hat{\ }) = \\
&= (-iu(w')J^{-1}z)^{1/2}\Theta(J^{-1}z; [\varphi \circ \alpha^{-1} \times \mathrm{e}(\tfrac{1}{2}wv^2)]\hat{\ } \times \mathrm{e}(\tfrac{1}{2}w'v^2)) = \\
&= (-iu(w')J^{-1}z)^{1/2}(-iz)^{1/2}\Theta(z; \{[\varphi \circ \alpha^{-1} \times \mathrm{e}(\tfrac{1}{2}wv^2)]\hat{\ } \times \mathrm{e}(\tfrac{1}{2}w'v^2))\}\hat{\ }).
\end{aligned}$$

Finally, replace v by $-v$ in the definition of the theta series on the righthand side, to obtain

$$\begin{aligned}
&\Theta(\gamma(z); \varphi) = \\
&\quad = (-iu(w')J^{-1}z)^{1/2}(-iz)^{1/2}\Theta(z; \{[\varphi \circ \alpha^{-1} \times \mathrm{e}(\tfrac{1}{2}wv^2)]\hat{\ } \times \mathrm{e}(\tfrac{1}{2}w'v^2))\}\check{\ }),
\end{aligned}$$

since, for any finite-adele Schwartz function ψ, $\hat{\psi}(-v) = \check{\psi}(v)$. *This identity is the crux of the matter: We must explicate the nature of the two items*

$$(-iu(w')J^{-1}z)^{1/2}(-iz)^{1/2} \qquad \{[\varphi \circ \alpha^{-1} \times \mathrm{e}(\tfrac{1}{2}wv^2)]\hat{\ } \times \mathrm{e}(\tfrac{1}{2}w'v^2))\}\check{\ }$$

occurring above, *under the hypothesis that* $a = \alpha(1 - ww')$ is prime to $2\mathfrak{o}^{*-1}$, $b = \alpha w \equiv 0 \bmod 2\mathfrak{o}^{*-1}$, $c = -w'\alpha^{-1} \equiv 0 \bmod 2\mathfrak{o}^{*-1}$, and $d = \alpha^{-1}$ is prime to $2\mathfrak{o}^{*-1}$.

For each finite prime we will compute the $\mathfrak{p}$-adic Fourier transform (in v)

$$\{[\varphi_{\mathfrak{p}} \circ \alpha^{-1} \times \tau_{\mathfrak{p}}(-\tfrac{1}{2}wv^2)]\hat{\ } \times \tau_{\mathfrak{p}}(-\tfrac{1}{2}w'v^2)\}\check{\ }$$

where $\tau_\mathfrak{p}$ is the standard character on $F_\mathfrak{p}$. Take $\mathfrak{p}$ dividing $2\mathfrak{o}^{*-1}$. Under our hypotheses, for $v \in \mathfrak{o}_\mathfrak{p}$ we have $\tau_\mathfrak{p}(-\frac{1}{2}wv^2) = 1$ and $\varphi_\mathfrak{p} \circ \alpha^{-1} = \varphi_\mathfrak{p}$. Therefore, we have

$$\{[\varphi_\mathfrak{p} \circ \alpha^{-1} \times \tau_\mathfrak{p}(-\tfrac{1}{2}wv^2)]\hat{\ } \times \tau_\mathfrak{p}(-\tfrac{1}{2}w'v^2)\}\check{\ }(v) = \{[\varphi_\mathfrak{p} \circ \alpha^{-1} \times 1]\hat{\ } \times 1\}\check{\ }(v) =$$

$$= (\hat{\varphi}_\mathfrak{p})\check{\ }(v) = \varphi_\mathfrak{p}(v),$$

by Fourier inversion.

Now suppose that $\mathfrak{p}$ does not divide $2\mathfrak{o}^{*-1}$. By changing variables in the integral, we obtain

$$[\varphi_\mathfrak{p} \circ \alpha^{-1} \times \tau_\mathfrak{p}(-\tfrac{1}{2}wv^2)]\hat{\ }(v) = \|\alpha\|_\mathfrak{p} \int_{\mathbb{Z}_\mathfrak{p}} \tau_\mathfrak{p}(-\tfrac{1}{2}w(\alpha x)^2 - \alpha x v)\, dx.$$

LEMMA

$$\int_{\mathfrak{o}_\mathfrak{p}} \tau_\mathfrak{p}(-\tfrac{1}{2}w(\alpha x)^2 - \alpha x v)\, dx \times \tau_\mathfrak{p}(-\tfrac{1}{2}w'v^2) = \int_{\mathfrak{o}_\mathfrak{p}} \tau_\mathfrak{p}(-\tfrac{1}{2}w(\alpha x)^2)\, dx \quad (\text{if } v \in \mathfrak{o}_\mathfrak{p}),$$

$$= 0 \qquad (\text{otherwise}).$$

Proof Upon replacing x by $x + \alpha^{-1}v'$ for $v' \in \mathfrak{o}_\mathfrak{p}$,

$$\int_{\mathfrak{o}_\mathfrak{p}} \tau_\mathfrak{p}(-\tfrac{1}{2}w(\alpha x)^2 - \alpha x v)\, dx = \int_{\mathfrak{o}_\mathfrak{p}} \tau_\mathfrak{p}(-\tfrac{1}{2}w(\alpha x + v')^2 - \alpha x v - v'v)\, dx =$$

$$= \tau_\mathfrak{p}(-v'v) \int_{\mathfrak{o}_\mathfrak{p}} \tau_\mathfrak{p}(-\tfrac{1}{2}w(\alpha x + v')^2 - \alpha x v)\, dx.$$

Now

$$\tfrac{1}{2}w(\alpha x + v')^2 = \tfrac{1}{2}w(\alpha x)^2 + \alpha w x v' + \alpha^{-1}(\tfrac{1}{2}\alpha w)v'^2 \in \tfrac{1}{2}w(\alpha x)^2 + \mathfrak{o}_\mathfrak{p}$$

since xv', $\frac{1}{2}\alpha w$, α^{-1}, and v'^2 are all in $\mathfrak{o}_\mathfrak{p}$. Therefore,

$$\tau_\mathfrak{p}(-\tfrac{1}{2}w(\alpha x + v')^2) = \tau_\mathfrak{p}(-\tfrac{1}{2}w(\alpha x)^2).$$

Then

$$\int_{\mathfrak{o}_\mathfrak{p}} \tau_\mathfrak{p}(-\tfrac{1}{2}w(\alpha x)^2 - \alpha x v)\, dx = \tau_\mathfrak{p}(-v'v) \int_{\mathfrak{o}_\mathfrak{p}} \tau_\mathfrak{p}(-\tfrac{1}{2}w(\alpha x)^2 - \alpha x v)\, dx.$$

As $\mathfrak{p}$ is absolutely unramified, if $v \notin \mathfrak{o}_\mathfrak{p}$ then there is $v' \in \mathfrak{o}_\mathfrak{p}$ so that $\mathrm{Tr}\, v'v \notin \mathfrak{o}_\mathfrak{p}$, i.e., we would have $\tau_\mathfrak{p}(-v'v) \neq 1$. This shows that *the integral vanishes if* $v \notin \mathfrak{o}_\mathfrak{p}$.

Now $\alpha(1-ww')$, αw, α^{-1}, and $\alpha^{-1}w'$ all are in $\mathfrak{o}_\mathfrak{p}$. Therefore, for $v \in \mathfrak{o}_\mathfrak{p}$, $x \in \mathfrak{o}_\mathfrak{p}$,

$$\begin{aligned}\tau_\mathfrak{p}(-\tfrac{1}{2}w(\alpha x)^2 - \alpha xv) &= \tau_\mathfrak{p}(-\tfrac{1}{2}w(\alpha x)^2 - \alpha ww'xv) = \\ &= \tau_\mathfrak{p}(-\tfrac{1}{2}w\alpha^2\{x^2 + \alpha^{-1}w'xv\}) = \\ &= \tau_\mathfrak{p}(-\tfrac{1}{2}w\alpha^2\{(x+\tfrac{1}{2}\alpha^{-1}w'v)^2 - (\alpha^{-1}w'v)^2\}).\end{aligned}$$

Further, since $\alpha^{-1}w' \in 2\mathfrak{o}_\mathfrak{p}$ and $\alpha - \alpha ww' \in \mathfrak{o}_\mathfrak{p}$, for $v \in \mathfrak{o}_\mathfrak{p}$ we have

$$\tau_\mathfrak{p}(\tfrac{1}{2}ww'^2v^2) = \tau_\mathfrak{p}(\tfrac{1}{2}(\alpha ww')(\alpha^{-1}w')v^2) = \tau_\mathfrak{p}(\tfrac{1}{2}\alpha(\alpha^{-1}w')v^2) = \tau_\mathfrak{p}(\tfrac{1}{2}w'v^2).$$

This cancels the factor $\tau_\mathfrak{p}(-\tfrac{1}{2}w'v^2)$ outside the integral. Since $\alpha^{-1}w' \in 2\mathfrak{o}_\mathfrak{p}$, $\tfrac{1}{2}(\alpha^{-1}w')v$ is in $\mathfrak{o}_\mathfrak{p}$, so we can replace x by $x-(\alpha^{-1}w')v$ in the integral. Thus, for v in $\mathfrak{o}_\mathfrak{p}$,

$$\int_{\mathfrak{o}_\mathfrak{p}} \tau_\mathfrak{p}(-\tfrac{1}{2}w(\alpha x)^2 - \alpha xv)\,dx \times \tau_\mathfrak{p}(-\tfrac{1}{2}w'v^2) = \int_{\mathfrak{o}_\mathfrak{p}} \tau_\mathfrak{p}(-\tfrac{1}{2}w(\alpha x)^2)\,dx,$$

and the integral is 0 otherwise. ∎

COROLLARY Let $\tau_0 = \prod_\mathfrak{p} \tau_\mathfrak{p}$ (over finite primes). Under our hypotheses on α, w, w' we have

$$\begin{aligned}&\{[\varphi \circ \alpha^{-1} \times \tau_0(-\pi iwv^2)]^\wedge \times \tau_0(-\pi iw'v^2)\}^\vee = \\ &= \varphi(v) \times \prod_{\mathfrak{p}\in X}\left\{\|\alpha\|_p \int_{\mathfrak{o}_\mathfrak{p}} \tau_\mathfrak{p}(-\tfrac{1}{2}w(\alpha x)^2)\,dx\right\},\end{aligned}$$

where X is the set of primes $\mathfrak{p}$ (not dividing $2\mathfrak{o}^{*-1}$) where α is not a $\mathfrak{p}$-unit.

Proof This just summarizes our computations of the $\mathfrak{p}$-adic Fourier transforms. ∎

Therefore, we have

$$\theta(\gamma(z)) = \theta(z)(-iu(w')J^{-1}z)^{1/2}(-iz)^{1/2} \prod_{\mathfrak{p}\in X}\left\{\|\alpha\|_\mathfrak{p} \int_{\mathfrak{o}_\mathfrak{p}} \tau_\mathfrak{p}(-\tfrac{1}{2}w(\alpha x)^2)\,dx\right\}.$$

LEMMA Take $0 \neq \beta \in \mathfrak{o}$, and let $\mathfrak{p}$ be a prime of $\mathfrak{o}$ not dividing $2\mathfrak{o}^{*-1}$. Suppose that $\eta \in \mathfrak{o}$, and that $\eta\mathfrak{o}$ is relatively prime to $\mathfrak{p}$. Let $e = \mathrm{ord}_\mathfrak{p}\,\beta$. Then

$$\int_{x\in\mathfrak{o}_\mathfrak{p}} \tau_\mathfrak{p}(\eta x^2/\beta)\,dx = (\eta/\mathfrak{p}^e)_2 \times \int_{x\in\mathfrak{o}_\mathfrak{p}} \tau_\mathfrak{p}(x^2/\beta)\,dx.$$

Proof The proof of this is essentially identical to the classical case, taking into account that $\mathfrak{p}$ is absolutely unramified and does not divide 2. ∎

PROPOSITION Use notation as above. We have

$$\prod_{\mathfrak{p}\in X} \|\alpha\|_{\mathfrak{p}} \int_{\mathfrak{o}_{\mathfrak{p}}} \tau_{\mathfrak{p}}(-\tfrac{1}{2}w(\alpha x)^2)\,dx = \varepsilon(\alpha^{-1})(\tfrac{1}{2}\alpha w/\alpha^{-1})_2|\alpha^{-1}|^{1/2}.$$

Proof Again, X is the set of primes dividing $\alpha^{-1}\mathfrak{o}$, and $\mathfrak{p} \in X$ implies that $\mathfrak{p}$ does not divide $2\mathfrak{o}^{*-1}$. In each integral

$$\int_{\mathfrak{o}_{\mathfrak{p}}} \tau_{\mathfrak{p}}(-\tfrac{1}{2}w(\alpha x)^2)\,dx = \int_{\mathfrak{o}_{\mathfrak{p}}} \tau_{\mathfrak{p}}(-\tfrac{1}{2}\alpha w \alpha x^2)\,dx$$

we have $b = \alpha w \in \mathfrak{o}_{\mathfrak{p}}^{\times}$, since $\gamma \in SL(2, \mathfrak{o})$. From the facts above about quadratic symbols, if $\mathrm{ord}_{\mathfrak{p}}(\alpha^{-1}) = e$, then

$$\int_{\mathfrak{o}_{\mathfrak{p}}} \tau_{\mathfrak{p}}(-\tfrac{1}{2}\alpha w x^2/\alpha^{-1})\,dx = (\tfrac{1}{2}\alpha w/p^e)_2 \int_{\mathfrak{o}_{\mathfrak{p}}} \tau_{\mathfrak{p}}(-x^2/\alpha^{-1})\,dx.$$

From the definitions, using the Approximation Theorem,

$$\prod_{\mathfrak{p}} \int_{\mathfrak{o}_{\mathfrak{p}}} \tau_{\mathfrak{p}}(-x^2/\alpha^{-1})\,dx = [\mathfrak{o} : \alpha^{-1}\mathfrak{o}]^{-1} \sum_{x \in \mathfrak{o}/\alpha^{-1}\mathfrak{o}} \mathrm{e}(x^2/\alpha^{-1}).$$

From the Reciprocity Law and its corollaries, we obtain

$$\prod_{\mathfrak{p}} \int_{\mathfrak{o}_{\mathfrak{p}}} \tau_{\mathfrak{p}}(-x^2/\alpha^{-1})\,dx = |\alpha^{-1}|^{-1/2}\varepsilon(\alpha^{-1}).$$

Therefore, by the product formula,

$$\prod_{p\in X} \|\alpha\|_p \int_{\mathfrak{o}_{\mathfrak{p}}} \tau_{\mathfrak{p}}(-\tfrac{1}{2}\alpha w \alpha x^2)\,dx = |\alpha^{-1}|^{1/2}(\tfrac{1}{2}\alpha w/\alpha^{-1})_2\varepsilon(\alpha^{-1}). \qquad \blacksquare$$

Therefore, we have

$$\theta(\gamma(z)) = \theta(z) \times (-iu(w')J^{-1}z)^{1/2}(-iz)^{1/2}\varepsilon(\alpha^{-1})(\tfrac{1}{2}\alpha w/\alpha^{-1})_2|\alpha^{-1}|^{1/2} =$$

$$= \theta(z) \times (-i(-cd^{-1} - z^{-1}))^{1/2}(-iz)^{1/2}\varepsilon(d)(\tfrac{1}{2}b/d)_2|d|^{1/2}.$$

Now

$$(-i(-cd^{-1} - z^{-1}))^{1/2}(-iz)^{1/2} = (cd^{-1}z + 1)^{1/2},$$

taking the square root lying in the right half-plane, since all the components of both $(-i(-cd^{-1} - z^{-1}))^{1/2}$ and $(-iz)^{1/2}$ (by convention...) have arguments between $-\pi/4$ and $\pi/4$. Further,

$$(cd^{-1}z + 1)^{1/2}|d|^{1/2} = \tilde{\varepsilon}(d)(cz + d)^{1/2}.$$

Also, as $ad - bc = 1$, it must be that $(-\frac{1}{2}c/d)_2 = (\frac{1}{2}b/d)_2$. Therefore,

$$\theta(\gamma(z)) = \theta(z) \times \tilde{\varepsilon}(d)\varepsilon(d)(cz + d)^{1/2}(-\tfrac{1}{2}c/d)_2.$$

The facts concerning θ immediately give the results regarding θ^2. ∎

5.2 Hecke's Identity

Let $\langle\ ,\ \rangle$ be the usual inner product on $\mathbb{R}^n$, and let $f \to \hat{f}$ be the Fourier transform

$$\hat{f}(x) = \int_{\mathbb{R}^n} f(\xi)\exp(-2\pi i\langle x, \xi\rangle)\,d\xi.$$

Let Δ be the usual Laplacian $\sum_i \partial^2/\partial x_i^2$ on $\mathbb{R}^n$. Let P be a homogeneous polynomial of degree d on $\mathbb{R}^n$ annihilated by Δ: P is a *homogeneous harmonic polynomial* of degree d.

THEOREM The Fourier transform of $P(x)\exp(-\pi\langle x, x\rangle)$ is $i^{-d}P(x)\exp(-\pi\langle x, x\rangle)$.

Proof Extend $\langle\ ,\ \rangle$ to a $\mathbb{C}$-bilinear function on $\mathbb{C}^n \times \mathbb{C}^n$. Then

$$\{P(x)\exp(-\pi\langle x, x\rangle)\}\hat{\ }(x) = \int_{\mathbb{R}^n} P(\xi)\exp(-\pi\langle \xi, \xi\rangle)\exp(-2\pi i\langle x, \xi\rangle)\,d\xi =$$

$$= \exp(-\pi\langle x, x\rangle)\int_{\mathbb{R}^n} P(\xi - ix)\exp(-\pi\langle \xi, \xi\rangle)\,d\xi$$

by Cauchy's Theorem in one variable, replacing each of the components ξ_j of the variable ξ by $\xi_j - ix_j$. Using the homogeneity of P, we have

$$\{P(x)\exp(-\pi\langle x, x\rangle)\}\hat{\ }(x) = i^{-d}\exp(-\pi\langle x, x\rangle)\int_{\mathbb{R}^n} P(x + i\xi)\exp(-\pi\langle \xi, \xi\rangle)\,d\xi.$$

What remains to be shown is that the integral is just $P(x)$.

Let P_j be the partial derivative of P_j with respect to its j-th argument. We have

$$0 = \int_{\mathbb{R}^n} \sum_j (\partial/\partial x_j)^2 P(x + i\xi)\exp(-\pi\langle \xi, \xi\rangle)\,d\xi =$$

$$= \int_{\mathbb{R}^n} \sum_j P_{jj}(x + i\xi)\exp(-\pi\langle \xi, \xi\rangle)\,d\xi =$$

$$= i^{-2d}\int_{\mathbb{R}^n} \exp(-\pi\langle \xi, \xi\rangle)\sum_j (\partial/\partial \xi_j)^2 P(x + i\xi)\,d\xi =$$

$$= i^{-2d}(-1)^n \int_{\mathbb{R}^n} \sum_j (\partial \exp(-\pi\langle \xi, \xi\rangle)/\partial \xi_j \times \partial P(x + i\xi)/\partial \xi_j)\,d\xi$$

by integrating by parts. Thus,

$$0 = \int_{\mathbb{R}^n} \exp(-\pi\langle\xi,\xi\rangle) \sum_j i\xi_j P_j(x + i\xi)\, d\xi =$$

$$= \int_{\mathbb{R}^n} \exp(-\pi\langle\xi,\xi\rangle) \sum_j [(x_j + i\xi_j) - x_j] P_j(x + i\xi)\, d\xi =$$

$$= d\int_{\mathbb{R}^n} P(x + i\xi)\exp(-\pi\langle\xi,\xi\rangle)\, d\xi - \sum_j x_j \int_{\mathbb{R}^n} \exp(-\pi\langle\xi,\xi\rangle) P_j(x + i\xi)\, d\xi$$

by Euler's identity for homogeneous polynomials. Now P_j is still homogeneous and harmonic, of degree one less than the degree d of P, so by induction on the (total) degree of P

$$\int_{\mathbb{R}^n} \exp(-\pi\langle\xi,\xi\rangle) P_j(x + i\xi)\, d\xi = P_j(x).$$

When $d = 0$, the assertion is elementary. Therefore, by induction we obtain

$$0 = d\int_{\mathbb{R}^n} P(x + i\xi)\exp(-\pi\langle\xi,\xi\rangle)\, d\xi - \sum_j x_j P_j(x) =$$

$$= d\int_{\mathbb{R}^n} P(x + i\xi)\exp(-\pi\langle\xi,\xi\rangle)\, d\xi - dP(x)$$

again by Euler's identity. Therefore,

$$\int_{\mathbb{R}^n} P(x + i\xi)\exp(-\pi\langle\xi,\xi\rangle)\, d\xi = P(x).$$ ■

5.3 Pluriharmonic Theta Series

We prove that harmonic theta series are Hilbert modular forms for some explicitly determinable congruence subgroups of $SL(2, \mathfrak{o})$. The "estimate" given here for the congruence subgroup is not the best possible; some improved estimates in special cases will be given subsequently.

Definition Let Q be a quadratic form on F^n, so that $Q[v] = Q(v, v) \in F$ is totally positive for $v \neq 0$. Let $\mathbb{A}$ be the adeles of F, and $\mathbb{A}_0$ the finite adeles of F. Let φ be a Schwartz function on $\mathbb{A}_0^n$. For $z \in \mathfrak{H}^m$, *the theta series* attached to Q and φ is defined to be

$$\Theta(z; Q, \varphi) = \sum_{v \in F^n} \varphi(v) \mathrm{e}(\tfrac{1}{2} z Q[v]).$$

Definition Let Q and φ be as above. Let $\{\sigma_j\}$ be the real imbeddings of F; let σ_j also denote the induced imbedding of F^n into $\mathbb{R}^n$, or of matrices over F to real matrices; extend these maps $\mathbb{R}$-linearly to $F^n \otimes_{\mathbb{Q}} \mathbb{R} \approx \mathbb{R}^{mn}$, etc. Let $f(v)$ be a polynomial in $v \in F^n \otimes \mathbb{R}$ which is homogeneous of degree k_j in $\sigma_j(v)$, and which is *harmonic* in v_j for each j; i.e., let Δ_j be the usual Euclidean Laplacian on $\sigma_j(F^n \otimes \mathbb{R})$, and require that $\Delta_j f = 0$ for $j = 1, \ldots, m$. For $z \in \mathfrak{H}^n$, the *pluriharmonic theta series* attached to Q, f, and φ is

$$\Theta(z; Q, f, \varphi) = \sum_{v \in F^m} \varphi(v) f(Q^{1/2} v) \mathrm{e}(\tfrac{1}{2} z Q[v]),$$

where we define $Q^{1/2} v \in F^n \otimes \mathbb{R}$ by

$$\sigma_j(Q^{1/2} v) = \sigma_j(Q)^{1/2} \sigma_j(v),$$

where Q_j is the image of Q under σ_j, and we take the positive definite square root of $\sigma_j(Q)$. (Clearly the former sort of theta series is the special case $f(v) \equiv 1$ of the latter.)

Let $\mathfrak{f}$ be a non-zero ideal in $\mathfrak{o}$, and as usual define the principal congruence subgroup of $SL(2, \mathfrak{o})$ of level $\mathfrak{f}$ by

$$\Gamma(\mathfrak{f}) = \left\{ \begin{pmatrix} a & b \\ c & d \end{pmatrix} \in SL(2, \mathfrak{o}) \colon a \equiv d \equiv 1 \text{ modulo } \mathfrak{f},\ b \equiv c \equiv 0 \text{ modulo } \mathfrak{f} \right\}.$$

Definition Given an F-valued (positive-definite...) quadratic form Q on F^n and a finite-adele Schwartz function φ on $\mathbb{A}_0^n$ factoring over finite primes as $\varphi = \prod_{\mathfrak{p}} \varphi_{\mathfrak{p}}$, let $\mathfrak{f}$ be the largest integral ideal in $\mathfrak{o}$ so that all the following conditions are satisfied:

i. $\mathfrak{f} \subset 4\mathfrak{o}^{*-1}$.
ii. If Q is not in $GL(m, \mathfrak{o}_{\mathfrak{p}})$, then $\mathfrak{f} \subset \mathfrak{p}$.
iii. If $\varphi_{\mathfrak{p}}$ is not the characteristic function of $\mathfrak{o}_{\mathfrak{p}}^n$, then $\mathfrak{f} \subset \mathfrak{p}$.
iv. $Q[v] \subset 2\mathfrak{o}^* \mathfrak{f}^{-1}$ for all $v \in F$ in the support of φ.
v. $Q^{-1}[v] \subset 2\mathfrak{o}^* \mathfrak{f}^{-1}$ for all $v \in F$ in the support of φ.

Definition For $d \in \mathfrak{o} - \{0\}$, define

$$\varepsilon(d) = (i \operatorname{sgn} d)^{1/2} 2^{-m/2} D^{-1/2} \sum_{v \in \mathfrak{o}^*/2\mathfrak{o}} \mathrm{e}(-v^2 d/4)$$

where $\mathrm{sgn}(d)$ is $\{\pm 1\}^m$-valued, and $(i\,\mathrm{sgn}\,d)^{1/2}$ is meant in a multi-index sense. For $d \in (\mathbb{R}^\times)^m$, let s be the number of negative components, and define $\tilde{\varepsilon}(d) = i^s$.

THEOREM Given a positive quadratic form Q of dimension n, a pluriharmonic polynomial f of (multi-) degrees $(k_1, \ldots, k_m)$, and $\varphi = \prod_{\mathfrak{p}} \varphi_{\mathfrak{p}}$, let $\mathfrak{f}$ be as just above. Suppose that

$$\varphi(-v) = (-1)^{k_1 + \cdots + k_m}\varphi(v)$$

for $v \in \mathbb{A}_0^n$. Let $\gamma = \begin{pmatrix} a & b \\ c & d \end{pmatrix} \in \Gamma(\mathfrak{f})$. We have the functional equation

$$\Theta(\gamma z; Q, f, \varphi) = \Theta(z; Q, f, \varphi)\varepsilon^n(d)\tilde{\varepsilon}^n(d)(\det Q/d)_2(-\tfrac{1}{2}c/d)_2^n(cz + d)^{(k+n/2)}$$

That is, $\Theta(z; Q, f, \varphi)$ is a Hilbert modular form (possibly with a character) for $\Gamma(\mathfrak{f})$ of (possibly half-integral) weight

$$k + \tfrac{1}{2}n = (k_1 + \tfrac{1}{2}n, \ldots, k_m + \tfrac{1}{2}n) \in (\tfrac{1}{2}\mathbb{Z})^m.$$

Remarks For n odd, the quadratic symbol $(-\frac{1}{2}c/d)_2$ and the $\varepsilon\tilde{\varepsilon}$ make it impossible to view such a theta series as a restriction of an automorphic form on $GL(2, \mathbb{A})$.

As demonstrated in the previous section, $(\varepsilon\tilde{\varepsilon})^2$ is a character χ defined modulo $4\mathfrak{o}^{*-1}$, given by

$$\chi(d) = (\mathrm{sgn}\,d)^1(-1/d)_2$$

and $\chi^2 = 1$; in particular, χ is trivial on $1 + 4\mathfrak{o}^{*-1}$. Therefore, for $n \in 2\mathbb{Z}$ and $\gamma \in \Gamma(\mathfrak{f})$, we have

$$\Theta(\gamma(z); Q, f, \varphi) = \Theta(z; Q, f, \varphi) \times (\det Q/d)_2(cz + d)^{(k+n/2)}.$$

Further, when $\det Q$ is a square, this is simply

$$\Theta(\gamma(z); Q, f, \varphi) = \Theta(z; Q, f, \varphi)(cz + d)^{(k+n/2)}.$$

Proof

Definition Take $y \in F^\times$, $x \in F$. Put

$$t(y) = \begin{pmatrix} y & 0 \\ 0 & y^{-1} \end{pmatrix} \qquad u(x) = \begin{pmatrix} 1 & x \\ 0 & 1 \end{pmatrix} \qquad J = \begin{pmatrix} 0 & -1 \\ 1 & 0 \end{pmatrix}.$$

PROPOSITION For all $\alpha \in F^\times$ and $w \in F$, we have functional equations:

1. $\Theta(t(\alpha)z; Q, f, \varphi) = \Theta(z; Q, f, \psi)$, where $\psi(v) = \alpha^{-k}\varphi(\alpha^{-1}v)$.
2. $\Theta(u(w)z; Q, f, \varphi) = \Theta(z; Q, f, \psi)$, where $\psi(v) = \varphi(v)e(\frac{1}{2}wQ[v])$.
3. $\Theta(Jz; Q, f, \varphi) = (-iz)^{(n/2+k)}(\det Q)^{1/2}i^k\Theta(z; Q^{-1}, f, \hat{\varphi})$.

Proof The first assertion follows from the definition of the theta series by replacing v by $\alpha^{-1}v$ in the definition of $\Theta(t(\alpha); Q, f, \varphi)$. The second assertion follows immediately from

$$\varphi(v)f(Q^{1/2}v)e(\tfrac{1}{2}(z+w)Q[v]) =$$

$$= \varphi(v)f(Q^{1/2}v)e(\tfrac{1}{2}zQ[v]) \times e(\tfrac{1}{2}wQ[v]).$$

The third assertion follows from the adelic Poisson summation formula: For $v \in \mathbb{A}^n$, let

$$\Phi(v) = \varphi(v_0)f(Q^{1/2}v_\infty)e(\tfrac{1}{2}zQ[v_\infty]),$$

where v_0 and v_∞ are the finite-adele and infinite-adele parts of v, respectively. The Poisson summation formula asserts that

$$\Theta(z; Q, f, \varphi) = \sum_{v \in F^n} \Phi(v) = \sum_{v \in F^n} \hat{\Phi}(v).$$

From the definitions of adelic Fourier transforms,

$$\hat{\Phi}(v) = \hat{\varphi}(v_0) \times \{f(Q^{1/2}v_\infty)e(\tfrac{1}{2}zQ[v_\infty])\}^{\wedge},$$

(i.e., taking Fourier transforms on $\mathbb{A}_0^n$ and $\mathbb{A}_\infty^n$, respectively), since Φ factors into finite-adele and infinite-adele parts. We need only an explicit calculation of the infinite-adele part. Clearly it suffices to consider the case $F = \mathbb{Q}$, since the infinite-adele part factors over (infinite) primes.

Let f be a homogeneous harmonic polynomial of degree q on $\mathbb{R}^n$, and let R be a positive definite quadratic form on $\mathbb{R}^n$. First, *Hecke's identity* asserts that the Fourier transform of the function

$$f(v)\exp(-\pi\|v\|^2)$$

is just

$$i^{-q}f(v)\exp(-\pi\|v\|^2).$$

By standard formal properties of the Fourier transform on $\mathbb{R}^n$, we find that the Fourier transform of the function

$$f(R^{1/2}v)\exp(-\pi R[v]) = f(R^{1/2}v)\exp(-\pi\|R^{1/2}v\|^2)$$

is, similarly,

$$i^{-q}\det R^{-1/2}f(R^{-1/2}v)\exp(-\pi R^{-1}[v]) =$$

$$= i^{-q}\det R^{-1/2}f(R^{-1/2}v)\exp(-\pi\|R^{-1}v\|^2).$$

For $R = yQ$, with Q another positive definite quadratic form, and $y > 0$, we have

$$\{f((yQ)^{1/2}v)\exp(-\pi(yQ)[v])\}^{\wedge} =$$

$$= i^{-q}\det(yQ)^{-1/2}f((yQ)^{-1/2}v)\exp(-\pi(yQ)^{-1}[v]).$$

We can rewrite this as

$$\{f(Q^{1/2}v)\exp(-\pi yQ[v])\}^{\wedge} =$$

$$= i^{-q}y^{-(n/2+q)}\det(Q)^{-1/2}f(Q^{-1/2}v)\exp(-\pi y^{-1}Q^{-1}[v]).$$

Thus, we have computed the Fourier transform in the case $z = iy$. By analytic continuation in z, we have

$$\{f(Q^{1/2}v)\exp(\pi izQ[v])\}^{\wedge} =$$

$$= i^{-q}(-iz)^{-(n/2+q)}(\det Q)^{-1/2}f(Q^{-1/2}v)\exp(\pi i(-z^{-1})Q^{-1}[v]).$$

Returning to the general situation, we have determined that

$$\hat{\Phi}(v) = (-iz)^{-(n/2+k)}i^{-k}(\det Q)^{-1/2} \times f(Q^{-1/2}v)e(\tfrac{1}{2}(-z^{-1})Q^{-1}[v]).$$

But this expression summed over F^n is just

$$(-iz)^{-(n/2+k)}i^{-k}(\det Q)^{-1/2}\Theta(-z^{-1};Q^{-1},f,\hat{\varphi}),$$

which yields our conclusion. ∎

Let P be the subgroup consisting of upper triangular matrices in $SL(2,F)$, U the subgroup consisting of unipotent upper triangular matrices, and T the subgroup consisting of diagonal matrices.

PROPOSITION The congruence subgroup $\Gamma(\mathfrak{f})$ is contained in the set $PJUJ^{-1} \subset SL(2,F)$.

Proof It is readily seen that $PJUJ^{-1}$ is exactly the collection of elements $\left(\begin{smallmatrix} a & b \\ c & d \end{smallmatrix}\right)$ of $SL(2,F)$ which have $d \neq 0$. The congruence condition $d \equiv 1$ modulo $\mathfrak{f}$, for $\mathfrak{f}$ a *proper* ideal in $\mathfrak{o}$, certainly requires that $d \neq 0$. Thus, $\Gamma(\mathfrak{f})$ is contained in $PJUJ^{-1}$. ∎

Now take $\alpha \in F^{\times}$, and u, $w \in F$, and consider an element

$$\gamma = t(\alpha)u(w)Ju(w')J^{-1} = \begin{pmatrix} \alpha(1 - ww') & \alpha w \\ -w'/\alpha & \alpha^{-1} \end{pmatrix} = \begin{pmatrix} a & b \\ c & d \end{pmatrix} \in \Gamma(\mathfrak{f}).$$

We will compute $\Theta(\gamma z; Q, f, \varphi)$, using the functional equations in the proposition above. We have

$$\begin{aligned}
&\Theta(\gamma z; Q, f, \varphi) = \\
&\quad = \Theta(t(\alpha)u(w)Ju(w')J^{-1}z; Q, f, \varphi) = \\
&\quad = \alpha^{-k}\Theta(u(w)Ju(w')J^{-1}z; Q, f, \varphi \circ \alpha^{-1}) = \\
&\quad = \alpha^{-k}\Theta(Ju(w')J^{-1}z; Q, f, \varphi \circ \alpha^{-1} \times \mathrm{e}(\tfrac{1}{2}wQ[\ ])) = \\
&\quad = \alpha^{-k}(-iu(w')J^{-1}z)^{(n/2+k)}(\det Q)^{1/2} i^k \times \\
&\qquad \times \Theta(u(w')J^{-1}z; Q^{-1}, f, [\varphi \circ \alpha^{-1} \times \mathrm{e}(\tfrac{1}{2}wQ[\ ])]\hat{\ }) = \\
&\quad = \alpha^{-k}(-iu(w')J^{-1}z)^{(n/2+k)}(\det Q)^{1/2} i^k \times \\
&\qquad \times \Theta(J^{-1}z; Q^{-1}, f, [\varphi \circ \alpha^{-1} \times \mathrm{e}(\tfrac{1}{2}wQ[\ ])]\hat{\ } \times \mathrm{e}(\tfrac{1}{2}w'Q^{-1}[\ ])) = \\
&\quad = \alpha^{-k}(-iu(w')J^{-1}z)^{(n/2+k)}(\det Q)^{1/2} i^k \times (-iz)^{(m/2+k)}(\det Q^{-1})^{1/2} i^k \times \\
&\qquad \times \Theta(z; Q, f, \{[\varphi \circ \alpha^{-1} \times \mathrm{e}(\tfrac{1}{2}wQ[\ ])]\hat{\ } \times \mathrm{e}(\tfrac{1}{2}w'Q^{-1}[\ ])\}\hat{\ }) = \\
&\quad = (-1)^k\alpha^{-k}(-iu(w')J^{-1}z)^{(n/2+k)}(-iz)^{(m/2+k)} \times \\
&\qquad \times \Theta(z; Q, f, \{[\varphi \circ \alpha^{-1} \times \mathrm{e}(\tfrac{1}{2}wQ[\ ])]\hat{\ } \times \mathrm{e}(\tfrac{1}{2}w'Q^{-1}[\ ])\}\hat{\ }),
\end{aligned}$$

upon simplifying. *This is the crux of the matter.*

Thus, *we must explicate the nature of*

$$\begin{aligned}
&(-1)^k\alpha^{-k}(-iu(w')J^{-1}z)^{(n/2+k)}(-iz)^{(n/2+k)} \times \\
&\qquad \times \Theta(z; Q, f, \{[\varphi \circ \alpha^{-1} \times \mathrm{e}(\tfrac{1}{2}wQ[\ ])]\hat{\ } \times \mathrm{e}(\tfrac{1}{2}w'Q^{-1}[\ ])\}\hat{\ }),
\end{aligned}$$

under the hypothesis that $a = \alpha(1 - ww')$, $b = \alpha w$, $c = -w'\alpha^{-1}$, $d = \alpha^{-1}$, and that $\gamma = \left(\begin{smallmatrix} a & b \\ c & d \end{smallmatrix}\right) \in \Gamma(\mathfrak{f})$. In particular, for each finite-prime $\mathfrak{p}$ we want to compute the $\mathfrak{p}$-adic Fourier transforms

$$\{[\varphi_{\mathfrak{p}} \circ \alpha^{-1} \times \tau_{\mathfrak{p}}(-\tfrac{1}{2}wQ[\ ])]\hat{\ } \times \tau_{\mathfrak{p}}(-w'Q^{-1}[\ ])\}\hat{\ }.$$

Definition Say that a prime $\mathfrak{p}$ is *bad* (for φ and Q) if $\mathfrak{p}$ divides $\mathfrak{f}$. Otherwise say that $\mathfrak{p}$ is *good*.

At a *bad prime* $\mathfrak{p}$, Q has values in $2\mathfrak{o}^*\mathfrak{f}^{-1}$ on the support of $\varphi_{\mathfrak{p}}$, and Q^{-1} has values in $2\mathfrak{o}^*\mathfrak{f}^{-1}$ on the support of $\hat{\varphi}_{\mathfrak{p}}$. Therefore, we have

$$\begin{aligned}
&\{[\varphi_{\mathfrak{p}} \circ \alpha^{-1} \times \tau_{\mathfrak{p}}(-\tfrac{1}{2}wQ[\ ])]\hat{\ } \times \tau_{\mathfrak{p}}(-\tfrac{1}{2}w'Q^{-1}[\ ])\}\hat{\ }(v) = \\
&\qquad = \{[\varphi_{\mathfrak{p}} \circ \alpha^{-1} \times 1]\hat{\ } \times 1\}\hat{\ }(v) = (\hat{\varphi_{\mathfrak{p}}})\hat{\ }(v) = \varphi_{\mathfrak{p}}(-v),
\end{aligned}$$

by Fourier inversion.

Now suppose that $\mathfrak{p}$ is a (finite) *good prime* (i.e., not bad). Let $\langle\ ,\ \rangle$ be the standard $F_\mathfrak{p}$-valued bilinear form on $F_\mathfrak{p}^n$:

$$\langle(x_1,\ldots,x_n),(y_1,\ldots,y_n)\rangle = \sum_i x_i y_i.$$

Let $\Lambda(\mathfrak{p}) = \mathfrak{o}_\mathfrak{p}^n$. As $\mathfrak{p}$ is a good prime, $\varphi_\mathfrak{p}$ is the characteristic function of $\Lambda(\mathfrak{p})$, so we obtain

$$[\varphi_\mathfrak{p}\circ\alpha^{-1}\times\tau_\mathfrak{p}(-\tfrac{1}{2}wQ[\ ])]\hat{\ }(v) = \|\alpha\|_\mathfrak{p}^n\int_{\Lambda(\mathfrak{p})}\tau_\mathfrak{p}(-\tfrac{1}{2}wQ[\alpha x]-\langle\alpha x,v\rangle)\,dx.$$

LEMMA Let $\mathfrak{p}$ be a good prime. Then

$$\|\alpha\|_\mathfrak{p}^n\int_{\Lambda(\mathfrak{p})}\tau_\mathfrak{p}(-\tfrac{1}{2}wQ[\alpha x]-\langle\alpha x,v\rangle)\,dx\times\tau_\mathfrak{p}(-\tfrac{1}{2}w'Q^{-1}[\ ]) =$$

$$= \|\alpha\|_\mathfrak{p}^n\int_{\Lambda(\mathfrak{p})}\tau_\mathfrak{p}(-\tfrac{1}{2}wQ[\alpha x])\,dx \qquad (\text{if } v\in\Lambda(\mathfrak{p})),$$

$$= 0 \qquad (\text{otherwise}).$$

Proof Upon replacing x by $x+\alpha^{-1}v'$ for $v'\in\Lambda(\mathfrak{p})$, the integral is

$$\|\alpha\|_\mathfrak{p}^n\int_{\Lambda(\mathfrak{p})}\tau_\mathfrak{p}(-\tfrac{1}{2}wQ[\alpha x]-\langle\alpha x,v\rangle)\,dx =$$

$$= \|\alpha\|_\mathfrak{p}^n\int_{\Lambda(\mathfrak{p})}\tau_\mathfrak{p}(-\tfrac{1}{2}wQ[\alpha x+v']-\langle\alpha x,v\rangle-\langle v',v\rangle)\,dx =$$

$$= \tau_\mathfrak{p}(-\langle v',v\rangle)\|\alpha\|_\mathfrak{p}^n\int_{\Lambda(\mathfrak{p})}\tau_\mathfrak{p}(-\tfrac{1}{2}wQ[\alpha x+v']-\langle\alpha x,v\rangle)\,dx.$$

Now

$$\tfrac{1}{2}wQ[\alpha x+v'] = \tfrac{1}{2}wQ[\alpha x]+\alpha wQ(x,v')+\tfrac{1}{2}\alpha^{-1}\alpha wQ[v']\in\tfrac{1}{2}wQ[\alpha x]+\mathfrak{o}_\mathfrak{p}$$

since $Q(x,v')$, $\frac{1}{2}\alpha w$, α^{-1}, and $Q[v']$ are all in $\mathfrak{o}_\mathfrak{p}$. Therefore,

$$\tau_\mathfrak{p}(-\tfrac{1}{2}wQ[\alpha x+v']) = \tau_\mathfrak{p}(-\tfrac{1}{2}wQ[\alpha x]).$$

Then

$$\int_{\Lambda(\mathfrak{p})}\tau_\mathfrak{p}(-\tfrac{1}{2}wQ[\alpha x]-\langle\alpha x,v\rangle)\,dx =$$

$$= \tau_\mathfrak{p}(-\langle v',v\rangle)\int_{\Lambda(\mathfrak{p})}\tau_\mathfrak{p}(-\tfrac{1}{2}wQ[\alpha x]-\langle\alpha x,v\rangle)\,dx.$$

If $v \notin \Lambda(\mathfrak{p})$, there is $v' \in \Lambda(\mathfrak{p})$ so that $\langle v', v\rangle \notin \mathfrak{o}_\mathfrak{p}$; i.e., since $\mathfrak{p}$ is unramified, we would have $\tau_\mathfrak{p}(-\langle v', v\rangle) \neq 1$. This shows that *the integral vanishes if* $v \notin \Lambda(\mathfrak{p})$.

Now $a = \alpha(1 - ww')$, $b = \alpha w$, $d = \alpha^{-1}$, and $c = \alpha^{-1}w'$ all are in $\mathfrak{o}_\mathfrak{p}$. Therefore, for $v \in \Lambda(\mathfrak{p})$, $x \in \Lambda(\mathfrak{p})$,

$$\begin{aligned}\tau_\mathfrak{p}(-\tfrac{1}{2}wQ[\alpha x] - \langle \alpha x, v\rangle) &= \tau_\mathfrak{p}(-\tfrac{1}{2}wQ[\alpha x] - \langle \alpha ww'x, v\rangle) = \\ &= \tau_\mathfrak{p}(-\tfrac{1}{2}w\alpha^2\{Q[x] + \langle \alpha^{-1}w'x, v\rangle\}) = \\ &= \tau_\mathfrak{p}(-\tfrac{1}{2}w\alpha^2\{Q[x + Q^{-1}\alpha^{-1}w'v] - Q^{-1}[\alpha^{-1}w'v]\}).\end{aligned}$$

Further, since $\alpha^{-1}w' \in 2\mathfrak{o}_\mathfrak{p}$, $\alpha - \alpha ww' \in \mathfrak{o}_\mathfrak{p}$, and $Q \in GL(n, \mathfrak{o}_\mathfrak{p})$, for $v \in \Lambda(\mathfrak{p})$ we have

$$\begin{aligned}\tau_\mathfrak{p}(\tfrac{1}{2}ww'^2Q^{-1}[v]) &= \tau_\mathfrak{p}(\tfrac{1}{2}(\alpha ww')(\alpha^{-1}w')Q^{-1}[v]) = \tau_\mathfrak{p}(\tfrac{1}{2}\alpha(\alpha^{-1}w')Q^{-1}[v]) = \\ &= \tau_\mathfrak{p}(\tfrac{1}{2}w'Q^{-1}[v]).\end{aligned}$$

This cancels the factor outside the integral. Since $Q \in GL(n, \mathfrak{o}_\mathfrak{p})$ and $\alpha^{-1}w \in 2\mathfrak{o}_\mathfrak{p}$, $(\alpha^{-1}w)Q^{-1}v$ is in $\Lambda(\mathfrak{p})$, so we can replace x by $x - (\alpha^{-1}w')Q^{-1}v$ in the integral. Thus, for v in $\mathfrak{o}_\mathfrak{p}^n$,

$$\|\alpha\|_\mathfrak{p}^n \int_{\Lambda(\mathfrak{p})} \tau_\mathfrak{p}(-\tfrac{1}{2}w'Q[\alpha x] - \langle \alpha x, v\rangle)\,dx \times \tau_\mathfrak{p}(-\tfrac{1}{2}w'Q^{-1}[v]) = \|\alpha\|_\mathfrak{p}^n \int_{\Lambda(\mathfrak{p})} \tau_\mathfrak{p}(-\tfrac{1}{2}wQ[\alpha x])\,dx,$$

and the integral is 0 otherwise. ■

PROPOSITION Let $\tau_0 = \prod_\mathfrak{p} \tau_\mathfrak{p}$ (product over finite primes). When $d = \alpha^{-1}$, $b = \alpha w$, $c = \alpha^{-1}w'$, and $\gamma = \left(\begin{smallmatrix} a & b \\ c & d\end{smallmatrix}\right) \in \Gamma(\mathfrak{f})$, we have

$$\begin{aligned}&\{[\varphi \circ \alpha^{-1} \times \tau_0(-\tfrac{1}{2}wQ[\ ])]^\wedge \times \tau_0(-\tfrac{1}{2}w'Q^{-1}[\ ])\}^\wedge(v) = \\ &= \varphi(-v) \times \prod_{\mathfrak{p} \in X} \left\{\|\alpha\|_\mathfrak{p}^n \int_{\Lambda(\mathfrak{p})} \tau_\mathfrak{p}(-\tfrac{1}{2}wQ[\alpha x])\,dx\right\},\end{aligned}$$

where X is the set of good primes where α is not a $\mathfrak{p}$-unit.

Proof This just summarizes our computations of the $\mathfrak{p}$-adic Fourier transforms, noting that $\varphi_\mathfrak{p}(-v) = \varphi_\mathfrak{p}(v)$ at a good prime $\mathfrak{p}$. ■

Therefore, we have

$$\begin{aligned}\Theta(\gamma(z); Q, f, \varphi) &= \Theta(z; Q, f, \varphi) \times \alpha^{-k}(-iu(w')J^{-1}(z))^{(n/2+k)}(-iz)^{(n/2+k)} \times \\ &\times \prod_{\mathfrak{p} \in X} \left\{\|\alpha\|_\mathfrak{p}^n \int_{\Lambda(\mathfrak{p})} \tau_\mathfrak{p}(-\tfrac{1}{2}wQ[\alpha x])\,dx\right\},\end{aligned}$$

where $\Lambda(\mathfrak{p}) = \mathfrak{o}_\mathfrak{p}^n$ for each $\mathfrak{p}$, using $\varphi(-v) = (-1)^k\varphi(v)$.

PROPOSITION Use notation as above. We have

$$\prod_{\mathfrak{p} \in X} \|\alpha\|_{\mathfrak{p}}^{\frac{n}{2}} \int_{\Lambda(\mathfrak{p})} \tau_{\mathfrak{p}}(-\tfrac{1}{2}wQ[\alpha x])\,dx = (\tfrac{1}{2}\alpha w/\alpha^{-1})_2^n (\det Q/\alpha^{-1})_2 |\alpha|^{-n/2}.$$

Proof Again, X is the set of primes dividing $\alpha^{-1}\mathfrak{o}$, and $\mathfrak{p} \in X$ implies $\mathfrak{p}$ is unramified over $\mathbb{Z}$, and does not divide 2. In each integral

$$\int_{\Lambda(\mathfrak{p})} \tau_{\mathfrak{p}}(-\tfrac{1}{2}wQ[\alpha x])\,dx = \int_{\Lambda(\mathfrak{p})} \tau_{\mathfrak{p}}(-\tfrac{1}{2}\alpha w\alpha Q[x])\,dx$$

for $\mathfrak{p}$ a good prime dividing $\alpha^{-1}\mathfrak{o}$, we have $Q \in GL(n, \mathfrak{o}_{\mathfrak{p}})$; and we have $\alpha w \in \mathfrak{o}_{\mathfrak{p}}^{\times}$, since $\gamma \in SL(2, \mathfrak{o})$. Since $\mathfrak{p}$ does not divide 2, the quadratic form Q may be *diagonalized* over $\mathfrak{o}_{\mathfrak{p}}$: there is $A \in SL(n, \mathfrak{o}_{\mathfrak{p}})$ so that $Q = A^T\delta A$ with δ diagonal, having diagonal entries $\delta_j \in \mathfrak{o}_{\mathfrak{p}}^{\times}$. Then, by changing coordinates in the integral,

$$\int_{\Lambda(\mathfrak{p})} \tau_{\mathfrak{p}}(-\tfrac{1}{2}\alpha w\alpha Q[x])\,dx = \int_{\Lambda(\mathfrak{p})} \tau_{\mathfrak{p}}(-\tfrac{1}{2}\alpha w\alpha\delta[x])\,dx =$$

$$= \prod_j \int_{\mathfrak{o}_{\mathfrak{p}}} \tau_{\mathfrak{p}}(-\tfrac{1}{2}\alpha w\delta_j x^2/\alpha^{-1})\,dx.$$

From the earlier discussion of quadratic symbols, for each index j, if $\mathrm{ord}_{\mathfrak{p}}(\alpha^{-1}) = e$,

$$\int_{\mathfrak{o}_{\mathfrak{p}}} \tau_{\mathfrak{p}}(-\tfrac{1}{2}\alpha w\delta_j x^2/\alpha^{-1})\,dx = (\tfrac{1}{2}(\alpha w)\delta_j/\mathfrak{p}^e)_2 \int_{\mathfrak{o}_{\mathfrak{p}}} \tau_{\mathfrak{p}}(-x^2/\alpha^{-1})\,dx.$$

Almost immediately from the definitions of the $\mathfrak{p}$-adic integrals,

$$\prod_{\mathfrak{p} \in X} \int_{\mathfrak{o}_{\mathfrak{p}}} \tau_{\mathfrak{p}}(-x^2/\alpha^{-1})\,dx = [\mathfrak{o} : \alpha^{-1}\mathfrak{o}]^{-1} \sum_{x \in \mathfrak{o}/\alpha^{-1}\mathfrak{o}} \mathrm{e}(x^2/\alpha^{-1}).$$

Using Hecke's reciprocity law for quadratic exponential sums,

$$[\mathfrak{o} : \alpha^{-1}\mathfrak{o}]^{-1} \sum_{x \in \mathfrak{o}/\alpha^{-1}\mathfrak{o}} \mathrm{e}(x^2/\alpha^{-1}) = [\mathfrak{o} : \alpha^{-1}\mathfrak{o}]^{-1/2} \times \varepsilon(\alpha^{-1}).$$

Therefore,

$$\prod_{\mathfrak{p} \in X} \int_{\mathfrak{o}_{\mathfrak{p}}} \tau_{\mathfrak{p}}(-\tfrac{1}{2}\alpha w\delta_j x^2/\alpha^{-1})\,dx = [\mathfrak{o} : \alpha^{-1}\mathfrak{o}]^{-1/2}\varepsilon(\alpha^{-1}) \prod_{\mathfrak{p} \in X} (\tfrac{1}{2}(\alpha w)\delta_j/\mathfrak{p}^e)_2.$$

Then, using the product formula,

$$\prod_{\mathfrak{p} \in X} \|\alpha\|_{\mathfrak{p}}^{n} \int_{\Lambda(\mathfrak{p})} \tau_{\mathfrak{p}}(-\tfrac{1}{2}\alpha w \alpha Q[x])\,dx =$$

$$= |\alpha^{-1}|^{n} \prod_{\mathfrak{p} \in X} \int_{\Lambda(\mathfrak{p})} \tau_{\mathfrak{p}}(-\tfrac{1}{2}\alpha w \alpha Q[x])\,dx =$$

$$= |\alpha^{-1}|^{n} \prod_{j} \left\{ [\mathfrak{o} : \alpha^{-1}\mathfrak{o}]^{-1/2} \varepsilon(\alpha^{-1}) \prod_{\mathfrak{p} \in X} (\tfrac{1}{2}(\alpha w)\delta_j/\mathfrak{p}^{e})_2 \right\} =$$

$$= |\alpha^{-1}|^{n/2} \varepsilon(\alpha^{-1})^{n}((\tfrac{1}{2}\alpha w)^{n} \det Q/\alpha^{-1})_2.$$

■

Summarizing what we have so far:

$$\Theta(\gamma z; Q, f, \varphi) =$$

$$= \Theta(z; Q, f, \varphi) \times \mathrm{d}^{k}\varepsilon(d)^{n} \times (-i(-cd^{-1} - z^{-1}))^{(n/2+k)}(-iz)^{(n/2+k)}(\det Q/d)_2((\tfrac{1}{2}b)^{n}/d)_2|d|^{n/2}.$$

Now

$$(-i(-cd^{-1} - z^{-1}))^{1/2}(-iz)^{1/2} = (cd^{-1}z + 1)^{1/2},$$

taking the square root lying in the right half-plane, since all the components of both $(-i(-cd^{-1} - z^{-1}))^{1/2}$ and $(-iz)^{1/2}$ (by convention...) have arguments between $-\pi/4$ and $\pi/4$. Further,

$$(cd^{-1}z + 1)^{1/2}|d|^{1/2} = \tilde{\varepsilon}(d)(cz + d)^{1/2}.$$

Also, as $ad - bc = 1$, it must be that $(-\tfrac{1}{2}c/d)_2 = (\tfrac{1}{2}b/d)_2$. Therefore,

$$\Theta(\gamma z; Q, f, \varphi) =$$

$$= \Theta(z; Q, f, \varphi) \times \tilde{\varepsilon}(d)^{n}\varepsilon(d)^{n}(-\tfrac{1}{2}c/d)_2^{n}(\det Q/d)_2(cz + d)^{(n/2+k)}.$$

This finishes the proof of the theorem. ■

5.4 Some Theta Series of Level One

The remarks after the statement of the theorem of the previous section might hint at the possibility of the existence of (integral weight) theta series which are Hilbert modular forms for all of $SL(2, \mathfrak{o})$. For quadratic forms of dimensions not divisible by 4, this is obviously unlikely, because of the presence of $\varepsilon\tilde{\varepsilon}$. The subtler issue of existence of theta series which are Hilbert modular forms for all of $GL^{+}(2, \mathfrak{o})$ will not be addressed here. We will content ourselves merely with construction of some examples.

Definition Let $0 < n \in \mathbb{Z}$, and let Q be a positive quadratic form on F^n. Suppose that Q is $2\mathfrak{o}^*$-valued on a lattice $\Lambda \subset F^n$. Then say that Q is *even-integral on* Λ.

THEOREM Let Q be a quadratic form on F^n with $n \in 2\mathbb{Z}$, and suppose that $\det Q$ is a square in $F^\times$. Suppose that Q is even-integral on $\mathfrak{o}^n$, and that Q^{-1} is even-integral on $\mathfrak{o}^{*n}$. Let the finite-adele Schwartz function φ be the characteristic function of $\prod_{\mathfrak{p}\text{ finite}} \mathfrak{o}_\mathfrak{p}$, and take any pluriharmonic polynomial f with degrees $(k,\ldots,k)$ divisible by 4. Then the harmonic theta function $\Theta(z; Q, f, \varphi)$ is a Hilbert modular form (of integral weight) for $SL(2, \mathfrak{o})$, with trivial character.

Proof First, recall that the Fourier transform of φ is a constant times the characteristic function of $\mathfrak{o}^* \prod_\mathfrak{p} \mathfrak{o}_\mathfrak{p}$. The result of the previous section assures that $\Theta(z; Q, f, \varphi)$ is certainly a Hilbert modular form for $\Gamma(4\mathfrak{o}^{*-1})$ (without character). The hypothesis that Q is even-integral on $\mathfrak{o}^n$ is exactly what is needed to assure that $\Theta(z; Q, f, \varphi)$ is invariant under elements $u(w)$ with $w \in \mathfrak{o}$. From the elementary functional equations used in the proof that theta series are modular forms, one easily finds that the hypothesis that Q^{-1} is even-integral on $\mathfrak{o}^{*n}$ is exactly what is needed to assure that $\Theta(z; Q, f, \varphi)$ is invariant under elements $Ju(w)J^{-1}$ with $w \in \mathfrak{o}$. Also, for $\eta \in \mathfrak{o}^\times$ it is elementary that $\Theta(t(\eta)z; Q, f, \varphi) = \eta^k \Theta(z; Q, f, \varphi)$. Therefore, to prove the theorem it suffices to prove

LEMMA The whole group $SL(2, \mathfrak{o})$ is generated by the congruence subgroup $\Gamma(4\mathfrak{o}^{*-1})$ together with the elements $u(w)$ with $w \in \mathfrak{o}$, $u'(w) = Ju(w)J^{-1}$ with $w \in \mathfrak{o}$, and $t(\eta)$ with $\eta \in \mathfrak{o}^\times$.

Proof of Lemma Let $R = \mathfrak{o}/4\mathfrak{o}^{*-1}$. For the moment, identify the elements $u(w)$ and $u'(w)$ with their images in $SL(2, R)$. Take $g = \begin{pmatrix} a & b \\ c & d \end{pmatrix} \in SL(2, R)$. If $a \in R^\times$, then take $w_1 \equiv -ca^{-1}$ modulo $4\mathfrak{o}^{*-1}$, and then $u'(w_1)g$ is of the form $\begin{pmatrix} * & * \\ 0 & * \end{pmatrix}$. Now suppose that $a \notin R^\times$; since $g \in SL(2, R)$, $aR + cR = R$. Let $\mathfrak{m}_1, \ldots, \mathfrak{m}_n$ be the maximal ideals not containing a. By the Chinese Remainder Theorem, there is $x \in R$ such that $x \equiv 1$ modulo aR and $x \equiv 0$ modulo $\mathfrak{m}_1 \ldots \mathfrak{m}_n$. Then $r = a + xc$ is a unit in R: If r is in some $\mathfrak{m}_i$, then $a \in \mathfrak{m}_i$, contradiction; if r lies in a maximal ideal $\mathfrak{m}$ so that $a \in \mathfrak{m}$, then $c \in \mathfrak{m}$, contradiction. Thus, the top left entry of $u'(x)g$ is a unit in R, and we are reduced to the first case. That is, there are w_1, $w_2 \in \mathfrak{o}$ so that, in $SL(2, \mathfrak{o})$,

$$g' = u'(w_1)gu(w) \equiv \begin{pmatrix} * & * \\ 0 & * \end{pmatrix} \quad \text{modulo } 4\mathfrak{o}^{*-1}.$$

Given such an element g', the upper left entry g_{11} is a unit modulo $4\mathfrak{o}^{*-1}$ and the lower left entry is $g_{21} \in 4\mathfrak{o}^{*-1}$. Let $z \equiv g_{11}^{-1}$ modulo $4\mathfrak{o}^{*-1}$; then $u'(-zy) \in \Gamma(4\mathfrak{o}^{*-1})$, and $u'(-zy)$ is upper triangular in $SL(2, \mathfrak{o})$ and so is clearly a product of elements $t(\eta)$ and $u(w)$. ∎

Now we construct some families of quadratic forms Q satisfying the hypotheses of the previous theorem.

Definition For x and y in $\mathbb{R}$, let $E_j(x, y)$ denote the j-by-j matrix

$$\begin{pmatrix} 2x & y & 0 & 0 & 0 & 0\ldots0 \\ y & 2 & 1 & 0 & 0 & 0\ldots0 \\ 0 & 1 & 2 & 1 & 0 & 0\ldots0 \\ 0 & 0 & 1 & 2 & 1 & 0\ldots0 \\ & & & \cdots & & \\ 0\ldots0 & 0 & 1 & 2 & 1 & 0 \\ 0\ldots0 & 0 & 0 & 1 & 2 & 1 \\ 0\ldots0 & 0 & 0 & 0 & 1 & 2 \end{pmatrix},$$

and let $E_j = E_j(1, 1)$. Note that, if $x, y \in \mathbb{Z}$, then $E_j(x, y)$ is even-integral.

LEMMA The determinant of $E_j(x, y)$ is $2jx - (j-1)y^2$. For $x > 0$, $y \in \mathbb{R}$ so that $2nx - (n-1)y^2 > 0$, the matrix $E_n(x, y)$ is positive definite.

Proof First, it is easy to see that

$$\det E_j(x, y) = 2x \det E_{j-1} - y^2 \det E_{j-2}$$

by expanding by minors. Clearly we have $\det E_1 = 2$ and $\det E_2 = 3$. By induction, for $j > 2$,

$$\det E_j = 2 \det E_{j-1} - \det E_{j-2} = 2j - (j-1) = j + 1.$$

Then

$$\det E_j(x, y) = x \det E_{j-1} - y^2 \det E_{j-2} = 2jx - (j-1)y^2.$$

This gives the first assertion.

It is sufficient to show that the determinants of all the principal minors of $E_n(x, y)$ are positive. These determinants are positive multiples of

$$2jx - (j-1)y^2 \qquad \text{(with } 1 \le j \le m)$$

$$2j - (j-1) = j + 1 > 0 \qquad \text{(with } 1 \le j < m).$$

For $\xi \in \mathbb{R}$, $0 \le \xi \le m$ the linear function

$$f(\xi) = 2\xi x - (\xi - 1)y^2$$

has minimum on the interval $[1, m]$ occurring at an endpoint. But we have

$$2x - (1-1)y^2 = 2x > 0, \qquad 2mx - (m-1)y^2 > 0. \qquad \blacksquare$$

COROLLARY Take $F = \mathbb{Q}$. The matrix $Q = E_8(4,3)$ is positive definite, and is even-integral on $\mathbb{Z}^8$. Further, Q^{-1} is even-integral on Z^8. Therefore,

$$\sum_{v \in \mathbb{Z}^8} \mathrm{e}(\tfrac{1}{2}zQ[v])$$

is a modular form of weight 4 for $SL(2, \mathbb{Z})$.

Proof From the previous lemma, $\det Q = 1$. The only non-obvious point is that Q^{-1} is also even-integral on $\mathbb{Z}^8$. For $v \in \mathbb{Z}^8$, we have $Q^{-1}[v] = Q[Q^{-1}v]$. Since $Q \in GL(8, \mathbb{Z})$, $Q^{-1}v \in \mathbb{Z}^8$, and Q^{-1} is even-integral. The theorem gives the assertion. ∎

COROLLARY Let F be a totally real number field with different $\mathfrak{o}^{*-1} = \delta\mathfrak{o}$ for some totally positive element δ of F. Put $Q = \delta^{-1}E_8(4,3)$. Then

$$\sum_{v \in \mathfrak{o}^8} \mathrm{e}(\tfrac{1}{2}zQ[v])$$

is a modular form of weight $(4, \ldots, 4)$ for $SL(2, \mathfrak{o})$.

Proof It is easy to see that Q is even-integral on $\mathfrak{o}^8$, and the same argument as in the previous corollary shows that Q^{-1} is even-integral on $\mathfrak{o}^{*8}$. ∎

Remarks Such number fields may be obtained, for example, by taking totally real subfields of $K = \mathbb{Q}(\zeta)$ where ζ is a p-th root of unity with an odd prime p. Indeed, it is well-known that the different of K over $\mathbb{Q}$ is generated by $\tilde{\delta} = (1-\zeta)^{p-1}$. Every totally real subfield of K is contained in $\mathbb{Q}(\zeta + \zeta^{-1})$, whose different over $\mathbb{Q}$ is generated by the norm δ of $\tilde{\delta}$ from K to $\mathbb{Q}(\zeta + \zeta^{-1})$. That is,

$$\delta = (1-\zeta)(1-\zeta^{-1}) = 2 - (\zeta + \zeta^{-1}) = 2 - 2\cos(2\pi k/p) > 0$$

for some integer k prime to p. This δ is obviously totally positive. If F were any other totally real subfield of $\mathbb{Q}(\zeta + \zeta^{-1})$, then the different of F over $\mathbb{Q}$ is generated by the norm of δ from $\mathbb{Q}(\zeta + \zeta^{-1})$ to F, which must be totally positive.

5.5 Action of $GL^+(2, F)$ on Theta Series

Fix a positive definite quadratic form Q of *even* dimension n over F and a pluri-harmonic polynomial f of (multi-) degree $k = (k_1, \ldots, k_m)$.

THEOREM The collection of theta series

$$X(Q, f) = \{\Theta(z; Q, f, \varphi) \colon \varphi \text{ is a Schwartz functions on } \mathbb{A}_0^n\}$$

is stable under the action

$$\Theta(z; Q, f, \varphi) \to \Theta(z; Q, f, \varphi)|\gamma = \Theta(\gamma z; Q, f, \varphi)(cz + d)^{-(k+n/2)}$$

of $\gamma = \left(\begin{smallmatrix} a & b \\ c & d \end{smallmatrix}\right) \in SL(2, F)$. That is, such maps only have the effect of changing the finite-adele Schwartz function. The collection of theta series

$$\tilde{X}(Q, f) = \{\Theta(z; \lambda Q, f, \varphi)\colon \varphi \text{ is a Schwartz function on } \mathbb{A}_0^n \text{ and } \lambda \text{ is a totally positive element of } F\}$$

is stable under the action

$$\Theta(z; Q, f, \varphi) \to \Theta(z; Q, f, \varphi)|\gamma = \Theta(\gamma z; Q, f, \varphi)(cz + d)^{-(k+n/2)} \det(\gamma)^{(k+n/2)/2}$$

of $\gamma = \left(\begin{smallmatrix} a & b \\ c & d \end{smallmatrix}\right) \in GL^+(2, F)$. That is, such maps only have the effect of changing the finite-adele Schwartz function, and changing the quadratic form by a totally positive scalar.

Remarks For n odd, the maps above do not give a group action, because of the ambiguity of the square root. This difficulty cannot be overcome on $GL(2)$.

Proof For all $\alpha \in F^\times$ and $w \in F$, we have the elementary functional equations obtained previously:

1. $\Theta(t(\alpha)z; Q, f, \varphi) = \Theta(z; Q, f, \psi)$, where $\psi(v) = \alpha^{-k}\varphi(\alpha^{-1}v)$.
2. $\Theta(u(w)z; Q, f, \varphi) = \Theta(z; Q, f, \psi)$, where $\psi(v) = \varphi(v)\mathrm{e}(\frac{1}{2}wQ[v])$.
3. $\Theta(Jz; Q, f, \varphi) = (-iz)^{(n/2+k)}(\det Q)^{1/2} i^k \Theta(z; Q^{-1}, f, \hat{\varphi})$.

The last one may be rewritten as

3′. $\Theta(Jz; Q, f, \varphi) = (-iz)^{(n/2+k)}(\det Q)^{1/2} i^k \Theta(z; Q, f, \hat{\varphi} \circ Q)$

by replacing v by Qv in the sum defining the theta series, since

$$Q^{-1}[Qv] = Q[v] \qquad f(Q^{-1/2}Qv) = f(Q^{1/2}v).$$

That is, $X(Q, f)$ is stable under the maps given by the elements $t(\alpha)$, $u(w)$, J. These elements generate $SL(2, F)$, so we have stability of $X(Q, f)$ under $SL(2, F)$.

Since $GL^+(2, F)$ is generated by $SL(2, F)$ and elements $\tilde{t} = \left(\begin{smallmatrix} t & 0 \\ 0 & 1 \end{smallmatrix}\right)$ with $t \in F^\times$ totally positive, to obtain the second assertion of the theorem it suffices to consider just these elements $\tilde{t}$. Since each component of the polynomial f is homogeneous, the stability is clear. The action of such $\tilde{t}$ causes Q to be replaced by tQ; unless t is a square, we cannot turn this back to Q by a change of coordinates in the sum defining the theta series. ∎

Definition Let Q be a positive definite quadratic form of even dimension n, and f a pluriharmonic polynomial of multi-degree k. Let $\mathrm{Th}(Q, f)$ denote the collection of automorphic forms θ on $GL(2, \mathbb{A})$ such that for all $\alpha \in GL(2, \mathbb{A}_0)$ the function $R_\alpha\theta$ on $x + iy \in \mathfrak{H}^m$ given by

$$(R_\alpha\theta)(x + iy) = y^{-(k+n/2)/2}\theta\left(\begin{pmatrix} 1 & x \\ 0 & 1 \end{pmatrix}\begin{pmatrix} y & 0 \\ 0 & 1 \end{pmatrix}\alpha\right)$$

is in $\tilde{X}(Q, f)$.

Remarks It is immediate that $\mathrm{Th}(Q, f)$ is *stable under the action of the* (*spherical*) *Hecke operators at every finite prime*, since these operators are given by right convolutions over the groups $GL(2, F_{\mathfrak{p}})$. What is not so apparent is that $\mathrm{Th}(Q, f)$ is not empty. The previous theorem assures that this is so; for example, one finds that $\tilde{X}(Q, f)$ injects to $\mathrm{Th}(Q, f)$ via the "adelization" procedure discussed earlier in comparing the classical and adelic versions of automorphic forms. As a consequence, in any reasonable sense, $\tilde{X}(Q, f)$ *is stable under the action of Hecke operators.*

5.6 Cuspforms Obtained as Pluriharmonic Theta Series

THEOREM Let Q be a quadratic form of even dimension n and let f be a pluriharmonic polynomial of multi-degree $k = (k_1, \ldots, k_m)$. Unless $k_1 = \cdots = k_m = 0$, the theta series $\Theta(z; Q, f, \varphi)$ is a cuspform for any choice of finite-adele Schwartz function φ.

Proof Since Q is positive definite, the constant term in the (classical) Fourier expansion of a theta series

$$\Theta(z; Q, f, \varphi) = \sum_{v \in F^n} \varphi(v) f(Q^{1/2} v) \mathrm{e}(\tfrac{1}{2} z Q[v])$$

is obtained by the summand $v = 0$. That is, the constant term is just $\varphi(0) f(0)$. If f is not a constant, then $f(0) = 0$. The theorem of the previous section shows that for any $g = \left(\begin{smallmatrix} a & b \\ c & d \end{smallmatrix}\right) \in GL^+(2, F)$

$$\Theta(gz; Q, f, \varphi)(cz + d)^{-(k+n/2)} \det(g)^{(k+n/2)/2} = \Theta(z; Q', f, \varphi')$$

where Q' differs from Q by a totally positive scalar, for some finite-adele Schwartz function φ'. The previous argument shows that the constant term in the Fourier expansion of $\Theta(z; Q', f, \varphi')$ is also zero. Thus, we have shown that $\Theta(z; Q, f, \varphi)$ is a cuspform. ∎

Remarks A certain complication here is that it is not so easy to tell whether a pluriharmonic theta series is identically zero or not when the pluriharmonic polynomial is not constant. Nevertheless, in the simple example (with $z \in \mathfrak{H}$)

$$\varphi(z) = \sum_{m, n \in \mathbb{Z}} (m + in)^k \exp(\pi i z(m^2 + n^2))$$

it is easy to check that φ is not identically zero if and only if $k \equiv 0$ modulo 4.

5.7 *L*-Functions with Grossencharacters

The purpose of this section is to construct some pluriharmonic theta series on $\mathfrak{H}^m$ whose Mellin transforms are L-functions with grossencharacters attached to totally imaginary quadratic extensions K of the totally real field F. We will be interested in grossencharacter L-functions of K of the form

$$L_K(s, \chi) = \prod_{\mathfrak{p}} 1/(1 - \chi(\mathfrak{P}) N\mathfrak{P}^{-s}) \qquad \text{(product over primes of } K \text{ unramified for } \chi)$$

where the grossencharacter χ is of a special type at infinite primes: We require that there be choices $\tilde{\sigma}_i$ of extensions of the imbeddings σ_i: $F \to \mathbb{R}$ to imbeddings of K into $\mathbb{C}$, and an integer n, so that for $\alpha \in K^\times$

$$\chi(\alpha_\infty) = \left(\prod_i \tilde{\sigma}_i(\alpha)/|\tilde{\sigma}_i(\alpha)| \right)^n.$$

Note that this implies that χ is totally odd if n is odd, and is totally even if n is even. Given a grossencharacter of K of this type, we wish to find a holomorphic modular form θ on $GL(2, \mathbb{A})$ so that $L_K(s, \chi)$ is essentially a Mellin transform of θ. Further, we wish to assure that θ is right-invariant under a subgroup of the form

$$\left\{ g = \begin{pmatrix} a & b \\ c & d \end{pmatrix} \in GL(2, \hat{\mathfrak{o}}) \colon c \equiv 0 \text{ and } d \text{ is a unit modulo } \mathfrak{n} \right\}.$$

We will accomplish this by reconsidering the proof in section 5.3, being more careful about "bad" primes.

Let $\mathfrak{o}$ denote the ring of integers of K. Let $\mathfrak{m} \subset \mathfrak{O}$ be the finite-prime part of the conductor of χ. By Dirichlet's theorem on primes in an arithmetic progression, we may choose representatives $\mathfrak{a}_i$ for the absolute ideal class group of $\mathfrak{O}$ so that each $\mathfrak{a}_i$ is prime to $\mathfrak{m}$.

Definition For $z \in \mathfrak{H}^m$, define a pluriharmonic theta series

$$\theta(z) = c_0 + \sum_i \chi(\mathfrak{a}_i) N\mathfrak{a}_i^{n/2} \sum_\alpha \chi(\alpha_\infty) \|\alpha_\infty\|_K^{n/2} \exp(2\pi i N\mathfrak{a}_i \operatorname{Tr} z\alpha\alpha')$$

where $\alpha \to \alpha'$ is the non-trivial Galois automorphism of K over F, $N\mathfrak{a}_i$ is the ideal norm of $\mathfrak{a}_i$, α_∞ is the infinite-prime part of $\alpha \in K^\times \subset \mathbb{J}_K$, $\| \ \|_K$ is the idele norm on K, $c_0 = 0$ if χ is nontrivial and c_0 is the class number of $\mathfrak{O}$ if χ is trivial, and for fixed $\mathfrak{a}_i$ the inner sum is over

$$\{\alpha \in K^\times \colon \alpha\mathfrak{a}_i \subset \mathfrak{O}, \alpha \text{ prime to } \mathfrak{m}\}.$$

Also, here $\chi(\mathfrak{a}_i)$ denotes the value of χ at a finite idele η_i whose $\mathfrak{P}$-component is 1 at primes dividing $\mathfrak{m}$, and so that $\mathfrak{a}_i\hat{\mathfrak{O}} = \eta_i\hat{\mathfrak{O}}$.

Note that the normalization of norms is such that for $\alpha \in \mathbb{J}_K$

$$\|\alpha\|_K = \|\operatorname{Norm}_{K/F} \alpha\|_F.$$

In particular, at an infinite prime $\mathfrak{P}$ of K associated to an imbedding $\tilde{\sigma}$: $K \to \mathbb{C}$,

$$\|\alpha\|_{\mathfrak{p}} = |\tilde{\sigma}(\alpha)|^2$$

where $|\ |$ is the usual complex absolute value extending that on $\mathbb{R}$. The polynomial given by the infinite-prime part of $\chi(\alpha_\infty)\|\alpha_\infty\|^{n/2}$ is then obviously of (multi-) degree $(n,\ldots,n)$; the vectorspace over which α is summed is of dimension 2 over F. Therefore, from section 5.3, we know that θ is a modular form for some congruence subgroup, and is of weight $(n+1,\ldots,n+1)$. We need a better result.

THEOREM Let $\mathfrak{n}$ be a sufficiently small non-zero ideal of $\mathfrak{o}$. Let Δ be a totally negative element of $\mathfrak{o}$ so that K is obtained from F by adjoining the square root δ of Δ. The theta series θ is a modular form of weight $(n+1,\ldots,n+1)$ for the subgroup

$$\Gamma_0(\mathfrak{n}) = \left\{g \in SL(2,\mathfrak{o})\colon g = \begin{pmatrix} a & b \\ c & d \end{pmatrix} \text{ with } c \equiv 0 \bmod \mathfrak{n}, \text{ and so that } d \text{ is an } \mathfrak{n}\text{-unit}\right\}$$

with character

$$\begin{pmatrix} a & b \\ c & d \end{pmatrix} \to \chi(d)\,\mathrm{sgn}(d)^1(\Delta/d)_2$$

(using a multi-index notation).

Remarks From the construction, it is clear that the *classical* Mellin transform of θ (as treated in Chapter 1) is essentially the desired L-function.

Note that the character

$$d \to \chi(d)\,\mathrm{sgn}(d)^1(\Delta/d)_2$$

is of parity *opposite* to that of χ; therefore, this character is indeed totally odd if $n+1$ is odd, and is totally even if $n+1$ is even.

In fact, the proof shows that each subsum corresponding to an ideal $\mathfrak{a}_i$ has the same property. Also, one can estimate how small the ideal $\mathfrak{n}$ really must be. In view of the discussion of section 3.1 comparing classical and adelic modular forms, in order to understand the "adelization" of a theta series θ (or any other modular form), it is desirable to have θ be a modular form for a congruence subgroup of the form of $\Gamma_0(\mathfrak{n})$ rather than merely for the smaller subgroup

$$\left\{\gamma = \begin{pmatrix} a & b \\ c & d \end{pmatrix} \in SL(2,\mathfrak{o})\colon c \equiv b \equiv 0 \text{ and } a \equiv d \equiv 1 \text{ modulo } \mathfrak{n}\right\}.$$

This is one motivation for the present discussion.

Proof We only give the proof for $\mathfrak{n}$ proper. The theorem of section 5.3 shows that the subsum of θ corresponding to one of the ideals $\mathfrak{a}_i$ is a modular form for

$$\Gamma(\mathfrak{n}) = \left\{g \in SL(2,\mathfrak{o})\colon g = \begin{pmatrix} a & b \\ c & d \end{pmatrix} \text{ with } b, c \equiv 0 \bmod \mathfrak{n}, \text{ and } a \equiv d \equiv 1 \bmod \mathfrak{n}\right\}$$

for *small enough* $\mathfrak{n}$, with character

$$\begin{pmatrix} a & b \\ c & d \end{pmatrix} \to \operatorname{sgn}(d)^1(-\det Q_i/d)_2$$

on the subgroup $\Gamma(\mathfrak{n})$. The proof of the present theorem is the same, but for treatment of primes $\mathfrak{p}$ dividing $\mathfrak{n}$; now we can be more careful. In effect, the proof in section 5.3 inadvertently required that d_∞ be in the kernel of χ, so that χ would have made no explicit appearance in that proof.

We fix an isomorphism $K \approx F^2$ by $x + \delta y \to (x, y)$. Conforming with the notation of the proof of section 5.3, for $v \in K \approx F^2$, define a quadratic form by

$$Q_i[v] = N\mathfrak{a}_i \operatorname{Norm}_{K/F}(v)$$

and a finite-adele Schwartz function on $\mathbb{A}_K \approx \mathbb{A}_F^2$ by

$$\varphi_{0i}(v) = N\mathfrak{a}_i^{n/2}\chi(\mathfrak{a}_i)\tilde{\chi}(v)\operatorname{ch}(v; \mathfrak{a}_i^{-1}X)$$

where

$$X = \mathfrak{n}\text{-units in } \hat{\mathfrak{O}}$$

and where $\tilde{\chi} \equiv 1$ at primes not dividing $\mathfrak{n}$, and is the restriction of χ at primes dividing $\mathfrak{n}$. Take the pluriharmonic polynomial on $\mathbb{C}^m \approx (\mathbb{R}^2)^m$ given by

$$f(u, v) = \prod_k (u_k + iv_k)^n.$$

Then, in the notation of section 5.3,

$$\theta(z) = \sum_i \Theta(z; Q_i, f, \varphi_{0i}).$$

Certainly this Schwartz function factors over primes of F as $\varphi_{0i} = \prod_\mathfrak{p} \varphi_{\mathfrak{p}i}$. Take

$$\gamma = \begin{pmatrix} a & b \\ c & d \end{pmatrix} = \begin{pmatrix} \alpha(1 - ww') & \alpha w \\ -\alpha^{-1}w' & \alpha^{-1} \end{pmatrix} \in SL(2, \mathfrak{o})$$

with α an $\mathfrak{n}$-unit, $w' \in \mathfrak{n}$, $\alpha w \in \mathfrak{o}$, and $(1 - ww')$ an $\mathfrak{n}$-unit.

Then the *good prime* ($\mathfrak{p}$ not dividing $\mathfrak{n}$) calculation of section 5.3 may be repeated verbatim. The *bad prime* case ($\mathfrak{p}$ dividing $\mathfrak{n}$) may be improved: Take $\mathfrak{p}$ dividing $\mathfrak{n}$; in the notation of the proof of section 5.3, after some preparations, we find that we wish to compute

$$\{[\varphi_{\mathfrak{p}i} \circ \alpha^{-1} \times \tau_\mathfrak{p}(-\tfrac{1}{2}wQ_i[\ ])]^\wedge \times \tau_\mathfrak{p}(-\tfrac{1}{2}w'Q_i^{-1}[\ ])\}^\wedge(v).$$

Certainly

$$[\varphi_{\mathfrak{p}i} \circ \alpha^{-1} \times \tau_{\mathfrak{p}}(-\tfrac{1}{2}wQ_i[\ ])]^{\wedge}$$

is compactly supported, so for w' sufficiently divisible this is the same as

$$\{[\varphi_{\mathfrak{p}i} \circ \alpha^{-1} \times \tau_{\mathfrak{p}}(-\tfrac{1}{2}wQ_i[\ ])]^{\wedge}\}^{\wedge}(v) =$$

$$= [\varphi_{\mathfrak{p}i} \circ \alpha^{-1} \times \tau_{\mathfrak{p}}(-\tfrac{1}{2}wQ_i[\ ])](-v) =$$

$$= \bar{\chi}(\alpha)\chi(-v)\tau_{\mathfrak{p}}(-\tfrac{1}{2}wQ_i[-\alpha^{-1}v])\,\mathrm{ch}(-v; \mathfrak{a}_i^{-1}(\mathfrak{O} \otimes \mathfrak{o}_{\mathfrak{p}} - \mathfrak{n}\mathfrak{O} \otimes \mathfrak{o}_{\mathfrak{p}})).$$

Taking w to be $\mathfrak{n}$-integral and α an $\mathfrak{n}$-unit, the function

$$\tau_{\mathfrak{p}}(-\tfrac{1}{2}wQ_i[-\alpha^{-1}v])$$

is surely identically 1 on the support of

$$v \to \mathrm{ch}(-v; \mathfrak{a}_i^{-1}(\mathfrak{O} \otimes \mathfrak{o}_{\mathfrak{p}} - \mathfrak{n}\mathfrak{O} \otimes \mathfrak{o}_{\mathfrak{p}})).$$

Therefore, the desired expression is

$$\chi(\alpha^{-1}) \times \varphi_{\mathfrak{p}i}(-v) = \chi(d) \times \varphi_{\mathfrak{p}i}(-v).$$

Supplementing the proof of section 5.3 with this result strengthens the conclusion of that theorem to the assertion that

$$\theta|_{n+1}\gamma = \chi(d)\,\mathrm{sgn}(d)^1(-1/d)_2(\det Q_i/d)_2\theta.$$

It is an easy exercise to see that the determinants $\det Q_i$ differ from each other by squares in F^x; in particular, they differ from $-\Delta$ by squares (in F^x). This yields the assertion of the theorem. ■

6 Arithmetic of Hilbert Modular Forms

The first section of this chapter contains the most serious result of this book, what we refer to as the "Arithmetic Structure Theorem." This theorem has two parts: The first concerns rationality properties of Fourier coefficients of cuspforms and is a weak version of a theorem originally due to Shimura [Sh3]; the second concerns comparison of certain inner products of cuspforms and is apparently new. The proof of both parts of the theorem is given in Chapter 7; it is an application of the ideas of [G1], and is a new method. The remaining sections of this chapter apply the Arithmetic Structure Theorem to determination of the arithmetic nature of special values of various types of L-functions (following Klingen [K1], [K2] and Shimura [Sh6]). All these results are compatible with the very general conjectures of Deligne [De] concerning the arithmetic nature of special values of a wide class of L-functions.

The second section is a reprise of Klingen's results on special values (at suitable integers) of L-functions (with suitable characters) of totally real number fields. This already powerfully illustrates the relevance of modular forms to seemingly more primitive questions. We note that Klingen's proof of this result needs only the first half of the Arithmetic Structure Theorem over $\mathbb{Q}$, which has been known for almost one hundred years. There are other proofs of this result, but the present proof is perhaps the easiest to understand.

The third section gives a sample of Shimura's application of Rankin's integral representations to L-functions with certain grossencharacters of totally imaginary quadratic extensions of totally real number fields (due to Damerell [D] in case the base field is $\mathbb{Q}$). These L-functions arise as the standard L-functions attached to certain theta series (as in section 5.7), which gives the connection to automorphic forms. There is the further coincidence that the "Rankin product" L-function of two L-functions with grossencharacter (over the same field) is a product of two grossencharacter L-functions; this feature is obviously peculiar to these L-functions. The fourth section gives an easy formal result in this vein.

The last section requires a little more preparation: Essentially following Shimura, an adroit choice of an auxiliary Eisenstein series in the Rankin integral representation yields a product of two special values of the standard L-function attached to a cuspform. Here we treat only the case that the cuspform is of level one (i.e., is right $GL(2, \hat{\mathfrak{o}})$-invariant) with trivial central character. This avoids some technicalities without obscuring the point. It should be noted that even with these strong hypotheses the full strength of the theorem of section 6.1 is genuinely needed.

Apart from small details, the proofs of the results in sections 6.3–6.5 are all exactly the same in spirit: They are applications of sections 4.10 and 6.1 together. This hardly seems harmful given the importance of the idea. It is fair to say that Shimura's papers [Sh4–Sh6] first fully manifested an appreciation of the possibilities here.

In all these results, one underlying theme is that *there are relatively few new "transcendental" numbers arising in the description of the special values*—only π and certain Petersson inner products. Naively, there is certainly no reason to expect any explicable relation between different special values of the same L-function (as in section 6.5), nor of grossencharacter L-functions with different grossencharacters (as in section 6.3).

Since the results of section 6.5 give some relations among *products* and *quotients* of special values of standard L-functions, the question might arise as to why there is no treatment of the values themselves. One ingenuous but not entirely misleading answer is that the special values, being Mellin integrals, are already represented as well as possible, and that the surprise ought to be that there are relations among them. See Deligne [De] for a very general conjectural framework which includes these (and all known) examples.

It should be emphasized that the results of sections 6.3–6.5 are far from the most general possible (see [Sh6]). Also, Shimura [Sh3] proves many other types of results beyond the first half of our result of section 6.1—e.g., concerning generation of class fields, which seem to be definitely beyond the scope of the present methods. Nevertheless, we have given some results here which should suffice to illustrate some of the ideas, and which do suffice for many applications.

6.1 The Arithmetic Structure Theorem

The purpose of this section is to give the statement of a theorem which has two parts. The first roughly states that spaces of holomorphic modular forms have $\mathbb{C}$-bases consisting of modular forms "with Fourier coefficients in finite field extensions of $\mathbb{Q}$." The second part compares certain inner products of cuspforms. In a classical situation, the Fourier coefficients are numbers, so "rationality of Fourier coefficients" has an obvious sense. On $GL(2, \mathbb{A})$, there are several possibilities for the sense of this notion. In the following, we have made a choice which suffices for many applications (including those in subsequent sections). In particular, the present choice of meaning of "rationality of Fourier coefficients" is somewhat sharper than the obvious classical numerical version. The comparison of inner products is necessary for the applications to special values results; the result given here is stronger than previously known results and allows a circumvention of consideration of the rather delicate idea of "newforms" (see [M], [C], [A]). Granting these things, the corollaries are essentially immediate.

Although we discuss modular forms only with respect to a certain form of congruence subgroup, as in several previous discussions it is elementary to see that the general situation may be reduced to the cases considered here. Also, although we discuss modular forms only with weights of the form $(\kappa, \ldots, \kappa)$, the methods of Chapter 7 can be used to give a much more general result.

Fix a weight $(\kappa, \ldots, \kappa)$, abbreviated simply as κ. Fix a non-zero ideal $\mathfrak{n}$ in $\mathfrak{o}$, and define

$$\tilde{\mathbb{K}} = \left\{ g \in GL(2, \hat{\mathfrak{o}}) : g \equiv \begin{pmatrix} * & 0 \\ 0 & 1 \end{pmatrix} \text{ modulo } \mathfrak{n} \right\}.$$

For a finite prime $\mathfrak{p}$ not dividing $\mathfrak{n}$, let

$$H(\mathfrak{p}) = \{\text{left and right } GL(2, \mathfrak{o}_\mathfrak{p})\text{-invariant compactly supported functions on } GL(2, F_\mathfrak{p})\}$$

$$H_1(\mathfrak{p}) = \{\text{left and right } SL(2, \mathfrak{o}_\mathfrak{p})\text{-invariant compactly supported functions on } SL(2, F_\mathfrak{p})\}.$$

Put

$$\mathscr{H}(\mathfrak{n}) = \bigotimes H(\mathfrak{p}) \qquad \mathscr{H}_1(\mathfrak{n}) = \bigotimes H_1(\mathfrak{p}).$$

We use notation

$$\delta(y) = \begin{pmatrix} y & 0 \\ 0 & y^{-1} \end{pmatrix}$$

for $y \in F_\mathfrak{p}^x$ or $y \in \mathbb{J}$, and

$$\ell(y) = \begin{pmatrix} y & 0 \\ 0 & 1 \end{pmatrix}.$$

PROPOSITION Let $K = GL(2, \mathfrak{o}_\mathfrak{p})$ and $K_1 = SL(2, \mathfrak{o}_\mathfrak{p})$. The map $\alpha: H_1(\mathfrak{p}) \to H(\mathfrak{p})$ given by

$$\alpha: \text{characteristic function of } K_1 \delta(y) K_1 \to \text{characteristic function of } K \delta(y) K$$

is an injective convolution algebra homomorphism (where we give maximal compact subgroups measure one in either case). Let $\eta \in \mathscr{H}_1(\mathfrak{n})$ act on right $\tilde{\mathbb{K}}$-invariant functions f by

$$\eta: f \to f * \eta \qquad \text{(right convolution over } SL(2, F_\mathfrak{p})).$$

Then

$$\eta(f) = f * \alpha(\eta).$$

That is, the effect of right $SL(2, F_\mathfrak{p})$ convolution with η is the same as right $GL(2, F_\mathfrak{p})$ convolution with $\alpha(\eta)$. (Proof is easy and formal.) ■

Let $Z^+(F_\infty)$ be the connected component of the archimedean part of $Z(\mathbb{A})$. Define

$$\mathrm{Cfm}(\mathfrak{n}, \kappa) = \{\text{right } \tilde{\mathbb{K}}\text{-invariant holomorphic cuspforms of weight } \kappa \text{ on } GL(2, \mathbb{A})\}$$

$$\mathrm{Cfm}(\mathfrak{n}, \kappa, \omega) = \{\text{right } \tilde{\mathbb{K}}\text{-invariant holomorphic cuspforms of weight } \kappa \text{ with central character } \omega \text{ on } GL(2, \mathbb{A})\}.$$

We have the inner product $\langle\ ,\ \rangle$ on $\mathrm{Cfm}(\mathfrak{n}, \kappa, \omega)$ given by

$$\langle f_1, f_2 \rangle = \int_{Z(\mathbb{A})GL(2,F)\backslash GL(2,\mathbb{A})} f_1(g)\bar{f}_2(g)\,dg.$$

Let $\lambda\colon \mathscr{H}_1(\mathfrak{n}) \to \mathbb{C}$ be an algebra homomorphism; define

$$\mathrm{Cfm}(\mathfrak{n}, \kappa)_\lambda = \{f \in \mathrm{Cfm}(\mathfrak{n}, \kappa)\colon f * \alpha(\eta) = \lambda(\eta)f \text{ for all } \eta \in \mathscr{H}_1(\mathfrak{n})\}$$

$$\mathrm{Cfm}(\mathfrak{n}, \kappa, \omega)_\lambda = \{f \in \mathrm{Cfm}(\mathfrak{n}, \kappa, \omega)\colon f * \alpha(\eta) = \lambda(\eta)f \text{ for all } \eta \in \mathscr{H}_1(\mathfrak{n})\}.$$

Let $\lambda\colon \mathscr{H}(\mathfrak{n}) \to \mathbb{C}$ be an algebra homomorphism; define

$$\mathrm{Cfm}(\mathfrak{n}, \kappa, \omega, \lambda) = \{f \in \mathrm{Cfm}(\mathfrak{n}, \kappa, \omega)\colon f * \eta = \lambda(\eta)f \text{ for all } \eta \in \mathscr{H}(\mathfrak{n})\}.$$

Definition Let Ω be a subfield of $\mathbb{C}$. For either sort of algebra homomorphism λ, for a fixed weight κ, let $\Omega(\lambda)$ be the field generated over Ω by the set of values $\lambda(\eta_\mathfrak{p})$, where

$$\eta_\mathfrak{p} = N\mathfrak{p}^{\kappa/2} \times \text{characteristic function of } GL(2, \mathfrak{o}_\mathfrak{p})\ell(\pi^{-1})GL(2, \mathfrak{o}_\mathfrak{p}) \quad (\text{for } \mathscr{H}(\mathfrak{n}))$$

$$\eta_\mathfrak{p} = \text{characteristic function of } GL(2, \mathfrak{o}_\mathfrak{p})\delta(\pi)GL(2, \mathfrak{o}_\mathfrak{p}) \quad (\text{for } \mathscr{H}_1(\mathfrak{n}))$$

where π is a local parameter at $\mathfrak{p}$. (The normalizing constant in the $\mathscr{H}(\mathfrak{n})$ case, which depends on the weight, is necessary for preservation of rationality properties.)

For a character ω on $\mathbb{J}/F_\infty^+ F^\times$, and for a subfield Ω of $\mathbb{C}$, put

$$\Omega(\omega) = \text{field generated over } \mathbb{Q} \text{ by the values of } \omega.$$

Also, let $\Omega(\omega, \lambda) = \Omega(\omega)\Omega(\lambda)$, etc. For a $\mathbb{C}$-valued function ψ and an element σ of $\mathrm{Aut}(\mathbb{C}/\mathbb{Q})$, let ψ^σ be the function whose value $\psi^\sigma(x)$ at a point x in the domain of ψ is $\psi(x)^\sigma$.

Let $\mathbb{J}_\infty^+$ be the connected component of the identity in $\mathbb{J}_\infty$, and put $\mathbb{J}^+ = \mathbb{J}_\infty^+\mathbb{J}_0$. On elements $g \in GL(2, \mathbb{A})$ of the form

$$g = u(x)\ell(y) = \begin{pmatrix} 1 & x \\ 0 & 1 \end{pmatrix}\begin{pmatrix} y & 0 \\ 0 & 1 \end{pmatrix}$$

with $x \in \mathbb{A}$ and $y \in \mathbb{J}^+$, a holomorphic automorphic form f on $GL(2, \mathbb{A})$ has a *Fourier expansion*

$$f(g) = \|y\|^{\kappa/2} \sum_{\xi \in F} \tau(\xi x) V_\xi(y_0) \exp(-2\pi \operatorname{Tr} \xi_\infty y_\infty)$$

where $V_\xi(y_0)$ is a function of only the finite-prime parts y_0 and ξ_0 of y and ξ (see section 3.5).

Definition A holomorphic modular form f of weight κ is said to have *Fourier coefficients in a field* $\Omega \subset \mathbb{C}$ if $V_\xi(y_0) \in \Omega$ for all $\xi \in F$ and $y_0 \in \mathbb{J}_0$.

Definition On the ring of *formal* expressions

$$f(g) = \|y\|^{\kappa/2} \sum_{\xi \in F} \tau(\xi x) V_\xi(y_0) \exp(-2\pi \operatorname{Tr} \xi_\infty y_\infty)$$

with $x \in \mathbb{A}$ and $y \in \mathbb{J}^+$, $g = u(x)\ell(y)$, define an action of $\sigma \in \operatorname{Aut}(\mathbb{C}/\mathbb{Q})$ "*on Fourier coefficients*" by

$$f^\sigma(g) = \|y\|^{\kappa/2} \sum_{\xi \in F} \tau(\xi x) V_\xi(y_0)^\sigma \exp(-2\pi \operatorname{Tr} \xi_\infty y_\infty).$$

Remarks If $f \in \operatorname{Cfm}(\mathfrak{n}, \kappa)$ and if f^σ is the restriction to

$$\{u(x)\ell(y) \colon x \in \mathbb{A}, y \in \mathbb{J}^+\} \subset GL(2, \mathbb{A})$$

of an element φ of $\operatorname{Cfm}(\mathfrak{n}, \kappa)$, then φ is uniquely determined. This follows from the last results of section 3.5.

The proof in section 7.10 shows that the following theorem is valid for weight κ sufficiently large so that a Siegel's Eisenstein series on $Sp(2)$ is convergent; in section 7.2 we show that $\kappa \geq 7$ suffices; in fact, $\kappa \geq 4$ suffices. (See the remarks in section 7.2.)

THEOREM Let $\sigma \in \operatorname{Aut}(\mathbb{C}/\mathbb{Q})$.

i. $\operatorname{Cfm}(\mathfrak{n}, \kappa)$ is spanned by cuspforms having Fourier coefficients in $\mathbb{Q}$; therefore, $\operatorname{Cfm}(\mathfrak{n}, \kappa)$ is stable under the action of $\operatorname{Aut}(\mathbb{C}/\mathbb{Q})$ on Fourier coefficients.
ii. $\operatorname{Cfm}(\mathfrak{n}, \kappa, \omega)$ is spanned by cuspforms having Fourier coefficients in $\mathbb{Q}(\omega)$, and

$$\operatorname{Cfm}(\mathfrak{n}, \kappa, \omega)^\sigma = \operatorname{Cfm}(\mathfrak{n}, \kappa, \omega^\sigma).$$

iii. For each homomorphism $\lambda \colon \mathscr{H}_1(\mathfrak{n}) \to \mathbb{C}$, the space $\operatorname{Cfm}(\mathfrak{n}, \kappa, \omega)_\lambda$ is spanned by cuspforms having Fourier coefficients in $\mathbb{Q}(\omega, \lambda)$; $\mathbb{Q}(\omega, \lambda) \subset \bar{\mathbb{Q}}$ and $\mathbb{Q}(\omega, \lambda)$ is finitely generated over $\mathbb{Q}$. Further,

$$\mathrm{Cfm}(\mathfrak{n}, \kappa, \omega)_\lambda^\sigma = \mathrm{Cfm}(\mathfrak{n}, \kappa, \omega^\sigma)_{\lambda^\sigma}.$$

iv. If $f \in \mathrm{Cfm}(\mathfrak{n}, \kappa, \omega_1)_\lambda$ and $\varphi \in \mathrm{Cfm}(\mathfrak{n}, \kappa, \omega_2)_\lambda$ have Fourier coefficients in $\Omega(\omega_1, \omega_2, \lambda)$ for some field Ω, then

$$\langle f, f\rangle/\langle \varphi, \varphi\rangle \in \Omega(\omega_1, \omega_2, \lambda) \qquad \langle f, \varphi\rangle/\langle \varphi, \varphi\rangle \in \Omega(\omega_1, \omega_2, \lambda).$$

v. If f and φ are in $\mathrm{Cfm}(\mathfrak{n}, \kappa)$ and f is an eigenfunction for $\mathscr{H}_1(\mathfrak{n})$, then

$$\{\langle \varphi, f\rangle/\langle f, f\rangle\}^\sigma = \langle \varphi^\sigma, f^\sigma\rangle/\langle f^\sigma, f^\sigma\rangle.$$

vi. $\mathrm{Cfm}(\mathfrak{n}, \kappa, \omega, \lambda)$ is spanned by cuspforms having Fourier coefficients in $\mathbb{Q}(\omega, \lambda)$; $\mathbb{Q}(\omega, \lambda) \subset \bar{\mathbb{Q}}$ and $\mathbb{Q}(\omega, \lambda)$ is finitely generated over $\mathbb{Q}$. Further, for $\sigma \in \mathrm{Aut}(\mathbb{C}/\mathbb{Q})$,

$$\mathrm{Cfm}(\mathfrak{n}, \kappa, \omega, \lambda)^\sigma = \mathrm{Cfm}(\mathfrak{n}, \kappa, \omega^\sigma, \lambda^\sigma).$$

(Proof is given in Chapter 7, especially section 7.10.)

COROLLARY The space of right $\tilde{\mathbb{K}}$-invariant holomorphic modular forms of weight $(\kappa, \ldots, \kappa)$ is spanned by modular forms with Fourier coefficients in $\mathbb{Q}$. The space spanned by $\tilde{\mathbb{K}}$-invariant left holomorphic Eisenstein series of weight $(\kappa, \ldots, \kappa)$ is spanned by modular forms with Fourier coefficients in $\mathbb{Q}$.

Proof of Corollary From section 4.9, we know that this space of modular forms is a direct sum of the space of cuspforms and the space spanned by certain explicit Eisenstein series. By the theorem, it suffices to prove the second assertion of the corollary. Section 4.8 already shows that all but possibly the 0-th Fourier coefficients lie in a cyclotomic field Ω (depending on $\mathfrak{n}$) of finite degree over $\mathbb{Q}$, and have explicit Galois equivariance properties. When the based field F is $\mathbb{Q}$, the explicit calculation in section 4.8 and the results of Appendix A.5 on Dirichlet L-functions over $\mathbb{Q}$ show that the *0-th* Fourier coefficients have the same behavior, being special values of such L-functions. Then the Galois equivariance property asserts that these specific Eisenstein series are merely permuted among themselves by the action of elements of $\mathrm{Gal}(\Omega/\mathbb{Q})$. From this it is elementary that the space spanned by these Eisenstein series has a basis of modular forms with Fourier coefficients in $\mathbb{Q}$.

The same argument applies to arbitrary (totally real) base fields, granting the result of section 6.2 on L-functions over totally real number fields. ∎

Remarks We will use the theorem above with $F = \mathbb{Q}$ to prove the theorem of section 6.2. The special values result for $F = \mathbb{Q}$ is proven directly (i.e., without application of this theorem), as indicated in Appendix A.5. In particular, we note that there is no circularity in the argument.

COROLLARY Let φ be a holomorphic modular form of weight κ which is right $\tilde{\mathbb{K}}$-invariant with central character ω. Let f be in $\mathrm{Cfm}(\mathfrak{n}, \kappa, \omega)_\lambda$ for some λ: $\mathscr{H}_1(\mathfrak{n}) \to \mathbb{C}$.

Then

$$\{\langle \varphi, f\rangle / \langle f, f\rangle\}^\sigma = \langle \varphi^\sigma, f^\sigma\rangle / \langle f^\sigma, f^\sigma\rangle.$$

Proof This is a part of the theorem if φ is a cuspform. Since every holomorphic modular form is a sum of a cuspform and an Eisenstein series, and $\langle E, f\rangle = 0$ for any Eisenstein series E, it suffices to show that σ stabilizes the space of Eisenstein series. This follows from the result of section 6.2 (or Appendix A.5 for $F = \mathbb{Q}$) together with the explicit calculations in section 4.8. ■

Remarks Though we have chosen a particular cofinite family of congruence subgroups in the above discussion, this result may be adapted to give a corresponding result for several other families of congruence subgroups. It should be noted that the theorem is not at all true for arbitrary congruence subgroups; nevertheless, via some linear algebra, a general result can be inferred from the present theorem. What will happen is that there will be a basis of spaces of cuspforms with Fourier coefficients in a cyclotomic field depending upon the congruence subgroup.

Also, Shimura [Sh3] shows (in effect) that the space of cuspforms with cyclotomic Fourier coefficients is stable under the right translation action of $GL(2, \hat{\mathfrak{o}})$. We do not discuss this issue here, though the method of Chapter 7, combined with a variant of that method, can be made to yield this result. Likewise, the case that the weight is of a more general form than $(\kappa, \ldots, \kappa)$ can be dealt with by the method of Chapter 7. On the other hand, the case of low weight does not seem so tractable by that method, though this is completely treated by Shimura.

6.2 Special Values of *L*-Functions of Totally Real Number Fields

Historically, the result of this section was the first example of the use of an imbedding of a smaller group in a larger one and restriction of an automorphic form from the larger to the smaller. Subsequently, this mechanism has been seen to be very useful in a variety of settings.

By "holomorphic automorphic form" we mean only the class of automorphic forms of weights $(\kappa, \ldots, \kappa)$ considered in the previous section, and by "Fourier coefficient" we mean in the sense discussed there for holomorphic automorphic forms on $GL(2, \mathbb{A})$.

Let F be a totally real number field of degree $m > 1$ over $\mathbb{Q}$, with absolute discriminant D. Let ω be a finite-order Hecke character on F, and suppose that ω is either totally even or totally odd. Put $\delta = 0$ if ω is totally even, and $\delta = 1$ if ω is totally odd. Let $L(s, \omega)$ be the L-function attached to ω without the prime factors at which ω is ramified, and without infinite-prime factors. Let β be a finite idele so that $\beta\hat{\mathfrak{o}} = \mathfrak{n}\mathfrak{o}^{*-1}\hat{\mathfrak{o}}$, where $\mathfrak{n}$ is the finite-prime part of the conductor of ω, and define a Gauss sum by

$$\gamma(\omega) = \int \tau(\beta^{-1}x)\bar{\omega}(x)\,d^\times x \qquad \text{(integral over } \hat{\mathfrak{o}}^\times\text{)}$$

where $d^\times x$ denotes the standard additive Haar measure on $\mathbb{J}_0$. As usual, this integral is equal to a finite sum with the same argument, so certainly takes values in $\mathbb{Q}_{ab}$, the field generated over $\mathbb{Q}$ by roots of unity. Therefore, for $\sigma = \sigma_\beta$ in $\mathrm{Gal}(\mathbb{Q}_{ab}/\mathbb{Q})$ sending a root of unity ζ to ζ^α, we have the Galois equivariance property

$$\gamma(\omega)^\sigma = \omega^\sigma(\alpha)\gamma(\omega^\alpha).$$

Further, standard methods show that $\gamma(\omega)$ *is non-zero* (e.g., by computing its absolute value). Let $\mathbb{Q}(\omega)$ be the field generated over $\mathbb{Q}$ by the values of ω.

Remarks In the following theorem, we can truly claim only to have proven the result for $(2k + \delta)m \geq 7$, because in section 7.2 (in the proof of section 6.1) we prove only convergence of Siegel's Eisenstein series for weight $\kappa \geq 7$. As remarked in sections 6.1 and 7.2, the Siegel's Eisenstein series can be shown to be convergent for $\kappa \geq 4$. With this technical improvement, the following holds for $(2k + \delta)m \geq 4$; that is, for all $k > 0$.

THEOREM For $0 < k \in \mathbb{Z}$

$$L(2k + \delta, \omega)/i^{\kappa m}\pi^{(2k+\delta)m}\gamma(\omega) \in \mathbb{Q}(\omega)$$

and there is the Galois equivariance property

$$\{L(2k + \delta, \omega)/i^{\kappa m}\pi^{(2k+\delta)m}\gamma(\omega)\}^\sigma = L(2k + \delta, \omega^\sigma)/i^{\kappa m}\pi^{(2k+\delta)m}\gamma(\omega^\sigma).$$

Proof We use the notation of section 4.8 regarding the special Eisenstein series $E(g; \kappa, 0, 1, \mathfrak{n}; \tfrac{1}{2}\kappa, \bar{\omega}, \mathbb{1})$ introduced in that section. From the second theorem of that section, we have an explicit Fourier expansion of certain Eisenstein series: Let $g \in GL(2, \mathbb{A})$ be of the form

$$g = u(x)\delta(y) = \begin{pmatrix} 1 & x \\ 0 & 1 \end{pmatrix}\begin{pmatrix} y & 0 \\ 0 & 1 \end{pmatrix}$$

with $x \in \mathbb{A}$ and $y \in \mathbb{J}^+$. Put $\kappa = 2k + \delta$. Then (in slightly different notation than in section 4.8)

$$\|y\|^{-\kappa/2}E(g; \kappa, 0, 1, \mathfrak{n}; \tfrac{1}{2}\kappa, \bar{\omega}, \mathbb{1}) =$$

$$= c\pi^{-\kappa m}L(\kappa, \omega) + (-2i)^{\kappa m}D^{-1/2}\sum_{\xi} c(\xi_0 y_0; \omega)\tau(\xi x)\exp(-2\pi \operatorname{Tr} \xi_\infty y_\infty)$$

where $c \in \mathbb{Q}^\times$, $c(\xi_0 y_0; \bar{\omega}) \in \mathbb{Q}_{ab}$, and ξ is summed over totally positive elements of $F^\times$. Further, for $\sigma = \sigma_\alpha$ in $\mathrm{Gal}(\mathbb{Q}_{ab}/\mathbb{Q})$ given by $\sigma\colon \zeta \to \zeta^\alpha$ for a root of unity ζ, we have a Galois equivariance property of the higher Fourier coefficients:

$$c(\xi_0 y_0; \omega)^\sigma = \omega^\sigma(\alpha) c(\xi_0 y_0; \omega^\sigma).$$

We need a variant corollary of the results of the previous section:

PROPOSITION Let f be a holomorphic modular form on $GL(2, \mathbb{A})$ *over* $\mathbb{Q}$. Then the values of the 0-th Fourier coefficient of f lie in the field generated over $\mathbb{Q}$ by the values of the other Fourier coefficients.

Proof of Proposition Let $\sigma \in \mathrm{Aut}(\mathbb{C}/\mathbb{Q})$ be an automorphism which is trivial on the higher Fourier coefficients of f. From the results of the previous section *for base field* $\mathbb{Q}$, f^σ is still an automorphic form, where σ acts on the Fourier coefficients; then $f - f^\sigma$ is an automorphic form, and has all but possibly its 0-th Fourier coefficient identically zero. From section 3.5, this implies that the constant is 0, and therefore that the 0-th Fourier coefficient of $f - f^\sigma$ must vanish identically. ∎

Now consider the obvious imbedding ι: $GL(2, \mathbb{Q}) \to GL(2, F)$. This extends to an imbedding of the adele groups. Further, the pullback of a holomorphic automorphic form of weight $(\kappa, \ldots, \kappa)$ for $GL(2, F)$ is readily verified to be a holomorphic automorphic form of weight $m\kappa$ for $GL(2, \mathbb{Q})$.

PROPOSITION Let f be a holomorphic automorphic form for $GL(2, F)$, and let $\iota^* f$ be its pullback to an automorphic form for $GL(2, \mathbb{Q})$. Then the 0-th Fourier coefficient of $\iota^* f$ is the (restriction of) the 0-th Fourier coefficient of f, and the higher Fourier coefficients of $\iota^* f$ are finite sums of (restrictions of) higher Fourier coefficients of f.

Proof of Proposition In such an assertion, we consider only elements g of $GL(2, \mathbb{A}_\mathbb{Q})$ of the form

$$g = u(x)\ell(y) = \begin{pmatrix} 1 & x \\ 0 & 1 \end{pmatrix} \begin{pmatrix} y & 0 \\ 0 & 1 \end{pmatrix}$$

with $x \in \mathbb{A}_\mathbb{Q}$ and $y \in \mathbb{J}_\mathbb{Q}^+$. As has been remarked before, the Koecher principle (treated classically in section 1.4) applies f: The ξ-th Fourier coefficient is 0 unless $\xi = 0$ or ξ is totally positive.

Suppose that f is right $\mathbb{K}$-invariant where

$$\mathbb{K} = \{g \in GL(2, \hat{\mathfrak{o}}) \colon g \equiv 1 \text{ modulo } \mathfrak{n}\}.$$

Then, for fixed finite-prime part y_0 of y, for $v \in \mathfrak{n}\hat{\mathfrak{o}}$, $x \in \mathbb{A}_F$ and $y \in \mathbb{J}_F^+$ we have

$$f(u(x)\ell(y)u(v)) = f(u(x + y_0^{-1}v)\ell(y)).$$

This implies that (for fixed y_0) the only non-zero Fourier coefficient indices ξ must lie in $y_0^{-1}\mathfrak{n}^{-1}\mathfrak{o}^*$.

Let n be a non-negative rational number. Then the n-th Fourier coefficient of $\iota^* f$ is clearly the sum of (the restrictions of) the ξ-th Fourier coefficients of ξ as ξ runs over elements of F whose trace (from F to $\mathbb{Q}$) is n. For each fixed y_0 and n, the set

$$X(y_0, n) = \{\xi \in F\colon \mathrm{Trace}_{F/\mathbb{Q}}\, \xi = n,\ \xi \in y_0^{-1}\mathfrak{n}^{-1}\mathfrak{d}^*,\ \xi \text{ totally positive or } 0\}$$

is *finite*. Further, the only element of $X(y_0, 0)$ is obviously $\xi = 0$, and $\xi = 0$ does not lie in any other set $X(y_0, n)$. This gives the assertion of the proposition. ■

This discussion applies to

$$f_\omega(g) = i^{-\kappa m} D^{-1/2} E(g; \kappa, 0, 1, \mathfrak{n}; \tfrac{1}{2}\kappa, \bar{\omega}, \mathbb{1}),$$

which has 0-th Fourier coefficient (in our present sense)

$$L(\kappa, \omega)/i^{\kappa m} D^{1/2}$$

and has higher Fourier coefficients $c_\xi(y_0; \omega)$ with values lying in $\mathbb{Q}_{ab}$ and satisfying the Galois equivariance condition

$$c_\xi(y_0; \omega)^\sigma = \omega^\sigma(\alpha) c_\xi(y_0; \omega^\sigma),$$

where $\sigma = \sigma_\alpha \in \mathrm{Gal}(\mathbb{Q}_{ab}/\mathbb{Q})$ sends a root of unity ζ to ζ^α. Therefore, we find that

$$L(\kappa, \omega)/i^{\kappa m} D^{1/2} \in \mathbb{Q}_{ab}.$$

Since the Gauss sum $\gamma(\omega)$ has the same Galois equivariance property as $c_\xi(y_0; \omega)$, with $\sigma = \sigma_\alpha$ in $\mathrm{Gal}(\mathbb{Q}_{ab}/\mathbb{Q})$ we have

$$\{c_\xi(y_0; \omega)/\gamma(\omega)\}^\sigma = c_\xi(y_0; \omega^\sigma)/\gamma(\omega^\sigma).$$

Therefore,

$$c_\xi(y_0; \omega^\sigma)/\gamma(\omega^\sigma) \in \mathbb{Q}(\omega).$$

By the two propositions above, it follows that

$$L(\kappa, \omega)/i^{\kappa m} D^{1/2} \gamma(\omega) \in \mathbb{Q}(\omega).$$

This is the first assertion of the theorem.

For the equivariance assertion, consider the holomorphic automorphic form

$$(f_\omega)^\sigma - f_{\omega^\sigma}$$

for $\sigma \in \mathrm{Gal}(\mathbb{Q}_{ab}/\mathbb{Q})$, where the action is on Fourier coefficients. By the Galois equivariance of the higher Fourier coefficients, this automorphic form has at most one non-vanishing Fourier coefficient, the 0-th one. As in the proof of the first proposition, this implies that this 0-th Fourier coefficient is 0. The yields the Galois equivariance assertion of the theorem. ■

Now we consider a special case which is of importance for application to discussion of rationality of Fourier coefficients of Siegel's Eisenstein series (on symplectic groups), considered in sections 7.1–7.8. (Again, by our restriction on weights, for $m = [F:\mathbb{Q}] > 1$ we must take $k \geq 7/m$ in the following corollary if we insist upon having an altogether self-contained proof.)

COROLLARY Let $\alpha \in F^{\times}$ so that $\alpha\mathfrak{o}$ is prime to $2\mathfrak{o}^*$, and let ω be idele class group character corresponding to the quadratic Dirichlet character

$$\mathfrak{a} \to (\alpha/\mathfrak{a})_2 = \begin{cases} 1 & \text{if } \alpha \text{ is a unit modulo } (N\mathfrak{a})\mathfrak{o}, \text{ and is a square modulo } \mathfrak{a} \\ -1 & \text{if } \alpha \text{ is a unit modulo } (N\mathfrak{a})\mathfrak{o}, \text{ and is a non-square modulo } \mathfrak{a} \\ 0 & \text{if } \alpha \text{ is not a unit modulo } (N\mathfrak{a})\mathfrak{o}. \end{cases}$$

Then, if α is totally positive and k is even, or if α is totally negative and k is odd,

$$L(k,\omega)/\pi^k D^{1/2}(\mathrm{Norm}_{F/\mathbb{Q}}\,\alpha)^{1/2} \in \mathbb{Q}.$$

Proof By the previous theorem, all that is required is to show that

$$(\mathrm{Norm}_{F/\mathbb{Q}}\,\alpha)^{1/2}/\gamma(\omega) \in \mathbb{Q}.$$

There is the identity

$$\gamma(\omega)^{\sigma} = \omega^{\sigma}(\beta)\gamma(\omega^{\sigma}) = \omega(\beta)\gamma(\omega)$$

(with $\sigma = \sigma_\beta$ in $\mathrm{Gal}(\mathbb{Q}_{ab}/\mathbb{Q})$). We will show that $\mathrm{Norm}_{F/\mathbb{Q}}\,\alpha)^{1/2}$ has exactly the same Galois equivariance, so that the ratio is in $\mathbb{Q}$. Let ζ be a root of unity so that both $(\mathrm{Norm}_{F/\mathbb{Q}}\,\alpha)^{1/2}$ and $\gamma(\omega)$ lie in $\mathbb{Q}(\zeta)$. By Dirichlet's theorem on primes in arithmetic progressions, it suffices to consider only elements of $\mathrm{Gal}(\mathbb{Q}_{ab}/\mathbb{Q})$ given by

$$\zeta \to \zeta^p$$

where p is an odd rational prime which is prime to $\alpha\mathfrak{o}$ and to the conductor of ω. We claim that

$$(\alpha/p)_{2,F} = (\mathrm{Norm}_{F/\mathbb{Q}}\,\alpha/p)_{2,\mathbb{Q}}$$

where $(*/*)_{2,\Omega}$ denotes the quadratic symbol for a field Ω. This would yield the assertion.

More generally, we claim that

$$(\alpha/\mathfrak{p})_{2,F} = (\mathrm{Norm}_{F/\Omega}\,\alpha/\mathfrak{p})_{2,\Omega}$$

for any finite extension F of a number field Ω, $\alpha \in F$, and a prime $\mathfrak{p}$ of Ω which is prime to α and does not ramify in F/Ω. First, consider the case that F is Galois over Ω. Let $\mathfrak{p}$ factor in F as $\prod_{i=1,\ldots,g} \mathfrak{P}_i$. Let $\{\sigma\}$ be a collection of representatives for $G_{\mathfrak{P}_1}\backslash G$ where G is the Galois group of F over Ω and $G_{\mathfrak{P}_1}$ is the decomposition group of $\mathfrak{P}_1$. Then

$$(\alpha/\mathfrak{p})_{2,F} = \prod_i (\alpha/\mathfrak{P}_i)_{2,F} = \prod_\sigma (\alpha^\sigma/\mathfrak{P}_1)_{2,F} \equiv \prod_\sigma (\alpha^\sigma)^{(N\mathfrak{P}_1-1)/2} \quad \text{modulo } \mathfrak{P}_1.$$

Therefore,

$$(\alpha/\mathfrak{p})_{2,F} \equiv \prod_\sigma ((\alpha^\sigma)^{(N\mathfrak{P}_1-1)/(N\mathfrak{p}-1)})^{(N\mathfrak{p}-1)/2} \quad \text{modulo } \mathfrak{P}_1.$$

Now

$$\prod_\sigma (\alpha^\sigma)^{(N\mathfrak{P}_1-1)/(N\mathfrak{p}-1)} \equiv \mathrm{Norm}_{F/\Omega}\,\alpha(\mathrm{mod}\ \mathfrak{P}_1) \equiv \mathrm{Norm}_{F/\Omega}\,\alpha(\mathrm{mod}\ \mathfrak{p}).$$

Therefore,

$$(\alpha/\mathfrak{p})_{2,F} = (\mathrm{Norm}_{F/\Omega}\,\alpha/\mathfrak{p})_{2,\Omega}$$

as asserted, when F/Ω is Galois.

When F/Ω is not Galois, let E be the Galois closure of F over Ω; let $\mathfrak{O}$ be the ring of integers of E. Let $\{\mathfrak{p}_i\}$ be the primes of F over the prime $\mathfrak{p}$ of Ω, let $\mathfrak{o}$ be the ring of integers in F, and let $\{\mathfrak{P}_{ij}\}$ be the primes of E over $\mathfrak{p}_i$. We wish to find $\tilde{\alpha}$ in $\mathfrak{O}$ so that

$$\mathrm{Norm}_{E/F}\,\tilde{\alpha} \equiv \alpha \text{ modulo } \prod_i \mathfrak{p}_i.$$

Let $\alpha_i \in \mathfrak{O}$ be such that the norm map $\mathfrak{O}/\mathfrak{P}_{i1} \to \mathfrak{o}/\mathfrak{p}_i$ sends α_i to α (using the surjectivity of such a norm map). Via the Chinese Remainder Theorem, we can choose α_i so that $\alpha_i \equiv 1$ modulo $\mathfrak{P}_{ij}$ for $j \neq i$. Then, by the Chinese Remainder Theorem, we may find $\tilde{\alpha}$ so that $\tilde{\alpha} \equiv \alpha_i$ modulo $\mathfrak{p}_i$, yielding the desired congruence. Using the result in the Galois extension case, we have

$$\begin{aligned}(\alpha/\mathfrak{p})_{2,F} &= \prod_i (\alpha/\mathfrak{p}_i)_{2,F} = \prod_i (\mathrm{Norm}_{E/F}\,\tilde{\alpha}/\mathfrak{p}_i)_{2,F} = \prod_i (\tilde{\alpha}/\mathfrak{p}_i)_{2,E} \\ &= \prod_{ij} (\tilde{\alpha}/\mathfrak{P}_{ij})_{2,E} = (\tilde{\alpha}/\mathfrak{p})_{2,E} = (\mathrm{Norm}_{E/\Omega}\,\tilde{\alpha}/\mathfrak{p})_{2,\Omega}.\end{aligned}$$

This finishes the proof of the claim, and so finishes the proof of the corollary. ■

6.3 Special Values of Grossencharacter L-Functions

Let K be a totally imaginary quadratic extension of the totally real number field F; let $\mathfrak{O}$ denote the ring of integers of K. We will be interested in grossencharacter L-functions of K of the form

$$L_K(s, \chi) = \prod_{\mathfrak{P}} 1/(1 - \chi(\mathfrak{P}) N\mathfrak{P}^{-s}) \qquad \text{(product over primes of } K \text{ unramified for } \chi)$$

where the grossencharacter χ *is of a special type*: We require that there is an integer n, and a choice of extensions $\tilde{\sigma}_i$ of the real imbeddings σ_i of F to complex imbeddings of K, so that

$$\chi(\alpha_\infty) = \left(\prod_i (\tilde{\sigma}_i(\alpha)/|\tilde{\sigma}_i(\alpha)|\right)^n$$

where α_∞ is the infinite-prime part of $\alpha \in K \subset \mathbb{J}_K$.

Further, to eliminate some technicalities which do not illustrate the main point, we *assume that* $\mathfrak{o}$ *has narrow class number* 1—i.e., that $\hat{\mathfrak{o}}^\times \backslash \mathbb{J}/F^\times F_\infty^+ \approx \{1\}$. This assumption is certainly not necessary.

For each such χ, let θ be the pluriharmonic theta series associated to χ as in section 5.7, as follows. Let $\mathfrak{m} \subset \mathfrak{O}$ be the finite-prime part of the conductor of χ. By Dirichlet's theorem on primes in an arithmetic progression, we may choose representative $\mathfrak{a}_i$ for the absolute ideal class group of $\mathfrak{O}$ so that each $\mathfrak{a}_i$ is prime to $\mathfrak{m}$. For $z \in \mathfrak{H}^m$, define a pluriharmonic theta series

$$\theta(z) = c_0 + \sum_i \chi(\mathfrak{a}_i) N\mathfrak{a}_i^{n/2} \sum_\alpha \chi(\alpha_\infty) \|\alpha_\infty\|_K^{n/2} \exp(2\pi i N\mathfrak{a}_i \operatorname{Tr} z\alpha\alpha')$$

where $\alpha \to \alpha'$ is the non-trivial Galois automorphism of K over F, $N\mathfrak{a}_i$ is the ideal norm of $\mathfrak{a}_i$, α_∞ is the infinite-prime part of $\alpha \in K^\times \subset \mathbb{J}_K$, $\| \ \|_K$ is the idele norm on K, $c_0 = 0$ if χ is non-trivial and c_0 is the class number of $\mathfrak{O}$ if χ is trivial, and for fixed $\mathfrak{a}_i$ the inner sum is over

$$\{\alpha \in K^\times: \alpha\mathfrak{a}_i \subset \mathfrak{O}, \alpha \text{ prime to } \mathfrak{m}\}.$$

We renormalize this slightly for present purposes. Let δ be a totally positive element of $F^\times$ so that δ generates the inverse different of $\mathfrak{o}$ over $\mathbb{Z}$.

Definition For χ as above and $z \in \mathfrak{H}^m$, put

$$\Theta(z; \chi) = \theta(\delta z).$$

PROPOSITION: For sufficiently small non-zero ideal $\mathfrak{n}$ of $\mathfrak{o}$, the theta series $\Theta(z; \chi)$ is a modular form of weight $(n + 1, \ldots, n + 1)$ for the congruence subgroup

$$\Gamma_0(\mathfrak{n}) = \left\{g = \begin{pmatrix} a & b \\ c & d \end{pmatrix} \in SL(2, \mathfrak{o}): c \equiv 0 \text{ modulo } \mathfrak{n}\right\}$$

with character

$$\begin{pmatrix} a & b \\ c & d \end{pmatrix} \to \chi(d)\,\mathrm{sgn}(d)^1(\Delta/d)_2$$

where K is generated by the square root of the totally negative element Δ of F. If χ is non-trivial, then $\Theta(z;\chi)$ is a cuspform.

Proof From the theorem of section 5.7 and the result of section 5.6, the assertions of this proposition hold for θ. It follows immediately that $\Theta(z;\chi)$ is a modular form for $\Gamma_0(\mathfrak{n})$ conjugated by

$$\begin{pmatrix} \delta & 0 \\ 0 & 1 \end{pmatrix},$$

with the same character. Now, for $x \in \mathfrak{o}$ and $\alpha\mathfrak{a}_i \subset \mathfrak{O}$, it is elementary that

$$\delta x N\mathfrak{a}_i \alpha\alpha' \in \mathfrak{o}^*$$

from which it follows that $\mathrm{Tr}(\delta x N\mathfrak{a}_i \alpha\alpha') \in \mathbb{Z}$. From this, it is immediate (from the definition) that Θ remains invariant under the group

$$\left\{\begin{pmatrix} 1 & x \\ 0 & 1 \end{pmatrix} : x \in \mathfrak{o}\right\}.$$

This group, together with the conjugate of $\Gamma_0(\mathfrak{n})$, certainly generates a congruence subgroup containing $\Gamma_0(\mathfrak{n})$. ■

Now we make use of the hypothesis that the narrow class number is 1. As pointed out in section 3.1, this allows us to define a holomorphic modular form $\tilde{\Theta}(g;\chi)$ of weight $n+1$ on $GL(2,\mathbb{A})$ by

$$\tilde{\Theta}(\gamma\zeta u(x)\delta(y)k_0) = \omega(\zeta)y^{(n+1)/2}\Theta(x+iy)$$

where

ω is the Hecke character on $\mathbb{J}_F$ associated to the $\alpha \to \chi(\alpha)\,\mathrm{sgn}(\alpha_\infty)^1(\Delta/\alpha_0)_2$

$$u(x) = \begin{pmatrix} 1 & x \\ 0 & 1 \end{pmatrix} \qquad \text{with } x \in F_\infty$$

$$\delta(y) = \begin{pmatrix} y^{1/2} & 0 \\ 0 & y^{-1/2} \end{pmatrix} \qquad \text{with } y \in F_\infty^+$$

$\gamma \in GL(2, F)$

$$k_0 = \begin{pmatrix} a & b \\ c & d \end{pmatrix} \in GL(2, \hat{\mathfrak{o}}) \qquad \text{with } c \equiv 0 \text{ and } d \equiv 1 \text{ modulo } \mathfrak{n}$$

$$\zeta \in Z(\mathbb{A}) \approx \mathbb{J}_F.$$

In particular, $\tilde{\Theta}$ is right-invariant under

$$\left\{ \begin{pmatrix} a & b \\ c & d \end{pmatrix} \in GL(2, \hat{\mathfrak{o}}) \colon c \equiv 0 \text{ and } d \equiv 1 \text{ modulo } \mathfrak{n} \right\}.$$

Note that ω is a *finite-order* character on $\mathbb{J}_F$, by the hypotheses on χ.

Remarks Given the imbeddings $\tilde{\sigma}_i$, there is an obvious divisibility condition on the integer n to avoid the possibility that Θ be identically 0, coming from the existence of roots of unity in $\mathfrak{O}$. For example, consideration of -1 imposes a non-trivial condition if $m = [F : \mathbb{Q}]$ is odd: n must be even.

For a character χ as above, let σ be the non-trivial Galois automorphism of K over F, and define

$$\chi^\sigma = \chi \circ (\text{the extension of } \sigma \text{ to } \mathbb{J}).$$

Let χ_1 and χ_2 be two characters as above, with associated integers n_1 and n_2 as above. Let $\mathfrak{n}_i$ be the finite-prime part of the conductor of χ_i, and let ω_i be associated to χ_i as just above. Suppose that $\mathfrak{n}_1$ and $\mathfrak{n}_2$ are relatively prime. Let $\Theta_i = \Theta(*; \chi_i)$ be as above. Let $\gamma(\omega)$ be the exponential sum attached to a character ω as in section 6.2. Let Ω be the field generated over $\mathbb{Q}$ by the values of χ_1 and χ_2 on the finite ideles of K. Let D be the absolute discriminant of $\mathfrak{o}$. Then we have the following theorem.

THEOREM Suppose that $n_2 - n_1 \geq 2$, and that $n_2 - n_1 > 2$ if $\omega_1 = \omega_2$ and $F = \mathbb{Q}$. Suppose that neither $\tilde{\Theta}_1$ nor $\tilde{\Theta}_2$ is identically 0. Then

$$A(\chi_1, \chi_2) = \pi^{-m(n_2 - n_1)} \pi^{-mn_2} D^{1/2} \gamma(\bar{\omega}_1 \omega_2)^{-1} i^{m(n_2 - n_1)} \times$$

$$\times L_K(*, \chi_1 \bar{\chi}_2) L_K(*, \chi_1 \bar{\chi}_2^\sigma) / \langle \tilde{\Theta}_2, \tilde{\Theta}_2 \rangle$$

lies in Ω. Further, for $\tau \in \mathrm{Aut}(\mathbb{C}/\mathbb{Q})$,

$$A(\chi_1, \chi_2)^\tau = A(\chi_1^\tau, \chi_2^\tau).$$

Remarks To have a complete proof of this via section 6.1, we must require that $n_2 + 1 \geq 7$, due (in effect) to our soft estimate in Chapter 7 concerning convergence of Siegel's Eisenstein series. A better estimate of that convergence allows the theorem of section 6.1 to apply when $n_2 + 1 \geq 4$. In Shimura's treatment of this, no such inequality is necessary. Further, this is far from the most general case Shimura considers, as mentioned earlier.

Proof This is application of the integral representation of section 4.10 together with the theorem of section 6.1, further playing upon a coincidence.

Let $\mathfrak{n}$ be the least common multiple of $\mathfrak{n}_1$ and $\mathfrak{n}_2$. Put

$$\mathbb{K} = \left\{ g \in GL(2, \mathfrak{o}) : g = \begin{pmatrix} a & b \\ c & d \end{pmatrix} \text{ with } c \equiv 0, d \equiv 1 \text{ modulo } \mathfrak{n} \right\}$$

$$\mathbb{K}_i = \left\{ g \in GL(2, \hat{\mathfrak{o}}) : g = \begin{pmatrix} a & b \\ c & d \end{pmatrix} \text{ with } c \equiv 0, d \equiv 1 \text{ modulo } \mathfrak{n}_i \right\}.$$

Both $\tilde{\Theta}_1$ and $\tilde{\Theta}_2$ are right-invariant by $\mathbb{K}$.

We certainly have not shown that $\tilde{\Theta}_1$ and $\tilde{\Theta}_2$ are holomorphic *normalized Hecke eigenforms* (in the sense of section 3.5). Nevertheless, by the remark adjacent to the theorem in section 4.10, all we really need to apply section 4.10 is right $\mathbb{K}$-invariance and a particular form of the Fourier coefficients on elements of the special form

$$\ell(y) = \begin{pmatrix} y & 0 \\ 0 & 1 \end{pmatrix}.$$

We will now see that the Fourier coefficients do have the required form.

What we must require is the following (as in section 4.10). In the Fourier expansions

$$\tilde{\Theta}_2(g) = \sum_{\xi \in F^\times} W_2(\ell(\xi)g)$$

$$\tilde{\Theta}_1(g) = (\text{0-th Fourier coefficient}) + \sum_{\xi \in F^\times} W_1(\ell(\xi)g)$$

with $y \in \mathbb{J}^+$ it must be that

$$W_i(\ell(y)) = \|y\|^{(n_i+1)/2} \exp(-2\pi \operatorname{Tr} y_\infty) V_i(y_o)$$

and that V_i factors over finite primes as

$$V_i(y_o) = \prod_{\mathfrak{p} < \infty} V_{i\mathfrak{p}}(y_\mathfrak{p})$$

and

$$V_{i\mathfrak{p}}(1) = 1.$$

Further, it must be that each factor

$$W_{i\mathfrak{p}}(\ell(y_\mathfrak{p})) = \|y_\mathfrak{p}\|^{(n_i+1)/2} V_{i\mathfrak{p}}(y_\mathfrak{p})$$

of "the" (non-zero) Fourier coefficient has an explicit form as in section 2.5. For $\mathfrak{p}$ not dividing $\mathfrak{n}_i$ it must be

$$W_{i\mathfrak{p}}(\ell(\delta_\mathfrak{p} y_\mathfrak{p})) = W_{i\mathfrak{p}}(\ell(\delta_\mathfrak{p}))(\alpha^{n+1} - \beta^{n+1})/(\alpha - \beta) \qquad (\text{if } y_\mathfrak{p}\mathfrak{o}_\mathfrak{p} = \mathfrak{p}^n\mathfrak{o}_\mathfrak{p} \text{ and } n \geq 0)$$

$$0 \qquad (\text{if } y_\mathfrak{p}\mathfrak{o}_\mathfrak{p} = \mathfrak{p}^n\mathfrak{o}_\mathfrak{p} \text{ and } n < 0)$$

for some complex numbers α and β depending on i and $\mathfrak{p}$ satisfying

$$\alpha\beta = N\mathfrak{p}^{-1}\omega_i(\mathfrak{p}).$$

For $\mathfrak{p}$ not dividing $\mathfrak{n}_i$ it must be

$$W_{i\mathfrak{p}}(\ell(\delta_\mathfrak{p} y_\mathfrak{p})) = W_{i\mathfrak{p}}(\ell(\delta_\mathfrak{p}))\alpha^{n+1} \qquad (\text{if } y_\mathfrak{p}\mathfrak{o}_\mathfrak{p} = \mathfrak{p}^n\mathfrak{o}_\mathfrak{p} \text{ and } n \geq 0)$$

$$0 \qquad (\text{if } y_\mathfrak{p}\mathfrak{o}_\mathfrak{p} = \mathfrak{p}^n\mathfrak{o}_\mathfrak{p} \text{ and } n < 0)$$

for some complex numbers α depending on i and $\mathfrak{p}$. In terms of the functions $V_{i\mathfrak{p}}$ *with the normalization* $V_{i\mathfrak{p}}(1) = 1$ this is as follows. For $\mathfrak{p}$ not dividing $\mathfrak{n}_i$:

$$V_{i\mathfrak{p}}(\delta_\mathfrak{p} y_\mathfrak{p}) = (\alpha_{i\mathfrak{p}}^{n+1} - \beta_{i\mathfrak{p}}^{n+1})/(\alpha_{i\mathfrak{p}} - \beta_{i\mathfrak{p}}) \qquad (\text{if } y_\mathfrak{p}\mathfrak{o}_\mathfrak{p} = \mathfrak{p}^n\mathfrak{o}_\mathfrak{p} \text{ and } n \geq 0)$$

$$0 \qquad (\text{if } y_\mathfrak{p}\mathfrak{o}_\mathfrak{p} = \mathfrak{p}^n\mathfrak{o}_\mathfrak{p} \text{ and } n < 0)$$

where

$$\alpha_{i\mathfrak{p}} = N\mathfrak{p}^{(n_i+1)/2}\alpha \qquad \beta_{i\mathfrak{p}} = N\mathfrak{p}^{(n_i+1)/2}\beta$$

in terms of the parameters α and β in $W_{i\mathfrak{p}}$. Therefore, we would have

$$\alpha_{i\mathfrak{p}}\beta_{i\mathfrak{p}} = \omega_i(\mathfrak{p})N\mathfrak{p}^{n_i}.$$

For $\mathfrak{p}$ dividing $\mathfrak{n}_i$:

$$V_{i\mathfrak{p}}(\delta_\mathfrak{p} y_\mathfrak{p}) = \alpha_{i\mathfrak{p}}^n \qquad (\text{if } y_\mathfrak{p}\mathfrak{o}_\mathfrak{p} = \mathfrak{p}^n\mathfrak{o}_\mathfrak{p} \text{ and } n \geq 0)$$

$$0 \qquad (\text{if } y_\mathfrak{p}\mathfrak{o}_\mathfrak{p} = \mathfrak{p}^n\mathfrak{o}_\mathfrak{p} \text{ and } n < 0)$$

where $\alpha_{i\mathfrak{p}} = N\mathfrak{p}^{(n_i+1)/2}\alpha$, in terms of the parameter α in $W_{i\mathfrak{p}}$.

It is almost immediate that these requirements are fulfilled: by construction of Θ_i, the classical Fourier coefficients have the right form; using the class number hypothesis assures that the adelic Fourier coefficients of $\tilde{\Theta}_i$ do as well, by the definition of $\tilde{\Theta}_i$. In particular, it is easy to verify directly that:

i. At $\mathfrak{p}$ not dividing $\mathfrak{n}_i$, with $\mathfrak{p}$ splitting as $\mathfrak{p}\mathfrak{O} = \mathfrak{P}_1\mathfrak{P}_2$ in the extension K over F,

$$(1 - \alpha_{i\mathfrak{p}}t)(1 - \beta_{i\mathfrak{p}}t) = (1 - \chi_i(\mathfrak{P}_1)N\mathfrak{p}^{n_i/2}t)(1 - \chi_i(\mathfrak{P}_2)N\mathfrak{p}^{n_i/2}t).$$

ii. At $\mathfrak{p}$ not dividing $\mathfrak{n}_i$, with $\mathfrak{p}$ inert ($\mathfrak{p}\mathfrak{O} = \mathfrak{P}$) in the extension K over F,

$$(1 - \alpha_{i\mathfrak{p}}t)(1 - \beta_{i\mathfrak{p}}t) = (1 - \chi_i(\mathfrak{P})N\mathfrak{p}^{n_i}t^2).$$

iii. At $\mathfrak{p}$ not dividing $\mathfrak{n}_i$, with $\mathfrak{p}$ ramified ($\mathfrak{p}\mathfrak{O} = \mathfrak{P}^2$) in the extension,

$$(1 - \alpha_{i\mathfrak{p}}t)(1 - \beta_{i\mathfrak{p}}t) = (1 - \chi_i(\mathfrak{P})N\mathfrak{p}^{n_i/2}t).$$

iv. at $\mathfrak{p}$ dividing $\mathfrak{n}_i$, in effect both $\alpha_{i\mathfrak{p}}$ and $\beta_{i\mathfrak{p}}$ are 0.

If a finite-prime $\mathfrak{p}$ does not divide $\mathfrak{n}$, say that $\mathfrak{p}$ is good; if $\mathfrak{p}$ does divide $\mathfrak{n}$, then say that $\mathfrak{p}$ is bad. Let $s' = s - 1 + \frac{1}{2}(n_1 + n_2 + 2)$, $\omega = \omega_2/\omega_1$, $\kappa = n_2 - n_1$. In the notation of section 4.10, we will make use of the Eisenstein series

$$E(g; s; \omega) = E^*(g; \kappa; 0, 1, \mathfrak{n}; s, \omega, \mathbb{1}).$$

With hypotheses as above, for $s \in \mathbb{C}$ away from poles of the Eisenstein series $E(*; s)$, by section 4.10 we conclude that

$$\langle \tilde{\Theta}_1 E(*; s; \omega), \tilde{\Theta}_2 \rangle$$

is a non-zero rational multiple of

$$\|\delta\|^{s'} \times (\pi^{-(s+\kappa/2)}\Gamma(s + \tfrac{1}{2}\kappa)(4\pi)^{-s'}\Gamma(s'))^m \times$$

$$\times \prod_{\mathfrak{p}\text{ good}} 1/(1 - \alpha_{1\mathfrak{p}}\bar{\alpha}_{2\mathfrak{p}}N\mathfrak{p}^{-s'})(1 - \alpha_{1\mathfrak{p}}\bar{\beta}_{2\mathfrak{p}}N\mathfrak{p}^{-s'})(1 - \beta_{1\mathfrak{p}}\bar{\alpha}_{2\mathfrak{p}}N\mathfrak{p}^{-s'})(1 - \beta_{1\mathfrak{p}}\bar{\beta}_{2\mathfrak{p}}N\mathfrak{p}^{-s'})$$

since in the present circumstances all the bad prime factors are just 1. The latter expression is immediately seen to be just

$$\|\delta\|^{s'} \times (\pi^{-(s+\kappa/2)}\Gamma(s + \tfrac{1}{2}\kappa)(4\pi)^{-s'}\Gamma(s'))^m \times$$

$$\times L_K(s' - \tfrac{1}{2}(n_1 + n_2), \chi_1\bar{\chi}_2)L_K(s' - \tfrac{1}{2}(n_1 + n_2), \chi_1\bar{\chi}_2^\sigma).$$

At $s = \frac{1}{2}\kappa = \frac{1}{2}(n_2 - n_1)$, we have

$$s' = s - 1 + \tfrac{1}{2}(n_1 + n_2 + 2) = \tfrac{1}{2}(n_2 - n_1) - 1 + \tfrac{1}{2}(n_1 + n_2 + 2) = n_2.$$

Therefore,

$$\langle \tilde{\Theta}_1 E(*; \tfrac{1}{2}\kappa; \omega), \tilde{\Theta}_2 \rangle$$

is a non-zero rational multiple of

$$\pi^{-m(n_2-n_1)}\pi^{-mn_2}L_K(\tfrac{1}{2}(n_2 - n_1), \chi_1\bar{\chi}_2)L_K(\tfrac{1}{2}(n_2 - n_1), \chi_1\bar{\chi}_2^\sigma).$$

At $s = \frac{1}{2}\kappa$ the Eisenstein series is holomorphic, and by sections 4.8 and 6.2

$$E(g; \omega) = E(g; \tfrac{1}{2}\kappa; \omega)/\{D^{1/2}i^{\kappa m}\gamma(\omega)\}$$

has Fourier coefficients in $\mathbb{Q}(\omega)$, and for $\tau \in \mathrm{Aut}(\mathbb{C}/\mathbb{Q})$

$$E(g;\omega)^{\tau} = E(g;\omega^{\tau})$$

under the action on Fourier coefficients. Therefore, dividing by $D^{1/2}i^{\kappa m}\gamma(\omega)$,

$$\langle \tilde{\Theta}_1 E(*;\omega), \tilde{\Theta}_2 \rangle$$

is a non-zero rational multiple of

$$\pi^{-m(n_2-n_1)}\pi^{-mn_2}D^{1/2}i^{\kappa m}\gamma(\omega)^{-1}L_K(\tfrac{1}{2}(n_2-n_1),\chi_1\bar{\chi}_2)L_K(\tfrac{1}{2}(n_2-n_1),\chi_1\bar{\chi}_2^{\sigma}).$$

Note that the rational constant of proportionality depends only on $\mathfrak{n}$; this constant is given explicitly in section 4.10 as a volume constant.

On the other hand, at primes not dividing $\mathfrak{n}_2$, the form of the $\mathfrak{p}$-part $W_{2\mathfrak{p}}$ of the Fourier coefficient of $\tilde{\Theta}_2$ on elements $\ell(y)$ determines $W_{2\mathfrak{p}}$ completely. Since the form is that of a Hecke eigenfunction, it must be that $\tilde{\Theta}_2$ is a Hecke eigenfunction at primes not dividing $\mathfrak{n}_2$. (The same applies to $\tilde{\Theta}_1$.) Therefore, we can apply the theorem of section 6.1: for $\tau \in \mathrm{Aut}(\mathbb{C}/\mathbb{Q})$

$$\{\langle \tilde{\Theta}_1 E(*;\omega), \tilde{\Theta}_2 \rangle / \langle \tilde{\Theta}_2, \tilde{\Theta}_2 \rangle\}^{\tau} = \langle \tilde{\Theta}_1^{\tau} E(*;\omega)^{\tau}, \tilde{\Theta}_2^{\tau} \rangle / \langle \tilde{\Theta}_2^{\tau}, \tilde{\Theta}_2^{\tau} \rangle.$$

Also by section 6.1, the map $\tilde{\Theta}_i \to \tilde{\Theta}_1^{\tau}$ merely has the effect of replacing χ_i by χ_i^{τ}, and $E(*;\omega)^{\tau} = E(*;\omega^{\tau})$. This gives the equivariance assertion of the theorem. Further, for $\tau \in \mathrm{Aut}(\mathbb{C}/\Omega)$, the equivariance assertion obviously implies that

$$A(\chi_1, \chi_2) \in \Omega. \qquad \blacksquare$$

6.4 Special Values of Product L-Functions

The results of the previous section play upon a coincidence which does not arise generally. Nevertheless, there is a general result which is essentially an immediate corollary of sections 4.10 and 6.1 in combination.

Let δ be a finite idele so that $\delta\hat{\mathfrak{o}} = \mathfrak{d}^*\hat{\mathfrak{o}}$. Let $\mathfrak{n}_1$ and $\mathfrak{n}_2$ be two ideals whose least common multiple is $\mathfrak{n}$. Put

$$\mathbb{K}_i = \left\{ g \in GL(2,\hat{\mathfrak{o}}) \colon g = \begin{pmatrix} a & b \\ c & d \end{pmatrix} \text{ with } c \equiv 0, d \equiv 1 \text{ modulo } \mathfrak{n}_i \right\}.$$

Let f and ψ be holomorphic cuspforms so that f is right-invariant by $\mathbb{K}_2$ and that ψ is right-invariant by $\mathbb{K}_1$. Let ω_2 be the central character of f, and $\omega_1 = \omega_2/\omega$ the central character of ψ. Suppose that f and ψ are *normalized Hecke eigenforms* (in the sense of section 3.5) of weights κ_2 and κ_1, respectively; take κ so that $\kappa_1 + \kappa = \kappa_2$. This requires that in the Fourier expansions

$$f(g) = \sum_{\xi \in F^\times} W_2(\ell(\xi)g)$$

$$\psi(g) = \sum_{\xi \in F^\times} W_1(\ell(\xi)g)$$

with $y \in \mathbb{J}^+$ we have

$$W_i(\ell(y)) = \|y\|^{\kappa_i/2} \exp(-2\pi \operatorname{Tr} y_\infty) V_i(y_0)$$

and that V_i factors over finite primes as

$$V_i(y_0) = \prod_{\mathfrak{p} < \infty} V_{i\mathfrak{p}}(y_\mathfrak{p})$$

and

$$V_{i\mathfrak{p}}(1) = 1.$$

Further, the Hecke eigenfunction property implies (see section 2.5) that each factor

$$W_{i\mathfrak{p}}(\ell(y_\mathfrak{p})) = \|y_\mathfrak{p}\|^{\kappa_i/2} V_{i\mathfrak{p}}(y_\mathfrak{p})$$

of "the" (non-zero) Fourier coefficient has an explicit form, from section 2.5. For $\mathfrak{p}$ not dividing $\mathfrak{n}_i$ it is

$$W_{i\mathfrak{p}}(\ell(\delta_\mathfrak{p} y_\mathfrak{p})) = W_{i\mathfrak{p}}(\ell(\delta_\mathfrak{p}))(\alpha^{n+1} - \beta^{n+1})/(\alpha - \beta) \qquad (\text{if } y_\mathfrak{p}\mathfrak{o}_\mathfrak{p} = \mathfrak{p}^n\mathfrak{o}_\mathfrak{p} \text{ and } n \geq 0)$$

$$0 \qquad (\text{if } y_\mathfrak{p}\mathfrak{o}_\mathfrak{p} = \mathfrak{p}^n\mathfrak{o}_\mathfrak{p} \text{ and } n < 0)$$

for some complex numbers α and β depending on i and $\mathfrak{p}$ satisfying

$$\alpha\beta = N\mathfrak{p}^{-1}\omega_i(\mathfrak{p}).$$

For $\mathfrak{p}$ dividing $\mathfrak{n}_i$ it is

$$W_{i\mathfrak{p}}(\ell(\delta_\mathfrak{p} y_\mathfrak{p})) = W_{i\mathfrak{p}}(\ell(\delta_\mathfrak{p}))\alpha^{n+1} \qquad (\text{if } y_\mathfrak{p}\mathfrak{o}_\mathfrak{p} = \mathfrak{p}^n\mathfrak{o}_\mathfrak{p} \text{ and } n \geq 0)$$

$$0 \qquad (\text{if } y_\mathfrak{p}\mathfrak{o}_\mathfrak{p} = \mathfrak{p}^n\mathfrak{o}_\mathfrak{p} \text{ and } n < 0)$$

for some complex numbers α depending on i and $\mathfrak{p}$. In terms of the functons V_{ip} *with the present normalization* $V_{i\mathfrak{p}}(1) = 1$ this is as follows. For $\mathfrak{p}$ not dividing $\mathfrak{n}_i$:

$$V_{i\mathfrak{p}}(\delta_\mathfrak{p} y_\mathfrak{p}) = (\alpha_{i\mathfrak{p}}^{n+1} - \beta_{i\mathfrak{p}}^{n+1})/(\alpha_{i\mathfrak{p}} - \beta_{i\mathfrak{p}}) \qquad (\text{if } y_\mathfrak{p}\mathfrak{o}_\mathfrak{p} = \mathfrak{p}^n\mathfrak{o}_\mathfrak{p} \text{ and } n \geq 0)$$

$$0 \qquad (\text{if } y_\mathfrak{p}\mathfrak{o}_\mathfrak{p} = \mathfrak{p}^n\mathfrak{o}_\mathfrak{p} \text{ and } n < 0)$$

where

$$\alpha_{i\mathfrak{p}} = N\mathfrak{p}^{\kappa_i/2}\alpha \qquad \beta_{i\mathfrak{p}} = N\mathfrak{p}^{\kappa_i/2}\beta$$

in terms of the parameters α and β in $W_{i\mathfrak{p}}$. Therefore,

$$\alpha_{i\mathfrak{p}}\beta_{i\mathfrak{p}} = \omega_i(\mathfrak{p})N\mathfrak{p}^{\kappa_i - 1}.$$

For $\mathfrak{p}$ dividing $\mathfrak{n}_i$:

$$V_{i\mathfrak{p}}(\delta_\mathfrak{p} y_\mathfrak{p}) = \begin{cases} \alpha_{i\mathfrak{p}}^n & (\text{if } y_\mathfrak{p}\mathfrak{o}_\mathfrak{p} = \mathfrak{p}^n\mathfrak{o}_\mathfrak{p} \text{ and } n \geq 0) \\ 0 & (\text{if } y_\mathfrak{p}\mathfrak{o}_\mathfrak{p} = \mathfrak{p}^n\mathfrak{o}_\mathfrak{p} \text{ and } n < 0) \end{cases}$$

where $\alpha_{i\mathfrak{p}} = N\mathfrak{p}^{\kappa_i/2}\alpha$, in terms of the parameter α in $W_{i\mathfrak{p}}$. If a finite-prime $\mathfrak{p}$ does not divide $\mathfrak{n}_1$ but does divide $\mathfrak{n}_2$, then (temporarily) say that $\mathfrak{p}$ is good$_1$; if $\mathfrak{p}$ does divide $\mathfrak{n}_1$ but does not divide $\mathfrak{n}_2$, then say that $\mathfrak{p}$ is good$_2$; if $\mathfrak{p}$ does not divide $\mathfrak{n}$, then say that $\mathfrak{p}$ is good; if $\mathfrak{p}$ divides both $\mathfrak{n}_1$ and $\mathfrak{n}_2$, then say that $\mathfrak{p}$ is bad.

THEOREM Let f and ψ have Fourier coefficients in a field Ω containing $\mathbb{Q}(\lambda_1, \lambda_2, \omega_1, \omega_2)$, supposing that $f \in \mathrm{Cfm}(\mathfrak{n}, \kappa, \omega_2, \lambda_2)$ and $\psi \in \mathrm{Cfm}(\mathfrak{n}, \kappa, \omega_1, \lambda_1)$. The ratio

$$L(\psi \otimes \bar{f}, \tfrac{1}{2}\kappa_2 - \tfrac{1}{2}\kappa_1 - 1)/\{\gamma(\omega)i^{m(\kappa_2-\kappa_1)}D^{1/2}\langle f, f\rangle\}$$

lies in Ω. Further, for $\sigma \in \mathrm{Aut}(\mathbb{C}/\mathbb{Q})$,

$$\{L(\psi \otimes \bar{f}, \tfrac{1}{2}\kappa_2 - \tfrac{1}{2}\kappa_1 - 1)/\{\gamma(\omega)i^{m(\kappa_2-\kappa_1)}D^{1/2}\langle f, f\rangle\}\}^\sigma =$$
$$= L(\psi^\sigma \otimes \bar{f}^\sigma, \tfrac{1}{2}\kappa_2 - \tfrac{1}{2}\kappa_1 - 1)/\{\gamma(\omega^\sigma)i^{m(\kappa_2-\kappa_1)}D^{1/2}\langle f^\sigma, f^\sigma\rangle\}.$$

Proof Let $s' = s - 1 + \tfrac{1}{2}(\kappa_1 + \kappa_2)$, $\kappa = \kappa_2 - \kappa_1$. With hypotheses as above, for $s \in \mathbb{C}$ away from poles of the Eisenstein series

$$E(*; s) = E^*(*; \kappa; 0, 1, \mathfrak{n}; s, \omega, \mathbb{1}),$$

by section 4.10 the inner product $\langle \psi E(*; s), f\rangle$ is

$$c\|\delta\|^{s'} \times (\pi^{-(s+\kappa/2)}\Gamma(s + \tfrac{1}{2}\kappa)(4\pi)^{-s'}\Gamma(s'))^m \times$$
$$\times \prod_{\mathfrak{p}\text{ good}} 1/(1 - \alpha_{1\mathfrak{p}}\bar{\alpha}_{2\mathfrak{p}}N\mathfrak{p}^{-s'})(1 - \alpha_{1\mathfrak{p}}\bar{\beta}_{2\mathfrak{p}}N\mathfrak{p}^{-s'})(1 - \beta_{1\mathfrak{p}}\bar{\alpha}_{2\mathfrak{p}}N\mathfrak{p}^{-s'}) \times$$
$$\times (1 - \beta_{1\mathfrak{p}}\bar{\beta}_{2\mathfrak{p}}N\mathfrak{p}^{-s'}) \prod_{\mathfrak{p}\text{ good}_1} 1/(1 - \alpha_{1\mathfrak{p}}\bar{\alpha}_{2\mathfrak{p}}N\mathfrak{p}^{-s'}) \times$$
$$\times (1 - \beta_{1\mathfrak{p}}\bar{\alpha}_{2\mathfrak{p}}N\mathfrak{p}^{-s'}) \prod_{\mathfrak{p}\text{ good}_2} 1/(1 - \alpha_{1\mathfrak{p}}\bar{\alpha}_{2\mathfrak{p}}N\mathfrak{p}^{-s'})(1 - \alpha_{1\mathfrak{p}}\bar{\beta}_{2\mathfrak{p}}N\mathfrak{p}^{-s'}) \times$$
$$\times \prod_{\mathfrak{p}\text{ bad}} 1/(1 - \alpha_{1\mathfrak{p}}\bar{\alpha}_{2\mathfrak{p}}N\mathfrak{p}^{-s'}) =$$
$$= \|\delta\|^{s'} \times (\pi^{-(s+\kappa/2)}\Gamma(s + \tfrac{1}{2}\kappa)(4\pi)^{-s'}\Gamma(s'))^m \times L_0(\psi \otimes \bar{f}, s' - \tfrac{1}{2}(\kappa_1 + \kappa_2))$$

for some non-zero rational number c depending only upon $\mathfrak{n}$. As mentioned in section 6.3, by sections 6.2 and 4.8 the holomorphic automorphic form

$$E_\omega(*) = E(*; \tfrac{1}{2}\kappa)/\{i^{\kappa m}\gamma(\omega)D^{1/2}\}$$

has Fourier coefficients in $\mathbb{Q}(\omega)$, and $E_\omega^\sigma = E_{\omega^\sigma}$. Taking $s = \frac{1}{2}\kappa$ makes $s' = \kappa_2 - 1$, and we find that

$$\langle \psi E_\omega, f \rangle$$

is a rational multiple of

$$D^{1/2} i^{\kappa m} \gamma(\omega)^{-1} \pi^{-m(\kappa_2 - \kappa_1)} \pi^{-m(\kappa_2 - 1)} L_0(\psi \otimes \bar{f}, \tfrac{1}{2}\kappa_2 - \tfrac{1}{2}\kappa_1 - 1).$$

On the other hand, by section 6.1,

$$\langle \psi E_\omega, f \rangle \in \langle f, f \rangle \Omega.$$

This gives the first assertion. For the second assertion, we use the result

$$\{\langle \psi E_\omega, f \rangle / \langle f, f \rangle\}^\sigma = \langle (\psi E_\omega)^\sigma, f^\sigma \rangle / \langle f^\sigma, f^\sigma \rangle$$

from section 6.1. We obtain

$$\{\langle \psi E_\omega, f \rangle / \langle f, f \rangle\}^\sigma = \langle \psi^\sigma E_{\omega^\sigma}, f^\sigma \rangle / \langle f^\sigma, f^\sigma \rangle$$

which is the *same* rational multiple of

$$D^{1/2} i^{\kappa m} \gamma(\omega^\sigma)^{-1} \pi^{-m(\kappa_2 - \kappa_1)} \pi^{-m(\kappa_2 - 1)} L_0(\psi^\sigma \otimes (f^\sigma)^-, \tfrac{1}{2}\kappa_2 - \tfrac{1}{2}\kappa_1 - 1).$$

This gives the theorem. ■

6.5 Special Values of Standard L-Functions Attached to Cuspforms

Now we return to the integral representation of section 4.10, applying the results of section 6.1 to obtain a comparison of different special values of (standard) L-functions attached to a holomorphic cuspform. We give two versions of the result: The first requires little preparation beyond previous results; the second (and more general) result requires construction of an auxiliary Eisenstein series. Also, the second theorem introduces the notion of "twist" of a cuspform, which merits some discussion.

These certainly are not the most general versions which can be proven (see [Sh4], [Sh6]), although the present methods (essentially those of [Sh4] and [Sh6]) do give the most general result. On the other hand, there is a technical simplification in the present argument: Because of the second part of section 6.1, regarding comparisons of inner products, we need no discussion of the idea of "newform" (as necessary in [Sh4], [Sh6]).

Let δ be a finite idele so that $\delta\hat{\mathfrak{o}} = \mathfrak{o}^*\hat{\mathfrak{o}}$, and let D be the discriminant of $\mathfrak{o}$ over $\mathbb{Z}$. Use notation $\gamma(\omega)$ as in section 6.2 for "Gauss sums" attached to a character ω. Let $\mathfrak{n}$ be an ideal of $\mathfrak{o}$. Put

$$\mathbb{K} = \left\{g \in GL(2, \hat{\mathfrak{o}}) \colon g = \begin{pmatrix} a & b \\ c & d \end{pmatrix} \text{ with } c \equiv 0, d \equiv 1 \text{ modulo } \mathfrak{n}\right\}.$$

Throughout this section, suppose that f is *right-invariant by* $GL(2, \hat{\mathfrak{o}})$, with *trivial central character*. Suppose that f is a holomorphic *normalized Hecke eigenform* (in the sense of section 3.5), of weight κ_2. First of all, this requires that in the Fourier expansion

$$f(g) = \sum_{\xi \in F^\times} W_2(\ell(\xi)g)$$

with $y \in \mathbb{J}^+$ we have

$$W_2(\ell(y)) = \|y\|^{\kappa_2/2} \exp(-2\pi \operatorname{Tr} y_\infty) V_2(y_o)$$

and that V_2 factors over finite primes as

$$V_2(y_o) = \prod_{\mathfrak{p} < \infty} V_{2\mathfrak{p}}(y_\mathfrak{p})$$

and

$$V_{2\mathfrak{p}}(1) = 1.$$

Further, the Hecke eigenfunction property implies (see section 2.5) that each factor

$$W_{2\mathfrak{p}}(\ell(y_\mathfrak{p})) = \|y_\mathfrak{p}\|^{\kappa_2/2} V_{2\mathfrak{p}}(y_\mathfrak{p})$$

of "the" (non-zero) Fourier coefficient has an explicit form. It is

$$\begin{aligned} W_{2\mathfrak{p}}(\ell(\delta_\mathfrak{p} y_\mathfrak{p})) &= W_{2\mathfrak{p}}(\ell(\delta_\mathfrak{p}))(\alpha_\mathfrak{p}^{n+1} - \beta_\mathfrak{p}^{n+1})/(\alpha_\mathfrak{p} - \beta_\mathfrak{p}) && (\text{if } y_\mathfrak{p}\mathfrak{o}_\mathfrak{p} = \mathfrak{p}^n\mathfrak{o}_\mathfrak{p} \text{ and } n \geq 0) \\ &\qquad 0 && (\text{if } y_\mathfrak{p}\mathfrak{o}_\mathfrak{p} = \mathfrak{p}^n\mathfrak{o}_\mathfrak{p} \text{ and } n < 0) \end{aligned}$$

for some complex numbers $\alpha_\mathfrak{p}$ and $\beta_\mathfrak{p}$ depending on $\mathfrak{p}$, and

$$\alpha_\mathfrak{p}\beta_\mathfrak{p} = N\mathfrak{p}^{-1}.$$

In terms of the functions $V_{2\mathfrak{p}}$ *with the normalization* $V_{2\mathfrak{p}}(1) = 1$ this becomes

$$V_{2\mathfrak{p}}(\delta_\mathfrak{p} y_\mathfrak{p}) = (\alpha_{2\mathfrak{p}}^{n+1} - \beta_{2\mathfrak{p}}^{n+1})/(\alpha_{2\mathfrak{p}} - \beta_{2\mathfrak{p}}) \qquad (\text{if } y_\mathfrak{p}\mathfrak{o}_\mathfrak{p} = \mathfrak{p}^n\mathfrak{o}_\mathfrak{p} \text{ and } n \geq 0)$$

$$0 \qquad (\text{if } y_\mathfrak{p}\mathfrak{o}_\mathfrak{p} = \mathfrak{p}^n\mathfrak{o}_\mathfrak{p} \text{ and } n < 0)$$

where

$$\alpha_{2\mathfrak{p}} = N\mathfrak{p}^{\kappa_2/2}\alpha_\mathfrak{p} \qquad \beta_{2\mathfrak{p}} = N\mathfrak{p}^{\kappa_2/2}\beta_\mathfrak{p}$$

in terms of the parameters $\alpha_\mathfrak{p}$ and $\beta_\mathfrak{p}$ in $W_{2\mathfrak{p}}$. Therefore,

$$\alpha_{2\mathfrak{p}}\beta_{2\mathfrak{p}} = N\mathfrak{p}^{\kappa_2-1}.$$

Further, assume that f has Fourier coefficients in a field Ω containing $\mathbb{Q}(\lambda)$ (in the notation of section 6.1). For present purposes, we normalize the (finite-prime part of) the standard L-function of f as

$$L_0(f,s) = \prod_\mathfrak{p} 1/(1 - \alpha_\mathfrak{p} N\mathfrak{p}^{-s})(1 - \beta_\mathfrak{p} N\mathfrak{p}^{-s}).$$

Finally, we note that *the hypothesis that the central character of f is trivial implies that $\alpha_\mathfrak{p} + \beta_\mathfrak{p}$ is real*; this follows from the fact that on cuspforms with trivial central character the Hecke operator given by $\eta_\mathfrak{p}$ is *self-adjoint*, and $\alpha_\mathfrak{p} + \beta_\mathfrak{p}$ is a real multiple of the $\eta_\mathfrak{p}$-eigenvalue (see section 2.5).

Remarks Because of the restriction on the weights κ for which we have proven the results of section 6.1, we have a complete proof of the following theorems only for $\kappa_2 \geq 7$, etc. In the papers of Shimura mentioned earlier there is no such restriction.

THEOREM Let κ_1 be an even integer. We require that $\kappa_1 \geq 2$ and $\kappa_2 - \kappa_1 \geq 2$. If $F = \mathbb{Q}$, then we require that $\kappa_1 > 2$ and $\kappa_2 - \kappa_1 > 2$. Then

$$\{D^{1/2}\pi^{m(1-\kappa_2)}\pi^{-m(\kappa_2-\kappa_1)}\} \times L_0(f, \tfrac{1}{2}\kappa_2 - \kappa_1)L_0(f, \tfrac{1}{2}\kappa_2 - 1)$$

lies in

$$\langle f, f\rangle\Omega.$$

Further, for $\sigma \in \mathrm{Aut}(\mathbb{C}/\mathbb{Q})$, we have the equivariance property

$$\{D^{1/2}\pi^{m(1-\kappa_2)}\pi^{-m(\kappa_2-\kappa_1)}L_0(f, \tfrac{1}{2}\kappa_2 - \kappa_1)L_0(f, \tfrac{1}{2}\kappa_2 - 1)/\langle f, f\rangle\}^\sigma =$$

$$= D^{1/2}\pi^{m(1-\kappa_2)}\pi^{-m(\kappa_2-\kappa_1)}L_0(f^\sigma, \tfrac{1}{2}\kappa_2 - \kappa_1)L_0(f^\sigma, \tfrac{1}{2}\kappa_2 - 1)/\langle f^\sigma, f^\sigma\rangle.$$

Remarks The hypotheses on κ_1 are exactly those to assure that the Eisenstein series occurring in this proof are holomorphic. The idea of the proof is quite simple, in principle. For two Eisenstein series E_1 and E_2 consider the integral $\langle E_1 E_2, f\rangle$. On one hand, if the Fourier coefficients of both E_1 and E_2 are in a field $\Omega(\omega)$, then a corollary of the theorem of section 6.1 shows that

$$\langle E_1 E_2, f\rangle \in \langle f, f\rangle \Omega(\omega).$$

Further, from section 6.1

$$\{\langle E_1 E_2, f\rangle / \langle f, f\rangle\}^\sigma = \langle E_1^\sigma E_2^\sigma, f^\sigma\rangle / \langle f^\sigma, f^\sigma\rangle.$$

On the other hand, one of these Eisenstein series can be unwound; by the concrete nature of the Fourier coefficients of the (adroitly chosen) E_1, following section 4.10 this unwound integral will be essentially a product of two values of the standard L-function attached to f. *The only real tasks are to determine the precise normalization of the Eisenstein series and to see exactly which values of the L-function come out.* In the proof of the present theorem, both Eisenstein series are exactly of a form previously considered. In the second (more general) theorem, some thought is necessary to make a suitable choice of Eisenstein series.

Proof We use the notation $E(g; \kappa; c_0, d_0, \mathfrak{n}; s, \omega, \chi)$ from Chapter 4, especially from section 4.8. Let $\kappa = \kappa_2 - \kappa_1$. Abbreviate

$$E_1(g; s) = E(g; \kappa_1; *, *, \mathfrak{o}; s, \mathbb{1}, \mathbb{1})$$

$$E_2(g; s) = E(g; \kappa; *, *, \mathfrak{o}; s, \mathbb{1}, \mathbb{1})$$

$$E_1(g) = E_1(g; \tfrac{1}{2}\kappa_1)/D^{1/2}$$

$$E_2(g) = E_1(g; \tfrac{1}{2}\kappa)/D^{1/2}.$$

From section 6.2 (and from section 4.8) the holomorphic automorphic forms $E_1(g)$ and $E_2(g)$ have Fourier coefficients in $\mathbb{Q}$. Therefore,

$$\langle E_1 E_2, f\rangle \in \langle f, f\rangle \Omega$$

by section 6.1.

On the other hand, we can unwind this inner product as in section 4.10. To apply the calculation there, we need to explicate the Fourier coefficient

$$\int_{\mathbb{A}/F} \bar{\tau}(x) E_1(u(x)\ell(y); \tfrac{1}{2}\kappa_1)\, d^+x.$$

This was computed explicitly (without the present renormalization) in the theorem in section 4.8; up to a rational constant, it is

$$\|y\|^{\kappa_1/2} \exp(-2\pi \operatorname{Tr} y_\infty) \times (-2i)^{\kappa_1 m} D^{-1/2} \sum_{\mathfrak{a}} N\mathfrak{a}^{\kappa_1 - 1}$$

where $\mathfrak{a}$ is summed over ideals so that $y\hat{\mathfrak{o}} \subset \mathfrak{a}\hat{\mathfrak{o}} \subset \mathfrak{o}^*\hat{\mathfrak{o}}$. Therefore, the sum over $\mathfrak{a}$ factors over primes as

$$\sum_{\mathfrak{a}} N\mathfrak{a}^{\kappa_1 - 1} = \prod_{\mathfrak{p} < \infty} \Psi_{\mathfrak{p}}(\ell(y))$$

where

$$\Psi_{\mathfrak{p}}(\ell(\delta_{\mathfrak{p}}\pi^n)) = \begin{cases} (\|\pi^{n+1}\|^{1-\kappa_1} - 1)/(\|\pi\|^{1-\kappa_1} - 1) & n \geq 0 \\ 0 & n < 0, \end{cases}$$

where π denotes a local parameter. That is, in the sense of section 3.5,

$$(-2i)^{-\kappa_1 m} D^{1/2} E_1(*; \tfrac{1}{2}\kappa_1)$$

is a *normalized Hecke eigenform.* In the notation of the theorem in section 4.10,

$$\alpha_{1\mathfrak{p}} = N\mathfrak{p}^{\kappa_1 - 1} \qquad \beta_{1\mathfrak{p}} = 1.$$

Let $s' = s - 1 + \frac{1}{2}(\kappa_1 + \kappa_2)$. By the theorem of section 4.10,

$$\langle E_1 E_2(*; s), f\rangle = (-2i)^{-\kappa_1 m} D^{1/2} \langle E_1(*; \tfrac{1}{2}\kappa_1) E_2(*; s), f\rangle$$

is a non-zero *rational* multiple of

$$\|\delta\|^{s'} \times (\pi^{-(s+\kappa/2)}\Gamma(s + \tfrac{1}{2}\kappa)(4\pi)^{-s'}\Gamma(s'))^m \times$$

$$\times \prod_{\mathfrak{p}} 1/(1 - \alpha_{1\mathfrak{p}}\bar{\alpha}_{2\mathfrak{p}} N\mathfrak{p}^{-s'})(1 - \alpha_{1\mathfrak{p}}\bar{\beta}_{2\mathfrak{p}} N\mathfrak{p}^{-s'})(1 - \beta_{1\mathfrak{p}}\bar{\alpha}_{2\mathfrak{p}} N\mathfrak{p}^{-s'})(1 - \beta_{1\mathfrak{p}}\bar{\beta}_{2\mathfrak{p}} N\mathfrak{p}^{-s'}) =$$

$$= \|\delta\|^{s'}(\pi^{-(s+\kappa/2)}\Gamma(s + \tfrac{1}{2}\kappa)(4\pi)^{-s'}\Gamma(s'))^m \times$$

$$\times \prod_{\mathfrak{p}} 1/(1 - \bar{\alpha}_{2\mathfrak{p}} N\mathfrak{p}^{\kappa_1 - 1 - s'})(1 - \bar{\beta}_{2\mathfrak{p}} N\mathfrak{p}^{\kappa_1 - 1 - s'})(1 - \bar{\alpha}_{2\mathfrak{p}} N\mathfrak{p}^{-s'})(1 - \bar{\beta}_{2\mathfrak{p}} N\mathfrak{p}^{-s'}) =$$

$$= \|\delta\|^{s'}(\pi^{-(s+\kappa/2)}\Gamma(s + \tfrac{1}{2}\kappa)(4\pi)^{-s'}\Gamma(s'))^m \times L_0(\bar{f}, s' + 1 - \kappa_1 - \tfrac{1}{2}\kappa_2) L_0(\bar{f}, s' - \tfrac{1}{2}\kappa_2)$$

using the relations

$$\alpha_{2\mathfrak{p}} = \alpha_{\mathfrak{p}} N\mathfrak{p}^{\kappa_2/2} \qquad \beta_{2\mathfrak{p}} = \beta_{\mathfrak{p}} N\mathfrak{p}^{\kappa_2/2}.$$

At $s = \frac{1}{2}\kappa$, we have

$$s' = \tfrac{1}{2}\kappa - 1 + \tfrac{1}{2}(\kappa_1 + \kappa_2) = \tfrac{1}{2}(\kappa_2 - \kappa_1) - 1 + \tfrac{1}{2}(\kappa_1 + \kappa_2) = \kappa_2 - 1.$$

Therefore,

$$\langle E_1(*)E_2(*;\tfrac{1}{2}\kappa),f\rangle = (-2i)^{-\kappa_1 m}D^{1/2}\langle E_1(*;\tfrac{1}{2}\kappa_1)E_2(*;\tfrac{1}{2}\kappa),f\rangle$$

is a non-zero rational multiple of

$$\|\delta\|^{\kappa_2-1} \times (\pi^{-\kappa}\Gamma(\kappa)(4\pi)^{1-\kappa_2}\Gamma(\kappa_2-1))^m L_0(\bar{f},\tfrac{1}{2}\kappa_2-\kappa_1)L_0(\bar{f},\tfrac{1}{2}\kappa_2-1)$$

which is a rational multiple of

$$(\pi^{1-\kappa_2}\pi^{-(\kappa_2-\kappa_1)})^m L_0(\bar{f},\tfrac{1}{2}\kappa_2-\kappa_1)L_0(\bar{f},\tfrac{1}{2}\kappa_2-1).$$

Dividing this relation by

$$i^{m\kappa}D^{1/2}$$

shows that

$$\langle E_1E_2,f\rangle$$

is a non-zero rational multiple of

$$\{D^{1/2}\pi^{m(1-\kappa_2)}\pi^{-m(\kappa_2-\kappa_1)}\} \times L_0(\bar{f},\tfrac{1}{2}\kappa_2-\kappa_1)L_0(\bar{f},\tfrac{1}{2}\kappa_2-1).$$

Combining this calculation with the first, we find that

$$\{D^{1/2}\pi^{m(1-\kappa_2)}\pi^{-m(\kappa_2-\kappa_1)}\} \times L_0(\bar{f},\tfrac{1}{2}\kappa_2-\kappa_1)L_0(\bar{f},\tfrac{1}{2}\kappa_2-1)$$

lies in

$$\Omega\langle f,f\rangle.$$

Since $\alpha_{\mathfrak{p}}+\beta_{\mathfrak{p}}$ is *real* for all finite primes $\mathfrak{p}$, we see that for $s\in\mathbb{R}$

$$L_0(\bar{f},s) = L_0(f,s).$$

This yields the first assertion of the theorem.

For the second assertion of the theorem, we use the property

$$\{\langle\psi,f\rangle/\langle f,f\rangle\}^\sigma = \langle\psi^\sigma,f^\sigma\rangle/\langle f^\sigma,f^\sigma\rangle$$

for $\sigma\in\mathrm{Aut}(\mathbb{C}/\mathbb{Q})$ from section 6.1. The normalized Eisenstein series E_1 and E_2 have Fourier coefficients in $\mathbb{Q}$, so $E_1^\sigma = E_1$ and $E_2^\sigma = E_2$. Therefore,

$$\{\langle E_1E_2,f\rangle/\langle f,f\rangle\}^\sigma = \langle E_1E_2,f^\sigma\rangle/\langle f^\sigma,f^\sigma\rangle.$$

From the first part of the proof, noting that the non-zero proportionality constants occurring throughout depend only upon $\mathfrak{n}$, this implies that

$$\{D^{1/2}\pi^{m(1-\kappa_2)}\pi^{-m(\kappa_2-\kappa_1)}L_0(f,\tfrac{1}{2}\kappa_2-\kappa_1)L_0(f,\tfrac{1}{2}\kappa_2-1)/\langle f,f\rangle\}^\sigma =$$

$$= D^{1/2}\pi^{m(1-\kappa_2)}\pi^{-m(\kappa_2-\kappa_1)}L_0(f^\sigma,\tfrac{1}{2}\kappa_2-\kappa_1)L_0(f^\sigma,\tfrac{1}{2}\kappa_2-1)/\langle f^\sigma,f^\sigma\rangle.$$

This is the assertion of this theorem. ■

Definition Let χ be a (finite-order) Hecke character, with conductor $\mathfrak{n}$. Define the (finite-prime part of the) L-function of f *twisted by* χ to be

$$L_0(f\otimes\chi,s) = \prod_{\mathfrak{p}} 1/(1-\chi(\mathfrak{p})\alpha_{\mathfrak{p}}N\mathfrak{p}^{-s})(1-\chi(\mathfrak{p})\beta_{\mathfrak{p}}N\mathfrak{p}^{-s})$$

where the product is over primes $\mathfrak{p}$ not dividing $\mathfrak{n}$.

Remarks It is an elementary exercise to show that there is a holomorphic cuspform (often denoted by $f\otimes\chi$) so that this twisted L-function is (essentially) the standard L-function attached to $f\otimes\chi$. The idea is to obtain $f\otimes\chi$ as a finite-linear combination of right translates of f by unipotent matrices in $GL(2,\mathbb{A}_0)$. The classical version of this is even easier to appreciate, and illustrates the idea very well: Let

$$f(z) = \sum_{n>0} c(n)\exp(2\pi inz)$$

be a cuspform on $\mathfrak{H}$ with respect to $SL(2,\mathbb{Z})$. What we want is to show that

$$\sum_{n>0} c(n)\chi(n)\exp(2\pi inz)$$

is also a cuspform, where χ has conductor N. We have

$$\sum_{b \bmod N}\bar{\chi}(b)f(z+N^{-1}b) = \sum_{b \bmod N}\bar{\chi}(b)\sum_{n>0} c(n)\exp(2\pi in(z+N^{-1}b)) =$$

$$= \sum_{n>0} c(n)\chi(n)\exp(2\pi inz)\sum_{b \bmod N}\bar{\chi}(b)\exp(2\pi inb/N).$$

The inner sum is 0 unless n is prime to N, in which case it is

$$\chi(n)\sum_{b \bmod N}\bar{\chi}(b)\exp(2\pi ib/N).$$

Therefore, the whole is

$$\sum_{b \bmod N}\bar{\chi}(b)\exp(2\pi ib/N)\times\sum_{n>0} c(n)\chi(n)\exp(2\pi inz),$$

as desired. The outer sum is non-zero. Via the Strong Approximation Theorem, one may easily ascertain the adelic version of this. The precise properties of such a twist are readily ascertained.

Actually, we do not use the fact that these twists $f \otimes \chi$ are cuspforms in the proof of the following theorem.

THEOREM Let κ_1 be an integer; let ω be a finite-order character which is totally even if κ_1 is even, is totally odd if κ_1 is odd, and so that the finite-prime part of the conductor of ω is $\mathfrak{n}$. We require that $\kappa_1 \geq 2$ and $\kappa_2 - \kappa_1 \geq 2$. If $F = \mathbb{Q}$ and $\omega = \mathbb{1}$, then we require that $\kappa_1 > 2$ and $\kappa_2 - \kappa_1 > 2$. Let

$$C(\omega) = i^{m(\kappa_2-\kappa_1)} D^{1/2} \gamma(\bar{\omega})^{-1} \pi^{m(1-\kappa_2)} \pi^{-m(\kappa_2-\kappa_1)}.$$

Then

$$C(\omega) L(f \otimes \omega, \tfrac{1}{2}\kappa_2 - \kappa_1) L(f, \tfrac{1}{2}\kappa_2 - 1) \in \langle f, f \rangle \Omega(\omega).$$

Further, for $\sigma \in \mathrm{Aut}(\mathbb{C}/\mathbb{Q})$, we have the equivariance property

$$\{C(\omega) L(f \otimes \omega, \tfrac{1}{2}\kappa_2 - \kappa_1) L(f, \tfrac{1}{2}\kappa_2 - 1)/\langle f, f \rangle\}^\sigma =$$

$$= C(\omega^\sigma) L(f^\sigma \otimes \omega^\sigma, \tfrac{1}{2}\kappa_2 - \kappa_1) L(f^\sigma, \tfrac{1}{2}\kappa_2 - 1)/\langle f^\sigma, f^\sigma \rangle.$$

Proof In spirit, this proof is identical to the proof of the previous theorem. Once again, all non-zero constants of proportionality depend only upon $\mathfrak{n}$.

Let $\kappa = \kappa_2 - \kappa_1$. Abbreviate

$$E_2(g; s) = E(g; \kappa; 0, 1, \mathfrak{n}; s, \bar{\omega}, \mathbb{1})$$

$$E_2(g) = E_2(g; \tfrac{1}{2}\kappa)/\{i^{m\kappa} D^{1/2} \gamma(\bar{\omega})\}.$$

From section 6.2 (and from section 4.8) the holomorphic automorphic form $E_2(g)$ has Fourier coefficients in $\mathbb{Q}(\omega)$. In the present circumstances, we must construct the appropriate substitute for the Eisenstein series denoted by E_1 in the proof of the previous theorem.

What we want is an Eisenstein series E_1 of weight κ_1 with central character ω which is right $\mathbb{K}$-invariant and whose Fourier coefficients have the form necessary to apply the theorem of section 4.10. As noted in the remark prior to the theorem of section 4.10, we need not know that E_1 is an eigenfunction for the Hecke operators at bad primes, but only that its Fourier coefficients have the desired form on elements $\ell(y)$ with $y \in \mathbb{J}^+$. Let η be a finite idele so that $\eta\hat{\mathfrak{o}} = \mathfrak{n}_1\hat{\mathfrak{o}}$, and let δ be a finite idele so that $\delta\hat{\mathfrak{o}} = \mathfrak{d}^*\hat{\mathfrak{o}}$. In the notation of section 4.1, let

$$E_1(g; s; \omega) = E^*(g; \kappa; \varphi_0; s, \omega, \mathbb{1}),$$

where the finite-adele Schwartz function φ_0 on $\mathbb{A}_0^2$ is described as follows. We require that φ_0 factors over finite primes as $\varphi_0 = \prod_\mathfrak{p} \varphi_\mathfrak{p}$, and that for $\mathfrak{p}$ not dividing $\mathfrak{n}$ the function $\varphi_\mathfrak{p}$ is the characteristic function of $(\mathfrak{o}_\mathfrak{p})^2$. At $\mathfrak{p}$ dividing $\mathfrak{n}$, we take

$$\varphi_\mathfrak{p}(u, v) = \alpha_\mathfrak{p}(u)\beta_\mathfrak{p}(v)$$

where $\alpha_\mathfrak{p}$ is the characteristic function of $\mathfrak{o}_\mathfrak{p}^*$, and the *Fourier transform* $\hat{\beta}_\mathfrak{p}$ of $\beta_\mathfrak{p}$ is

$$\hat{\beta}_\mathfrak{p}(v) = [\mathfrak{o}_\mathfrak{p}^* : \mathfrak{o}_\mathfrak{p}]^{-1/2} \|\eta_\mathfrak{p}\delta_\mathfrak{p}^{-1}\|^{1-\kappa_1} (\text{characteristic function of } \mathfrak{o}_\mathfrak{p}^\times)(\eta\delta^{-1}v) \times \bar{\omega}(v).$$

As in sections 4.4 and 4.8, for $g = u(x)\ell(y)$ with $x \in \mathbb{A}$ and $y \in \mathbb{J}^+$, we have a Fourier expansion

$$E_1(g; \tfrac{1}{2}\kappa_1; \omega) =$$

$$= \|y\|^{\kappa_1/2}\left\{(\text{0-th coefficient}) + (-2i)^{\kappa_1 m} D^{-1/2} \sum_{\xi \in F^\times} \tau(\xi x)V(\xi y; \omega)\exp(-2\pi \operatorname{Tr} \xi_\infty y_\infty)\right\}.$$

Further, by construction the coefficient $V(\xi y; \omega)$ factors over finite primes as

$$V(y; \omega) = \prod_\mathfrak{p} V_\mathfrak{p}(y_\mathfrak{p}; \omega_1),$$

and for $\mathfrak{p}$ not dividing $\mathfrak{n}$ we have already computed $V_\mathfrak{p}$; it is

$$V_\mathfrak{p}(y_\mathfrak{p}; \omega) = \sum_\mathfrak{a} \omega(\mathfrak{a}) N\mathfrak{a}^{\kappa_1 - 1}$$

where $\mathfrak{a}$ is summed over ideals of $\mathfrak{o}_\mathfrak{p}$ so that $\mathfrak{a} \subset \mathfrak{o}_\mathfrak{p}^*$ and $y\mathfrak{o}_\mathfrak{p} \subset \mathfrak{a}$. At $\mathfrak{p}$ dividing $\mathfrak{n}$, from the Fourier coefficient calculations of section 4.4 we obtain

$$V_\mathfrak{p}(y_\mathfrak{p}; \omega) = \int_{F_\mathfrak{p}^\times} \|t\|^{\kappa_1 - 1}\bar{\omega}(t)\alpha_\mathfrak{p}(ty)\hat{\beta}_\mathfrak{p}(t^{-1})\, d^\times t.$$

From our present definition of $\alpha_\mathfrak{p}$ and $\beta_\mathfrak{p}$, this is

$$V_\mathfrak{p}(y_\mathfrak{p}; \omega) = \|\eta_\mathfrak{p}\delta_\mathfrak{p}^{-1}\|^{1-\kappa_1} \int_{F_\mathfrak{p}^\times} \|t\|^{\kappa_1 - 1}\bar{\omega}(t)\operatorname{ch}(ty; \mathfrak{o}_\mathfrak{p}^*)\bar{\omega}(t^{-1})\operatorname{ch}(\eta\delta^{-1}t^{-1}; \mathfrak{o}_\mathfrak{p}^\times)\, d^\times t$$

where $\operatorname{ch}(*; X)$ is the characteristic function of a set X. Clearly this is just

$$V_\mathfrak{p}(y_\mathfrak{p}; \omega) = \operatorname{ch}(y; \mathfrak{o}_\mathfrak{p}^*) \int_{\mathfrak{o}_\mathfrak{p}^\times} d^\times t = \operatorname{ch}(y; \mathfrak{o}_\mathfrak{p}^*).$$

This is the desired form.

It is clear that $V(y_0;\omega)$ takes values in $\mathbb{Q}(\omega)$, and that

$$V(y_0;\omega)^\sigma = V(y_0;\omega^\sigma).$$

From the proposition in the proof of the theorem of section 6.2, this implies that the 0-th Fourier coefficient of

$$E_1(g;\omega) = (-2i)^{-\kappa_1 m} D^{1/2} E_1(g;\tfrac{1}{2}\kappa_1;\omega)$$

has the same property. Therefore, the normalization $E_1(g;\omega)$ has the property that its Fourier coefficients are in $\mathbb{Q}(\omega)$, and

$$E_1(g;\omega)^\sigma = E_1(g;\omega^\sigma).$$

Using the arithmetic properties of the Fourier coefficients, we can obtain half of the necessary conclusion. By the theorem of section 6.1,

$$\langle E_1(*;\omega)E_2(*;\bar{\omega}), f\rangle \in \langle f,f\rangle\Omega(\omega).$$

Further, by that theorem,

$$\{\langle E_1(*;\omega)E_2(*;\bar{\omega}), f\rangle/\langle f,f\rangle\}^\sigma = \langle E_1(*;\omega)^\sigma E_2(*;\bar{\omega})^\sigma, f^\sigma\rangle/\langle f^\sigma,f^\sigma\rangle =$$

$$= \langle E_1(*;\omega^\sigma)E_2(*;\bar{\omega}^\sigma), f^\sigma\rangle/\langle f^\sigma,f^\sigma\rangle.$$

On the other hand, we can unwind this inner product as in section 4.10. Say that a finite prime $\mathfrak{p}$ is good if $\mathfrak{p}$ does not divide $\mathfrak{n}$; say that $\mathfrak{p}$ is bad if $\mathfrak{p}$ does divide $\mathfrak{n}$. The previous calculations show that, in the sense of section 3.5,

$$E_1(*;\omega) = (-2i)^{-\kappa_1 m} D^{1/2} E_1(*;\tfrac{1}{2}\kappa_1;\omega)$$

is a *normalized Hecke eigenform.* In the notation of the theorem in section 4.10, at $\mathfrak{p}$ not dividing $\mathfrak{n}$, we have

$$\alpha_{1\mathfrak{p}} = \omega(\mathfrak{p})N\mathfrak{p}^{\kappa_1 - 1} \qquad \beta_{1\mathfrak{p}} = 1.$$

At $\mathfrak{p}$ dividing $\mathfrak{n}$,

$$\alpha_{1\mathfrak{p}} = 1 \qquad \beta_{1\mathfrak{p}} = 0.$$

Let $s' = s - 1 + \frac{1}{2}(\kappa_1 + \kappa_2)$. By the theorem of section 4.10,

$$\langle E_1(*;\omega)E_2(*;s;\bar{\omega}), f\rangle$$

is a non-zero *rational* multiple of

$$\|\delta\|^{s'} \times (\pi^{-(s+\kappa/2)}\Gamma(s + \tfrac{1}{2}\kappa)(4\pi)^{-s'}\Gamma(s'))^m \times$$

$$\times \prod_{\mathfrak{p}\text{ good}} 1/(1 - \alpha_{1\mathfrak{p}}\bar{\alpha}_{2\mathfrak{p}}N\mathfrak{p}^{-s'})(1 - \alpha_{1\mathfrak{p}}\bar{\beta}_{2\mathfrak{p}}N\mathfrak{p}^{-s'})(1 - \beta_{1\mathfrak{p}}\bar{\alpha}_{2\mathfrak{p}}N\mathfrak{p}^{-s'})(1 - \beta_{1\mathfrak{p}}\bar{\beta}_{2\mathfrak{p}}N\mathfrak{p}^{-s'}) \times$$

$$\times \prod_{\mathfrak{p}\text{ bad}} 1/(1 - \bar{\alpha}_{2\mathfrak{p}}N\mathfrak{p}^{-s'})(1 - \bar{\beta}_{2\mathfrak{p}}N\mathfrak{p}^{-s'}) =$$

$$= \|\delta\|^{s'} \times (\pi^{-(s+\kappa/2)}\Gamma(s + \tfrac{1}{2}\kappa)(4\pi)^{-s'}\Gamma(s'))^m \times$$

$$\times \prod_{\mathfrak{p}\text{ good}} 1/(1 - \omega(\mathfrak{p})\bar{\alpha}_{2\mathfrak{p}}N\mathfrak{p}^{\kappa_1-1-s'})(1 - \omega(\mathfrak{p})\bar{\beta}_{2\mathfrak{p}}N\mathfrak{p}^{\kappa_1-1-s'})(1 - \bar{\alpha}_{2\mathfrak{p}}N\mathfrak{p}^{-s'})(1 - \bar{\beta}_{2\mathfrak{p}}N\mathfrak{p}^{-s'}) \times$$

$$\times \prod_{\mathfrak{p}\text{ bad}} 1/(1 - \bar{\alpha}_{2\mathfrak{p}}N\mathfrak{p}^{-s'})(1 - \bar{\beta}_{2\mathfrak{p}}N\mathfrak{p}^{-s'}) =$$

$$= \|\delta\|^{s'} \times (\pi^{-(s+\kappa/2)}\Gamma(s + \tfrac{1}{2}\kappa)(4\pi)^{-s'}\Gamma(s'))^m L(\bar{f} \otimes \omega, s' + 1 - \kappa_1 - \tfrac{1}{2}\kappa_2)L(\bar{f}, s' - \tfrac{1}{2}\kappa_2)$$

using the relations

$$\alpha_{2\mathfrak{p}} = \alpha_{\mathfrak{p}}N\mathfrak{p}^{\kappa_2/2} \qquad \beta_{2\mathfrak{p}} = \beta_{\mathfrak{p}}N\mathfrak{p}^{\kappa_2/2}.$$

At $s = \frac{1}{2}\kappa$, we have

$$s' = \tfrac{1}{2}\kappa - 1 + \tfrac{1}{2}(\kappa_1 + \kappa_2) = \tfrac{1}{2}(\kappa_2 - \kappa_1) - 1 + \tfrac{1}{2}(\kappa_1 + \kappa_2) = \kappa_2 - 1.$$

Therefore,

$$\langle E_1(*;\omega)E_2(*;\tfrac{1}{2}\kappa;\bar{\omega}), f\rangle$$

is a non-zero rational multiple of

$$\|\delta\|^{\kappa_2-1} \times (\pi^{-\kappa}\Gamma(\kappa)(4\pi)^{1-\kappa_2}\Gamma(\kappa_2 - 1))^m L(\bar{f} \otimes \omega, \tfrac{1}{2}\kappa_2 - \kappa_1)L(\bar{f}, \tfrac{1}{2}\kappa_2 - 1)$$

which is a rational multiple of

$$(\pi^{1-\kappa_2}\pi^{-(\kappa_2-\kappa_1)})^m L(\bar{f} \otimes \omega, \tfrac{1}{2}\kappa_2 - \kappa_1)L(\bar{f}, \tfrac{1}{2}\kappa_2 - 1).$$

As in the proof of the first theorem, for $s \in \mathbb{R}$ we have

$$L(\bar{f}, s) = L(f, s) \qquad L(\bar{f} \otimes \omega, s) = L(f \otimes \omega, s),$$

so that

$$\langle E_1(*;\omega)E_2(*;\tfrac{1}{2}\kappa;\bar{\omega}), f\rangle$$

is a non-zero rational multiple of

$$(\pi^{1-\kappa_2}\pi^{-(\kappa_2-\kappa_1)})^m L(f\otimes\omega, \tfrac{1}{2}\kappa_2-\kappa_1)L(f, \tfrac{1}{2}\kappa_2-1).$$

Now

$$\langle E_1(*;\omega)E_2(*;\bar{\omega}), f\rangle = \langle E_1(*;\omega)E_2(*;\tfrac{1}{2}\kappa;\bar{\omega}), f\rangle\{i^{m\kappa}D^{1/2}\gamma(\bar{\omega})\}^{-1}$$

which is a non-zero rational multiple of

$$\{i^{m(\kappa_2-\kappa_1)}D^{1/2}\gamma(\bar{\omega})\}^{-1}(\pi^{1-\kappa_2}\pi^{-(\kappa_2-\kappa_1)})^m L(f\otimes\omega, \tfrac{1}{2}\kappa_2-\kappa_1)L(f, \tfrac{1}{2}\kappa_2-1).$$

Combining this calculation with the first, we find that

$$\{i^{m(\kappa_2-\kappa_1)}D^{1/2}\gamma(\bar{\omega})^{-1}\pi^{m(1-\kappa_2)}\pi^{-m(\kappa_2-\kappa_1)}\}L(f\otimes\omega_1, \tfrac{1}{2}\kappa_2-\kappa_1)L(f, \tfrac{1}{2}\kappa_2-1)$$

lies in $\langle f, f\rangle\Omega$. This is the first assertion of the present theorem.

The second assertion follows from the property as remarked before, using the $\mathrm{Aut}(\mathbb{C}/\mathbb{Q})$-equivariance properties of the Eisenstein series, and using section 6.1. This finishes the proof of this theorem. ■

7 Proof of the Arithmetic Structure Theorem

This chapter gives the proof of the theorem of section 6.1. The first part of that theorem was first obtained by Shimura [Sh3] as part of a much more general result; the present proof is completely different. The present proof also yields the second part of the theorem of section 6.1 at the same time. The seeming lengthiness of the argument should not be begrudged.

The idea is that arithmetic aspects of the Fourier coefficients of cuspforms can be deduced from the arithmetic properties of the Fourier coefficients of a holomorphic Eisenstein series *on a larger group* (a symplectic group of rank two). Following Siegel [Si2], the Fourier coefficients of the latter are (products and ratios of) special values of L-functions of the underlying number field F; miraculously, all the special values needed are those which are algebraic multiples of powers of π (with explicit Galois properties), as in section 6.2. (For the present purpose, we need no general development of the theory of Siegel modular forms.) The "Main Formula" of section 7.9 gives the connection to Hilbert modular forms, from which the assertions of section 6.1 are obtained in section 7.10. Part of the bulk of section 7.10 is the relatively elementary conversion of results for $SL(2)$, which the previous considerations provide most naturally, into the corresponding results for $GL(2)$.

We note that the present method of proof can be extended to provide a similarly direct discussion of the arithmetic properties of cuspforms on many of the simpler classical groups.

7.1 Siegel's Holomorphic Eisenstein Series

Let

$$\mathrm{J} = \begin{pmatrix} 0 & 0 & -1 & 0 \\ 0 & 0 & 0 & -1 \\ 1 & 0 & 0 & 0 \\ 0 & 1 & 0 & 0 \end{pmatrix} \in GL(4).$$

For a commutative ring R, the *symplectic group* $Sp(2, R)$ is defined by

$$Sp(2, R) = \{g \in SL(4, R) \colon g^T J g = J\},$$

where g^T is the transpose of g. We will often use two-by-two block decompositions of matrices in $Sp(2, R)$. The group $Sp(2, R)$ contains some explicit subgroups which will be of interest subsequently; define

$$\tilde{L}(R) = \left\{\ell'(A) = \begin{pmatrix} A & 0 \\ 0 & A^{T-1} \end{pmatrix} \colon A \in GL(2, R)\right\}$$

$$\tilde{U}(R) = \left\{u'(x) = \begin{pmatrix} 1 & x \\ 0 & 1 \end{pmatrix} \colon x = x^T \text{ is a two-by-two matrix over } R\right\}$$

$$\tilde{P}(R) = \tilde{L}(R)\tilde{U}(R) = \tilde{U}(R)\tilde{L}(R).$$

Note that $\tilde{L}$ normalizes $\tilde{U}$ and that $\tilde{U}$ is abelian. Also, we have a compact subgroup $U^*(2)$ in $Sp(2, \mathbb{R})$ given by

$$\left\{\begin{pmatrix} A & B \\ -B & A \end{pmatrix} \in Sp(2, \mathbb{R}) \colon (A + iB)^*(A + iB) = 1_2\right\} \subset Sp(2, \mathbb{R}).$$

Let $V(R)$ be the R-module of two-by-four matrices over R. Let φ_0 be a Schwartz function on the space $V(\mathbb{A})$. Let Tr denote trace of a matrix. *Fix an integer* κ; define a Schwartz function ψ on $v = (c \quad d) \in V(\mathbb{R})$ (in two-by-two blocks) by

$$\psi(v) = \det(-ic + d)^\kappa \exp(-\pi \operatorname{Trace}(c^T c + d^T d)).$$

For each infinite prime $\mathfrak{p}$ of F, let $\varphi_\mathfrak{p} = \psi$, and define a Schwartz function on $V(\mathbb{A})$ by

$$\varphi = \varphi_0 \prod_{\mathfrak{p} \text{ infinite}} \varphi_\mathfrak{p}.$$

Definition If κ is even, take $\chi = 1$. For κ odd, let χ be the idele class character associated to the ideal class character

$$\mathfrak{a} \to (-1/\mathfrak{a})_2 \qquad (\mathfrak{a} \text{ prime to } 2\mathfrak{o}^{*-1})$$

of F. That is, for an ideal $\mathfrak{a}$ in $\mathfrak{o}$ prime to 2 and prime to the different $\mathfrak{o}^{*-1}$, $(-1/\mathfrak{a})_2$ is 1 if -1 is a square mod $\mathfrak{a}$, and is -1 if not.

Remarks From section 5.1 on reciprocity laws, this ideal class character has conductor dividing

$$\left(\prod_{\mathfrak{p}\text{ infinite}} \mathfrak{p}\right) 4\mathfrak{o}^{*-1}.$$

In particular, every infinite prime does divide the conductor: It is *totally odd.*

Definition For $g \in Sp(2, \mathbb{A})$, put

$$q(g) = q(g; \varphi_0) = \int_{GL(2,\mathbb{A})} \|\det t\|^{\kappa} \chi(\det t) \varphi(t(0_2 1_2)g)\, dt$$

where dt is a Haar measure on $GL(2, \mathbb{A})$, and 0_2 (resp., 1_2) is the two-by-two zero (respectively, identity) matrix. Define an adelic *Siegel's Eisenstein series* (of weight κ) by

$$\tilde{E}(g; \varphi_0) = \sum_{\gamma} q(\gamma g) \qquad (\gamma \in \tilde{P}(F)\backslash Sp(2, F)).$$

PROPOSITION Granting uniform absolute convergence of $\tilde{E}(g; \varphi_0)$ for g in compacta in $Sp(2, \mathbb{A})$, the Eisenstein series $\tilde{E}(g; \varphi_0)$ is a left $Sp(2, F)$-invariant function of g, and is right $\mathbb{K}$-invariant for some open compact subgroup $\mathbb{K}$ of $Sp(2, \hat{\mathfrak{o}})$. Further, $\tilde{E}(g; \varphi_0)$ is right-equivariant by the compact subgroup $U^*(2)^m \subset Sp(2, \mathbb{R})^m \subset Sp(2, \mathbb{A})$ by

$$\tilde{E}\left(g\begin{pmatrix} A & B \\ -B & A \end{pmatrix}; \varphi_0\right) = \tilde{E}(g; \varphi_0) \det(A + iB)^{\kappa}.$$

Proof The left invariance is formal, from the definition. The right invariance for finite primes follows (as for $GL(2)$ Eisenstein series) from the definition of Schwartz functions. The definition of the Schwartz function at infinite primes was designed to assure the right equivariance at infinite primes. It suffices to prove this equivariance for the kernel which was "automorphized" by left translations to construct the Eisenstein series. Let $g = \begin{pmatrix} a & b \\ c & d \end{pmatrix}$ be in $Sp(2, \mathbb{R})$, and $k = \begin{pmatrix} A & B \\ -B & A \end{pmatrix}$ in $U^*(2)$. We wish to show that

$$\int_{GL(2,\mathbb{A})} \|\det t\|^{\kappa} \chi(\det t) \varphi(t(0_2 1_2)gk)\, dt =$$

$$= \int_{GL(2,\mathbb{A})} \|\det t\|^{\kappa} \chi(\det t) \varphi(t(0_2 1_2)g)\, dt \times \det(A + iB)^{\kappa}.$$

The definition of the infinite-prime Schwartz function is

$$\varphi((0_2 1_2)g) = \varphi(cd) = \det(-ic + d)^{\kappa} \exp(-\pi \operatorname{Tr}(cc^T + dd^T)) =$$

$$= \det(-ic + d)^{\kappa} \exp(-\pi \operatorname{Tr}(ic + d)(ic + d)^*)$$

since $cd^T = dc^T$. Noting that

$$ic + d = (c \quad d)\delta$$

with

$$\delta = \begin{pmatrix} -i \\ 1 \end{pmatrix}$$

one sees that

$$(c \quad d)k\delta = (c \quad d)\delta(A + iB).$$

Since

$$(A + iB)^*(A + iB) = (A + iB)(A + iB)^* = 1_2,$$

this gives the result. ■

Definition For $g \in Sp(2, \mathbb{A})$, define the ξ-*th Fourier coefficient* of $\tilde{E}(g; \varphi_0)$ by

$$\int \bar{\tau}(\mathrm{Tr}(\xi x))\tilde{E}(u'(x)g; \varphi_0)\, du \qquad (u \in \tilde{U}(F)\backslash\tilde{U}(\mathbb{A})).$$

This integral is well-defined since (granting convergence) $\tilde{E}(g; \varphi_0)$ is surely left $\tilde{U}(F)$-invariant.

Definition Fix a *proper* ideal $\mathfrak{n}$ of $\mathfrak{o}$. Define a special Schwartz function φ^* on $V(\mathbb{A}_0)$ by

$$\varphi^* = \text{characteristic function of} \left\{ v \equiv \begin{pmatrix} 1 & 1 & 0 & 0 \\ 0 & 0 & -1 & 1 \end{pmatrix} \text{modulo } \mathfrak{n}V(\hat{\mathfrak{o}}) \right\}.$$

For fixed $\mathfrak{n}$ and fixed κ as above, for $g \in Sp(2, \mathbb{A})$ define

$$E(g) = \tilde{E}(g; \varphi^*).$$

Remarks *The purpose of sections 7.2–7.8 is to determine the arithmetic properties of the Fourier coefficients of these Eisenstein series.*

7.2 Cell Decomposition

The purposes of this section are to develop a little linear algebra (which will also be used in the proof of the Main Formula in section 7.9), and to break the Eisenstein series into three pieces for further analysis. Each piece will have a Fourier expansion of its own, and the three pieces require different but similar treatments.

Let $w_2 = J \in Sp(2, F)$, $w_0 = 1 \in Sp(2, F)$, and let

$$w_1 = \begin{pmatrix} 0 & 0 & -1 & 0 \\ 0 & 1 & 0 & 0 \\ 1 & 0 & 0 & 0 \\ 0 & 0 & 0 & 1 \end{pmatrix} \in Sp(2, F).$$

LEMMA The double coset $\tilde{P}(F)w_i\tilde{P}(F)$ is exactly the collection of $g = \begin{pmatrix} a & b \\ c & d \end{pmatrix}$ in $Sp(2, F)$ so that the two-by-two block c is of rank i. In particular, we have a (Bruhat) decomposition

$$Sp(2, F) = \bigcup_{i=0,1,2} \tilde{P}(F)w_i\tilde{P}(F) \qquad \text{(disjoint union).}$$ ■

LEMMA Let

$$SCP(F) = \{\text{two-by-four matrices } (c \quad d) \text{ over } F \text{ of rank 2 so that } cd^T = dc^T\}.$$

The quotient $\tilde{P}(F)\backslash Sp(2, F)$ is in bijection with

$$GL(2, F)\backslash SCP(F)$$

with the left multiplication action of $GL(2, F)$ on $SCP(F)$, under the map

$$\rho\colon \tilde{P}(F)\begin{pmatrix} a & b \\ c & d \end{pmatrix} \to GL(2, F)(c \quad d).$$

Proof First, from $g^TJg = J$ the relation $gJg^T = J$ follows; from this, if $g = \begin{pmatrix} a & b \\ c & d \end{pmatrix}$, we have $cd^T = dc^T$. Clearly the left action of $\tilde{P}(F)$ can change only the lower half of an element of $Sp(2, F)$ by left multiplication by $GL(2, F)$. Thus, this map is well-defined.

If g in $Sp(2, F)$ has bottom half $(c \quad d)$ and h in $Sp(2, F)$ has bottom half $A(c \quad d)$ with $A \in GL(2, F)$, then from $g^TJg = J$ we have

$$g^{-1} = J^{-1}g^TJ = \begin{pmatrix} d^T & * \\ -c^T & * \end{pmatrix}.$$

Therefore,

$$hg^{-1} = \begin{pmatrix} * & * \\ 0 & * \end{pmatrix} \in \tilde{P}(F).$$

Thus, the map is injective.

To show the surjectivity, we need to see that for any $(c \quad d)$ of rank 2 with $cd^T = dc^T$ there is $g = \begin{pmatrix} a & b \\ c & d \end{pmatrix}$ in $Sp(2, F)$. Writing out the condition $gJg^T = J$ shows that this is equivalent to

$$ad^T - bc^T = 1 \qquad cd^T = dc^T \qquad ab^T = ba^T.$$

Since $(c \quad d)$ is of rank 2, there are two-by-two matrices x and y so that $xd^T - yc^T = 1$. We need to adjust x and y so that $xy^T = yx^T$. For any two-by-two matrix z, we may certainly replace x by $x + zc$ and y by $y + zd$, since $cd^T = dc^T$. We will solve the equation

$$(x + zc)(y + zd)^T = (y + zd)(x + zc)^T$$

for z. Writing this condition out, it is

$$xy^T + zcy^T + xd^Tz^T + zcd^Tz^T = yx^T + zdx^T + yc^Tz + zdc^Tz^T.$$

Rearranging, using $cd^T = dc^T$, this is

$$xy^T - yx^T = z(xd^T - yc^T)^T - (xd^T - yc^T)z^T.$$

Using $xd^T - yc^T = 1$, this is equivalent to

$$xy^T - yx^T = z - z^T.$$

Thus, evidently, taking $z = xy^T$ suffices to meet the condition for $\begin{pmatrix} a & b \\ c & d \end{pmatrix}$ to lie in $Sp(2, F)$. ■

LEMMA Let

$$Q = \left\{ \begin{pmatrix} * & 0 \\ * & * \end{pmatrix} \in GL(2, F) \right\}.$$

Then $\rho(\tilde{P}(F)\backslash Sp(2, F)) \approx GL(2, F)\backslash SCP(F)$ has irredundant representatives of three sorts:

$(0_2 \quad 1_2)$ (the "little cell")

$(1_2 \quad S)$ with S symmetric 2×2 (the "big cell")

$\begin{pmatrix} 1 & 0 & s & 0 \\ 0 & 0 & 0 & 1 \end{pmatrix} \begin{pmatrix} A & 0_2 \\ 0_2 & A^{T-1} \end{pmatrix}$ with $s \in F$, A in $Q(F)\backslash GL(2, F)$ (the "middle cell").

Proof From the image by ρ of the coset

$$\tilde{P}(F)\backslash \tilde{P}(F)w_0\tilde{P}(F) = \tilde{P}(F)\backslash \tilde{P}(F)$$

we take the sole representative $(0_2 1_2)$. Noting that

$$\tilde{P}(F) \cap w_2 \tilde{P}(F) w_2^{-1} = \tilde{L}(F),$$

we have

$$\tilde{P}(F) \backslash \tilde{P}(F) w_2 \tilde{P}(F) = \tilde{P}(F) \backslash \tilde{P}(F) w_2 \tilde{U}(F) \approx w_2 \tilde{U}(F).$$

Under ρ, $w_2 \tilde{U}(F)$ is sent to the representatives we have labelled as being in the "big cell." Finally, the image by ρ of the other double coset consists of $(c \quad d)$ in $SCP(F)$ with c of rank 1. By left multiplication by an element of $GL(2)$ we may make the lower row of c be 0. By right inverse multiplication by an element $\left(\begin{smallmatrix} A & 0 \\ 0 & A^{T-1} \end{smallmatrix}\right)$ of $Sp(2, F)$, we may make c be of the indicated form. That is, we may choose a representative of the form

$$\begin{pmatrix} 1 & 0 & * & * \\ 0 & 0 & * & * \end{pmatrix} \begin{pmatrix} A & 0_2 \\ 0_2 & A^{T-1} \end{pmatrix}.$$

From the requirement that $cd^T = dc^T$, this must actually be of the form

$$\begin{pmatrix} 1 & 0 & * & * \\ 0 & 0 & 0 & d' \end{pmatrix} \begin{pmatrix} A & 0_2 \\ 0_2 & A^{T-1} \end{pmatrix}$$

with $d' \in F^\times$. A further left multiplication by an upper triangular element of $GL(2, F)$ puts this into the form

$$\begin{pmatrix} 1 & 0 & s & 0 \\ 0 & 0 & 0 & 1 \end{pmatrix} \begin{pmatrix} A & 0_2 \\ 0_2 & A^{T-1} \end{pmatrix}.$$

It is then clear that left multiplying A by an element of $Q(F)$ does not affect the form (left modulo $GL(2, F)$). It is not hard to check that these representatives are irredundant. ∎

From the last result, we can break the sum defining the Eisenstein series into three parts, corresponding to the small, middle, and big cells, each of which has a Fourier expansion of its own:

COROLLARY

$$E(g) = E_0(g) + E_1(g) + E_2(g)$$

with

$$E_0(g) = \int_{GL(2,\mathbb{A})} \|\det t\|^\kappa \chi(\det t) \varphi^*(t(0_2 1_2)g)\, dt \qquad \text{(the small cell)}$$

$$E_1(g) = \sum_{s, A} \int_{GL(2,\mathbb{A})} \|\det t\|^\kappa \chi(\det t) \varphi^*(t Y(s) \ell'(A) g)\, dt \qquad \text{(the middle cell)}$$

$$E_2(g) = \sum_{S} \int_{GL(2,\mathbb{A})} \|\det t\|^\kappa \chi(\det t) \varphi^*(t(1_2 S)g)\, dt \qquad \text{(the big cell)}$$

where

$$Y(s) = \begin{pmatrix} 1 & 0 & s & 0 \\ 0 & 0 & 0 & 1 \end{pmatrix}$$

and the sums are as in the last lemma. Each $E_i(g)$ has a Fourier expansion.

Proof The first assertion is immediate from the previous discussion. That each E_i has a Fourier expansion is a formal consequence of the fact that each double coset space

$$\tilde{P}(F)\backslash\tilde{P}(F)w_i\tilde{P}(F)$$

has a right action of $\tilde{U}(F) \subset \tilde{P}(F)$ on it. ∎

7.3 Convergence

Here we give a proof of convergence of Siegel's Eisenstein series which has the virtue of simplicity and brevity, but suffers the vice of failing to give the best estimate possible. The present approach is also in harmony with the style of our present discussion.

THEOREM The defining expression for the Eisenstein series $\tilde{E}(g; \varphi_0)$ of weight κ is uniformly convergent for $\kappa > 6$ and for g in compacta in $Sp(2, \mathbb{A})$.

Remarks It is possible, though more taxing, to prove the convergence for $\kappa \geq 4$. See [Br1], [Br2] for a proof of this for $F = \mathbb{Q}$ and symplectic groups of arbitrary size.

Proof From results of the first part of the previous section, we may write

$$\tilde{E}(g; \varphi_0) = \int_{GL(2,\mathbb{A})} \|\det t\|^{\kappa}\chi(\det t) \sum_{c,d} \varphi(t(c \quad d)g)\, dt$$

where $(c \quad d)$ is summed over

$$GL(2, F)\backslash SCP(F) \subset GL(2, F)\backslash V(F).$$

To prove convergence, we may as well take $\chi = 1$ and arbitrary non-negative-valued φ. Then the expression for the Eisenstein series is dominated by

$$\int_{GL(2,\mathbb{A})} \|\det t\|^{\kappa} \sum_{v} \varphi(t(c \quad d)g)\, dt$$

where v is summed over

$$GL(2, F)\backslash\{\text{elements of } V(F) \text{ of rank } 2\}.$$

By integration theory, this is

$$\int_{GL(2,\mathbb{A})/SL(2,\mathbb{A})} \|\det t\|^\kappa \chi(\det t) \int_{SL(2,\mathbb{A})} \sum_v \varphi(tt'(c \quad d)g)\, dt'\, dt.$$

CLAIM The function Φ on $V(\mathbb{A})$ defined by

$$\Phi\colon v \to \int_{SL(2,\mathbb{A})} \varphi(t'v)\, dt'$$

yields a Schwartz function on

$$W(\mathbb{A}) = SL(2,\mathbb{A})\backslash V(\mathbb{A}) \approx \text{second exterior power } \textstyle\bigwedge^2 \mathbb{A}^4 \text{ of } \mathbb{A}^4 \approx \mathbb{A}^6$$

(where we view the 2 rows of an element of $V(\mathbb{A})$ as giving two elements in $\mathbb{A}^4$).
(Proof is clear after sufficient reflection upon the definitions!) ■

Note that $GL(2,\mathbb{A})/SL(2,\mathbb{A})$ and $g \in Sp(2,\mathbb{A})$ still act upon $\bigwedge^2 \mathbb{A}^4$. As usual, for g in a compact subset of $Sp(2,\mathbb{A})$, the Schwartz functions $\Phi(vg)$ on $\bigwedge^2 \mathbb{A}^4$ are uniformly dominated by a Schwartz function $\Phi'(v)$ (for all v in $\bigwedge^2 \mathbb{A}^4$). Choose representatives

$$\ell(t) = \begin{pmatrix} t & 0 \\ 0 & 1 \end{pmatrix}$$

for $GL(2,\mathbb{A})/SL(2,\mathbb{A})$. Then the integral is

$$\int_{\mathbb{J}} \|\det t\|^\kappa \sum_{0 \neq v \in W(F)/F^\times} \Phi'(tv)\, dt$$

since the condition that $(c \quad d)$ be of rank 2 assures that its image in W is non-zero, and because $GL(2,F)$ acts on W by $F^\times$. Note that our representatives $\ell(t)$ for $GL(2,\mathbb{A})/SL(2,\mathbb{A})$ act by t on $W(\mathbb{A})$. By the usual unwinding trick, this is just

$$\int_{\mathbb{J}/F^\times} \|\det t\|^\kappa \sum_{0 \neq v \in W(F)} \Phi'(tv)\, dt.$$

The arguments we have used before in proving convergence of such "Mellin transforms of theta series" show that for general $n \in \mathbb{Z}$ an expression

$$\int_{\mathbb{J}/F^\times} \|\det t\|^s \sum_{0 \neq v \in F^n} \Phi'(tv)\, dt$$

(with Φ' a Schwartz function on $\mathbb{A}^n$, $s \in \mathbb{C}$) is absolutely convergent for $\mathrm{Re}(s) > n$. Thus, we obtain convergence for $\kappa > 6$, as asserted. ■

Remarks This estimate surely is not sharp, as we have estimated a sum over $SCP(F)$ by a sum over two-by-four rational matrices: The former is a subset of the latter defined by two independent (scalar) polynomial relationships.

7.4 Two Archimedean Integrals

Here we determine two elementary integrals over $\mathbb{R}$, which will be used subsequently. Let $\mathfrak{p}$ be an infinite prime of F.

LEMMA For $s \in \mathbb{C}$ with $\mathrm{Re}(s) > \frac{1}{2}$, we have

$$\int_{GL(2,\mathbb{R})} \|\det t\|^{2s} \exp(-\pi \operatorname{Tr}(tt^T))\, dt = \pi^{-s}\pi^{-s+1/2}\Gamma(s)\Gamma(s - \tfrac{1}{2}).$$

Proof The integrand is left $SO(2)$-invariant, so the integral is equal to the corresponding integral over $\tilde{P}(F_\mathfrak{p})$ with right Haar measure on $\tilde{P}(F_\mathfrak{p})$. Using coordinates

$$\begin{pmatrix} a & b \\ 0 & d \end{pmatrix}$$

this is

$$\iiint |ad|^{2s} \exp(-\pi(a^2 + b^2 + d^2))|d|^{-1} d^\times a\, d^+ b\, d^\times d \qquad (a, d \in \mathbb{R}^\times, b \in \mathbb{R})$$

(with usual notation for the standard multiplicative and additive measures on $\mathbb{R}^\times$ and $\mathbb{R}$). It is elementary that this is

$$\pi^{-s}\pi^{-s+1/2}\Gamma(s)\Gamma(s - \tfrac{1}{2}). \qquad \blacksquare$$

LEMMA Let $(c \quad d) \in SCP(\mathbb{R})$. Then

$$\int_{GL(2,\mathbb{R})} \|\det t\|^{\kappa} \chi(\det t) \varphi_\kappa(t(c \quad d))\, dt = \pi^{-\kappa}\pi^{-(\kappa-1/2)}\Gamma(\kappa)\Gamma(\kappa - \tfrac{1}{2}) \det(ic + d)^{-\kappa}.$$

The two-by-two complex matrix $ic + d$ lies in $GL(2, \mathbb{C})$.

Proof Granting that the two-by-two complex matrix $ic + d$ lies in $GL(2, \mathbb{C})$, we compute the integral first. Replacing t by $t(cc^T + dd^T)^{-1/2}$, from the definition of φ_κ the integral is

$$\det(-ic + d)^\kappa \det(cc^T + dd^T)^{-\kappa} \int_{GL(2,\mathbb{R})} \|\det t\|^{2\kappa} \chi(\det t) \exp(-\pi \operatorname{Tr}(t))\, dt.$$

Since $cd^T = dc^T$,

$$\det(-ic+d)^\kappa \det(cc^T+dd^T)^{-\kappa} = \det(-ic+d)^\kappa \det((-ic+d)(ic^T+d^T))^{-\kappa} =$$

$$= \det(ic+d)^{-\kappa}.$$

As in the previous lemma, the remaining integrand is left $SO(2)$-invariant, so is equal to the corresponding integral over $\tilde{P}(F_\mathfrak{p})$ with right Haar measure on $\tilde{P}(F_\mathfrak{p})$. Using coordinates

$$\begin{pmatrix} a & b \\ 0 & d \end{pmatrix}$$

this is

$$\iiint |ad|K|ad|^\kappa \exp(-\pi(a^2+b^2+d^2))|d|^{-1}\, d^\times a\, d^+ b\, d^\times d \qquad (a, d \in \mathbb{R}^\times, b \in \mathbb{R})$$

(with usual notation for the standard multiplicative and additive measures on $\mathbb{R}^\times$ and $\mathbb{R}$), where we use the fact that χ is totally odd if κ is odd. It is elementary that this is

$$\pi^{-\kappa}\pi^{-(\kappa-1/2)}\Gamma(\kappa)\Gamma(\kappa - \tfrac{1}{2}).$$

To see that $ic + d$ is in $GL(2, \mathbb{C})$, using $cd^T = dc^T$ we have

$$(ic + d)(-ic + d) = (cc^T + dd^T).$$

The latter must be a positive definite symmetric matrix, since $(c \quad d)$ is of full rank (i.e., 2). In particular, $cc^T + dd^T$ is in $GL(2, \mathbb{R})$. Therefore, $ic + d$ (and $-ic + d$) must be in $GL(2, \mathbb{C})$. ∎

7.5 The Small-Cell Contribution

The small-cell integral is independent of x; with g of the special form

$$g = u'(x)\ell'(y)$$

it is

$$E_0(g) = \int_{GL(2,\mathbb{A})} \|\det t\|^\kappa \chi(\det t)\varphi^*(t(0_2 y^{T-1}))\, dt.$$

The integral in t factors over primes, by construction of φ^*. At a finite prime $\mathfrak{p}$ dividing $\mathfrak{n}$, there is no $t \in GL(2, F_\mathfrak{p})$ so that $t0_2$ meets the required congruence. That is, *when* $\mathfrak{n}$ *is a proper ideal of* $\mathfrak{o}$ *the small-cell contribution is zero.*

7.6 The Middle-Cell Contribution

Regarding the middle cell, let $W(R)$ denote the space of symmetric two-by-two matrices over a ring R. With $g \in Sp(2, \mathbb{A})$ of the special form

$$g = \ell'(y),$$

the ξ-th Fourier coefficient of the middle-cell part E_1 of the Eisenstein series (evaluated at g) is

$$\int \sum_{s,A} \int_{GL(2,\mathbb{A})} \|\det t\|^{\kappa} \chi(\det t) \varphi^*(tY(s)\ell'(A)u'(x)\ell'(y)) \bar{\tau}(\mathrm{Tr}(\xi x))\, dt\, d^+x$$

where x is integrated over $W(\mathbb{A})/W(F)$. Replacing x by $A^{-1}xA^{T-1}$ (noting that $A \in GL(2, F)$) turns this into

$$\int \sum_{s,A} \int_{GL(2,\mathbb{A})} \|\det t\|^{\kappa} \chi(\det t) \varphi^*(tY(s)u'(x)\ell'(Ay)) \bar{\tau}(\mathrm{Tr}(A^{T-1}\xi Ax))\, dt\, d^+x.$$

One readily verifies that

$$Y(s)u'(x) = pY(s)$$

with $p \in SL(2, F_{\mathfrak{p}})$ if x is of the form

$$x = \begin{pmatrix} 0 & * \\ * & * \end{pmatrix}.$$

Thus,

$$\sum_{s,A} \int_{GL(2,\mathbb{A})} \|\det t\|^{\kappa} \chi(\det t) \varphi^*(tY(s)u'(x)\ell'(Ay))\, dt$$

is invariant under translation by $u'(x)$ with such x. Therefore, this integral vanishes unless

$$A^{T-1}\xi A^{-1} = \begin{pmatrix} \xi_1 & 0 \\ 0 & 0 \end{pmatrix}.$$

In the latter case, we may replace the integral over $x \in W(\mathbb{A})/W(F)$ by an integral over

$$x = \begin{pmatrix} x_1 & 0 \\ 0 & 0 \end{pmatrix}$$

with $x_1 \in \mathbb{A}/F$. Then the sum over s may be "unwound," so that the Fourier transform is

$$\sum_A \int_{\mathbb{A}} \int_{GL(2,\mathbb{A})} \|\det t\|^\kappa \chi(\det t)\varphi^*(tY(x_1)\ell'(Ay))\bar{\tau}(\xi_1 x_1)\,dt\,d^+x.$$

Once again, for fixed $A \in Q(F)\backslash GL(2,F)$, the integral factors over primes.

First, we compute the archimedean integral factors. Let

$$Ay = \begin{pmatrix} a & b \\ c & d \end{pmatrix}$$

with determinant δ. One has

$$Y(x_1)\ell'(Ay) = \begin{pmatrix} a & b & x_1 d/\delta & -x_1 b/\delta \\ 0 & 0 & -c/\delta & a/\delta \end{pmatrix}.$$

Then the calculation of the archimedean integral (in section 7.4) yields

$$\int_{F_\mathfrak{p}} \int_{GL(2,F_\mathfrak{p})} \|\det t\|^\kappa \chi(\det t)\varphi_\kappa(tY(x_1)\ell'(Ay))\bar{\tau}(\xi_1 x_1)\,dt\,d^+x =$$

$$= \pi^{-\kappa}\pi^{-(\kappa-1/2)}\Gamma(\kappa)\Gamma(\kappa - \tfrac{1}{2})\delta^\kappa \det(x_1 + i(a^2 + b^2))^{-\kappa}.$$

Here we have used the fact that χ is of the same parity as κ, to avoid making this integral 0 for κ odd. By residues, the Fourier transform of $\det(x_1 + iu)^{-\kappa}$ for $u > 0$, evaluated at ξ_1, is

$$\begin{array}{ll} 0 & \text{if } \xi_1 \le 0 \\ (-2\pi i)^\kappa \Gamma(\kappa)^{-1}\xi_1^{\kappa-1}\exp(-2\pi\xi_1 u) & \text{if } \xi_1 > 0. \end{array}$$

(Here $\xi_1 > 0$ means at the prime $\mathfrak{p}\ldots$) In particular, this shows that the indices ξ contributing to E_1 are exactly those which are *totally positive semi-definite and of rank one.*

Note that we have made ξ_1 (which we now know to be in $F^\times$ for there to be a non-zero contribution) depend on A implicitly. To turn this around: It is easy to show that, given ξ of rank one, there is at most one $A \in Q(F)\backslash GL(2,F)$ and $\xi_1 \in F^\times$ so that

$$A^{T-1}\xi A^{-1} = \begin{pmatrix} \xi_1 & 0 \\ 0 & 0 \end{pmatrix}.$$

Therefore, we could unambiguously index the Fourier indices ξ which give a non-zero contribution to E_1 via $A \in Q(F)\backslash GL(2,F)$ and $\xi_1 \in F^\times$ by

$$\xi = A^T\begin{pmatrix}\xi_1 & 0\\ 0 & 0\end{pmatrix}A.$$

Therefore, the sum over A in the middle-cell contribution consists of just one summand (for fixed ξ totally positive semi-definite of rank 1), which factors over primes.

At a finite good prime $\mathfrak{p}$, as a function of $Ay \in GL(2, F_\mathfrak{p})$, the function

$$\int_{F_\mathfrak{p}}\int_{GL(2,F_\mathfrak{p})} \|\det t\|^\kappa \chi(\det t)\varphi^*(tY(x_1)\ell'(Ay))\bar{\tau}(\xi_1 x_1)\,dt\,d^+x$$

is right $GL(2, \mathfrak{o}_\mathfrak{p})$-invariant. Therefore, without loss of generality (by a $\mathfrak{p}$-adic Iwasawa decomposition), we may suppose that

$$(Ay)^{T-1} = \begin{pmatrix}a & b\\ 0 & d\end{pmatrix}.$$

Then

$$Y(x)\ell'(Ay) = \begin{pmatrix}a^{-1} & 0 & ax & b\\ 0 & 0 & 0 & d\end{pmatrix}.$$

Left multiplication of this on the left by an element of $SL(2, F_\mathfrak{p})$ puts it into the form

$$\begin{pmatrix}a^{-1} & 0 & ax_1 & 0\\ 0 & 0 & 0 & d\end{pmatrix}.$$

Thus, by changing variables in the integral, we may suppose that $b = 0$. Further, as the integrand is left $GL(2, \mathfrak{o}_\mathfrak{p})$-invariant, we may integrate over upper triangular t with respect to a suitable right Haar measure. Let $\mathrm{ch}(*; X)$ be the characteristic function of a set X; for brevity, let $\mathrm{ch}(*)$ denote the characteristic function of $\mathfrak{o}_\mathfrak{p}$. In coordinates

$$t = \begin{pmatrix}t_1 & t_2 t_3\\ 0 & t_3\end{pmatrix}$$

the inner integral over $GL(2, F_\mathfrak{p})$ becomes

$$\int \|t_1 t_3\|^\kappa \chi(t_1 t_3)\,\mathrm{ch}(t_1 a^{-1})\,\mathrm{ch}(t_1 a x_1)\,\mathrm{ch}(t_2 t_3 d)\,\mathrm{ch}(t_3 d)\,d^\times t_1\,d^+ t_2\,d^\times t_3.$$

The integral over t_2 is just $\|t_3 d\|_\mathfrak{p}^{-1}$ (with correct normalization of additive Haar measure). Thus, the integral over t_3 is a geometric series which sums to

$$\bar{\chi}(d)^{-1}\|d\|_\mathfrak{p}^{-\kappa}(1 - \chi(\mathfrak{p})N\mathfrak{p}^{1-\kappa})^{-1}.$$

Thus, the desired Fourier transform is

$$\bar{\chi}(d)^{-1}\|d\|_{\mathfrak{p}}^{-\kappa}(1-\chi(\mathfrak{p})N\mathfrak{p}^{1-\kappa})^{-1}\int_{F_{\mathfrak{p}}}\int_{F_{\mathfrak{p}}^{\times}}\|\det t_1\|^{\kappa}\chi(t_1)\operatorname{ch}(t_1a^{-1})\operatorname{ch}(t_1ax_1)\bar{\tau}(\xi_1x_1)\,d^{\times}t_1\,d^{+}x_1.$$

We can easily compute the indicated Fourier transform by interchanging the integration; it is

$$\|a\|_{\mathfrak{p}}^{-1}[\mathfrak{o}_{\mathfrak{p}}^{*}:\mathfrak{o}_{\mathfrak{p}}]^{-1/2}\int_{F_{\mathfrak{p}}^{\times}}\|\det t_1\|^{\kappa}\chi(t_1)\operatorname{ch}(t_1a^{-1})\operatorname{ch}(\xi_1/t_1a)\,d^{\times}t_1.$$

The remaining integral in t_1 is clearly a finite sum; for fixed a, d, ξ_1 *it is 1 for almost all primes* $\mathfrak{p}$, e.g., whenever a, d, ξ_1 are all $\mathfrak{p}$-units. That is, *for almost all finite primes not dividing* $\mathfrak{n}$, *the local integral is just*

$$[\mathfrak{o}_{\mathfrak{p}}^{*}:\mathfrak{o}_{\mathfrak{p}}]^{-1/2}/(1-\chi(\mathfrak{p})N\mathfrak{p}^{1-\kappa}).$$

In any case, at a finite prime $\mathfrak{p}$ *not dividing* $\mathfrak{n}$, *it is a rational multiple of the latter expression.*

Remarks One can use an argument identical to that just above to compute the higher Fourier coefficients of Eisenstein series for $GL(2)$, if all that were desired were to obtain the Galois equivariance property of the higher Fourier coefficients without knowing their precise nature.

Now consider a finite prime dividing $\mathfrak{n}$. The parameters a, d, b, ξ_1 are fixed. To evaluate

$$\int_{F_{\mathfrak{p}}}\int_{GL(2,F_{\mathfrak{p}})}\|\det t\|^{\kappa}\chi(\det t)\varphi^{*}(tY(x_1)\ell'(Ay))\bar{\tau}(\xi_1x_1)\,dt\,d^{+}x$$

let

$$(Ay)^{T-1}=\begin{pmatrix}a & b\\ c & d\end{pmatrix}$$

where $\det\begin{pmatrix}a & b\\ c & d\end{pmatrix}=\delta$. Let $t=\begin{pmatrix}A & B\\ C & D\end{pmatrix}$. Then

$$tY(x_1)\ell'(Ay)=\begin{pmatrix}Ad/\delta & -Ac/\delta & Aax_1+Bc & Abx+Bd\\ Cd/\delta & -Cc/\delta & Cax_1+Dc & Cbx+Dd\end{pmatrix}.$$

First, we claim that this integral is equal to a finite sum (of values of the same integrand) *multiplied by* $[\mathfrak{o}_{\mathfrak{p}}^{*}:\mathfrak{o}_{\mathfrak{p}}]^{-1/2}$, *by normalization of the additive Haar measure.* The support of φ^{*} consists of things congruent to

$$\begin{pmatrix}1 & 1 & 0 & 0\\ 0 & 0 & -1 & 1\end{pmatrix}$$

modulo $\mathfrak{n}$. Given t and x_1 so that this congruence holds, from observation of the left two columns of $tY(x_1)\ell'(Ay)$ it is clear that $\|A\|$ is determined (by a, b, c, d, δ) and $\|C\|$ has an upper bound. Given these observations on A and C, it is clear that there is a non-zero ideal $\mathfrak{a}$ in $\mathfrak{o}_\mathfrak{p}$ so that for *any* t and x_1 so that this congruence holds, t and $x_1 + \varepsilon$ also satisfy the condition, for all $\varepsilon \in \mathfrak{a}$. That is, $\varphi^*(tY(x_1)\ell'(Ay))$ is uniformly locally constant in x_1; since $\bar{\tau}(\xi_1 x_1)$ certainly has the same property, we have shown that the integrand is *uniformly locally constant in* x_1 (uniformly in t).

Toward proving that the integral is a finite sum, we next show that the integral is equal to an integral (with the same integrand) with compact support. For *arbitrary* x_1, if x_1 is replaced by $(1 + \eta)x_1$ with $\eta \in \mathfrak{n}\mathfrak{o}_\mathfrak{p}$ then by replacing A and C by $A/(1 + \eta)$ and $C/(1 + \eta)$ the value of the integral of $\varphi^*(tY(x_1)\ell'(Ay))$ over t is visibly unchanged. Note that this change of variables corresponds to right multiplying t by

$$\begin{pmatrix} (1+\eta)^{-1} & 0 \\ 0 & 1 \end{pmatrix} \in \mathbb{K}(\mathfrak{n}).$$

Fix a large positive integer n, and let

$$S_n = \{x_1 \in F_\mathfrak{p} \colon \|x_1\| = N\mathfrak{p}^n\}.$$

Then $H = \{1 + \eta \colon \eta \in \mathfrak{p}\mathfrak{o}_\mathfrak{p}\}$ acts on S_n by multiplication. Then for fixed $x_1 \in S_n$,

$$\int_{H(x_1)} \int_{GL(2,F\mathfrak{p})} \|\det t\|^\kappa \chi(\det t)\varphi^*(tY(x_1)\ell'(Ay))\bar{\tau}(\xi_1 x_1)\, dt\, d^+x =$$

$$= \text{constant} \times \int_{\mathfrak{p}\mathfrak{o}_\mathfrak{p}} \bar{\tau}(\xi_1 x_1 \eta)\, d^+\eta.$$

For n sufficiently large, the latter integral is 0. Since S_n is a disjoint union of orbits by H, this implies that

$$\int_Y \int_{GL(2,F_\mathfrak{p})} \|\det t\|^\kappa \chi(\det t)\varphi^*(tY(x_1)\ell'(Ay))\bar{\tau}(\xi_1 x_1)\, dt\, d^+x = 0$$

where

$$Y = \bigcup_{n \geq n_0} S_n$$

for n_0 sufficiently large. Therefore, we may as well integrate (the integral in t) in x over a sufficiently large *compact* set. Therefore, recalling the uniform local constancy in x, there is a *finite set* X of values of x_1 so that

$$\int_{F_\mathfrak{p}} \int_{GL(2,F_\mathfrak{p})} \|\det t\|^\kappa \chi(\det t)\varphi^*(tY(x_1)\ell'(Ay))\bar{\tau}(\xi_1 x_1)\, dt\, d^+x =$$

$$= [\mathfrak{o}_\mathfrak{p}^* : \mathfrak{o}_\mathfrak{p}]^{-1/2} \sum_{x_1 \in X} \int_{GL(2,F_\mathfrak{p})} \|\det t\|^\kappa \chi(\det t)\varphi^*(tY(x_1)\ell'(Ay))\bar{\tau}(\xi_1 x_1)\, dt.$$

Given x_1, it is immediate that, if there is t in $GL(2, F_\mathfrak{p})$ so that $tY(x_1)\ell'(Ay)$ is in the support of φ^*, then there is *exactly* one such t left modulo

$$\mathbb{K}(\mathfrak{n})_\mathfrak{p} = \{g \in GL(2, \mathfrak{o}_\mathfrak{p}) : g \equiv 1 \text{ modulo } \mathfrak{n}\}.$$

Certainly $\chi(\det t)$ is invariant under this group. Thus, for fixed x_1 the integral over t is given by a single value t_{x_1}. Therefore,

$$\int_{F_\mathfrak{p}} \int_{GL(2,F_\mathfrak{p})} \|\det t\|^\kappa \chi(\det t)\varphi^*(tY(x_1)\ell'(Ay))\overline{\tau}(\xi_1 x_1)\, dt\, d^+x =$$

$$= \text{measure}(\mathbb{K}(\mathfrak{n})_\mathfrak{p}) \times \sum_{x_1 \subset X} \|\det t_{x_1}\|^\kappa \chi(\det t_{x_1})\varphi^*(t_{x_1} Y(x_1)\ell'(Ay))\overline{\tau}(\xi_1 x_1).$$

This proves our claim.

Therefore, disregarding the factor of $[\mathfrak{o}_\mathfrak{p}^* : \mathfrak{o}_\mathfrak{p}]^{-1/2}$, the value of this bad-prime integral certainly lies in the field $\mathbb{Q}_{ab}$ generated by all roots of unity; indeed, it lies in a field Ω generated over $\mathbb{Q}$ by some $N\mathfrak{p}$-power roots of unity. Given an element $\sigma = \sigma_\eta$ of $\mathrm{Gal}(\Omega/\mathbb{Q})$, it can be described by $\zeta \to \zeta\eta^{-1}$ (ζ a root of unity) with $\eta \in \mathbb{Z}_p^x$ where p is the rational prime lying under $\mathfrak{p}$; for present purposes σ_η can be given by η of finite order. Therefore, we have

$$\sigma\left(\sum_{x_1 \in X} \|\det t_{x_1}\|^\kappa \chi(\det t_{x_1})\varphi^*(t_{x_1} Y(x_1)\ell'(Ay))\overline{\tau}(\xi_1 x_1) \right) =$$

$$= \sum_{x_1 \in X} \|\det t_{x_1}\|^\kappa \chi(\det t_{x_1})\varphi^*(t_{x_1} Y(x_1)\ell'(Ay))\overline{\tau}(\xi_1 x_1 \eta^{-1}) =$$

$$= \sum_{x_1 \in X} \|\det t_{\eta x_1}\|^\kappa \chi(\det t_{\eta x_1})\varphi^*(t_{\eta x_1} Y(\eta x_1)\ell'(Ay))\overline{\tau}(\xi_1 x_1),$$

since we can easily arrange the finite set X to be stable under multiplication by η. Further, if we have

$$t_{x_1} Y(x_1)\ell'(Ay) \equiv \begin{pmatrix} 1 & 1 & 0 & 0 \\ 0 & 0 & -1 & 1 \end{pmatrix} \quad \text{modulo } \mathfrak{n}\mathfrak{o}_\mathfrak{p}$$

then by replacing x_1 by ηx_1 and t_{x_1} by

$$t_{\eta x_1} = \begin{pmatrix} \eta & 0 \\ 0 & 1 \end{pmatrix} t \begin{pmatrix} \eta^{-1} & 0 \\ 0 & 1 \end{pmatrix}$$

we still have

$$t_{\eta x_1} Y(\eta x_1)\ell'(Ay) \equiv \begin{pmatrix} 1 & 1 & 0 & 0 \\ 0 & 0 & -1 & 1 \end{pmatrix} \quad \text{modulo } \mathfrak{n}\mathfrak{o}_\mathfrak{p}.$$

This certainly makes use of the specific nature of this congruence requirement. Therefore,

$$\sigma\left(\sum_{x_1\in X}\|\det t_{x_1}\|^\kappa\chi(\det t_{x_1})\bar\tau(\xi_1x_1)\right)=$$

$$=\sigma\left(\sum_{x_1\in X}\|\det t_{x_1}\|^\kappa\chi(\det t_{x_1})\varphi^*(t_{x_1}Y(x_1)\ell'(Ay))\bar\tau(\xi_1x_1)\right)=$$

$$=\sum_{x_1\in X}\|\det t_{\eta x_1}\|^\kappa\chi(\det t_{\eta x_1})\varphi^*(t_{\eta x_1}Y(\eta x_1)\ell'(Ay))\bar\tau(\xi_1x_1)=$$

$$=\sum_{x_1\in X}\|\det t_{\eta x_1}\|^\kappa\chi(\det t_{\eta x_1})\bar\tau(\xi_1x_1)=$$

$$=\sum_{x_1\in X}\|\det t_{x_1}\|^\kappa\chi(\det t_{x_1})\bar\tau(\xi_1x_1).$$

That is, the whole summation is unchanged. *Therefore, the local integral at a prime dividing* $\mathfrak{n}$ *is a rational multiple of* $[\mathfrak{o}_\mathfrak{p}^*:\mathfrak{o}_\mathfrak{p}]^{-1/2}$.

7.7 The Big-Cell Contribution

The Fourier expansion of the big-cell contribution E_2 is the most interesting. We have

$$E_2(g)=\sum_S\int_{GL(2,\mathbb{A})}\|\det t\|^\kappa\chi(\det t)\varphi^*(t(1_2S)g)\,dt.$$

Therefore, for an index ξ we have the ξ-th Fourier coefficient

$$\int_{W(F)\backslash W(\mathbb{A})}\sum_S\int_{GL(2,\mathbb{A})}\|\det t\|^\kappa\chi(\det t)\varphi^*(t(1_2S)u'(x)g)\bar\tau(\mathrm{Tr}(\xi x))\,dt\,d^+x=$$

$$=\int_{W(F)\backslash W(\mathbb{A})}\sum_S\int_{GL(2,\mathbb{A})}\|\det t\|^\kappa\chi(\det t)\varphi^*(t(1_2S+x)g)\bar\tau(\mathrm{Tr}(\xi x))\,dt\,d^+x=$$

$$=\int_{W(\mathbb{A})}\int_{GL(2,\mathbb{A})}\|\det t\|^\kappa\chi(\det t)\varphi^*(t(1_2x)g)\bar\tau(\mathrm{Tr}(\xi x))\,dt\,d^+x$$

by unwinding. *Once again, this is a product of local integrals.* As in the previous calculations, we have a fixed g of the form $g=\ell'(y)$ with $y\in GL(2,\mathbb{A})$.

From the archimedean calculation in section 7.4, for $F_\mathfrak{p}\approx\mathbb{R}$ we have

$$\int_{GL(2,\mathbb{R})}\|\det t\|^\kappa\chi(\det t)\varphi_\kappa(t(c\quad d))\,dt=\pi^{-\kappa}\pi^{-(\kappa-1/2)}\Gamma(\kappa)\Gamma(\kappa-\tfrac12)\det(ic+d)^{-\kappa}.$$

We must compute a Fourier transform along $W(\mathbb{R})$ of $\det(iM + x)^{-\kappa} = i^{-\kappa}\det(M - ix)^{-\kappa}$ with respect to the pairing given by the trace Tr, where $M = yy^T$. Define

$$\Gamma_2(s) = \int_{GL(2,\mathbb{R})} |\det t|^{2s} \exp(-\pi \operatorname{Tr}(t^T t))\, dt.$$

The first lemma in section 7.4 shows that

$$\Gamma_2(s) = \pi^{-s}\Gamma(s)\pi^{-s+1/2}\Gamma(s - \tfrac{1}{2}).$$

On the other hand, for a positive definite two-by-two real matrix M, consider

$$\Gamma_2(s; M) = \int_{GL(2,\mathbb{R})} |\det t|^{2s} \exp(-\pi \operatorname{Tr}(t^T t M))\, dt.$$

By replacing t by $tM^{-1/2}$ in the integral, we obtain the identity

$$\Gamma_2(s; M) = (\det M)^{-s}\Gamma_2(s)$$

first for $\operatorname{Re}(s) > \frac{1}{2}$, and then for $s \in \mathbb{C}$ by analytic continuation in s. For x in the space $W(\mathbb{R})$ of symmetric two-by-two real matrices, by analytic continuation inside the (contractible) set

$$\{M - ix \text{ in } W(\mathbb{C}): M \text{ is positive definite}\} \subset W(\mathbb{C})$$

we have

$$\int_{GL(2,\mathbb{R})} |\det t|^{2s} \exp(-\pi \operatorname{Tr}(t^T t(M - ix)))\, dt = \Gamma_2(s; M - ix) = \det(M - ix)^{-s}\Gamma_2(s).$$

Now the integrand is left $SO(2)$-invariant (in t), so we can rewrite this as an integral over

$$C_2 = \{\eta \text{ in } W(\mathbb{R}): \eta \text{ is positive definite}\} \approx SO(2)\backslash GL(2, \mathbb{R})$$

where the inverse of this map is

$$t \to \eta = t^T t$$

(with suitable choice of measure). Let $\eta = t^T t$; if $d^+\eta$ is the additive measure on $W(\mathbb{R})$ given by

$$d^+\begin{pmatrix}\eta_1 & \eta_2\\ \eta_2 & \eta_3\end{pmatrix} = 2d\eta_1\, d\eta_2\, d\eta_3,$$

then one may readily verify that

$$\det \eta^{-3/2}\, d^+\eta$$

is the compatible measure on C_2. The factor of 2 occurs because $SO(2)$ meets the subgroup $P(\mathbb{R})$ of upper triangular matrices in $\{\pm 1\}$, and because we defined the measure on $GL(2, \mathbb{R})$ via the Haar integral

$$f \to \int_{P(\mathbb{R})} \int_{SO(2)} f.$$

Therefore,

$$2 \int_{C_2} |\det \eta|^{s-3/2} \exp(-\pi \operatorname{Tr}(\eta(M - ix)))\, d^+\eta = \det(M - ix)^{-s}\Gamma_2(s).$$

Replacing η by 2η, we may write this as

$$2^{s+1} \int_{C_2} \{|\det \eta|^{s-3/2} \exp(-2\pi \operatorname{Tr} \eta M)\} \exp(2\pi i \operatorname{Tr}(\eta x))\, d^+\eta = \det(M - ix)^{-s}\Gamma_2(s).$$

That is, we are computing a certain normalization of an inverse Fourier transform on $W(\mathbb{R})$ of the function

$$\Phi(\eta) = \begin{cases} |\det \eta|^{s-3/2} \exp(-2\pi \operatorname{Tr} \eta M) & \text{if } \eta \in C_2 \\ 0 & \text{otherwise.} \end{cases}$$

It is important to note that

$$\operatorname{Tr}(\eta x) = \eta_1 x_1 + 2\eta_2 x_2 + \eta_3 x_3$$

where

$$\eta = \begin{pmatrix} \eta_1 & \eta_2 \\ \eta_2 & \eta_3 \end{pmatrix} \qquad x = \begin{pmatrix} x_1 & x_2 \\ x_2 & x_3 \end{pmatrix}.$$

Therefore, we write

$$2^{s+1} \int_{C_2} \{|\det \eta|^{s-3/2} \exp(-2\pi \operatorname{Tr} \eta M)\} \exp(2\pi i \operatorname{Tr}(\eta x))\, d^+\eta = \check{\Phi}(x_1, 2x_2, x_3).$$

Similarly, what we want to compute is the Fourier transform of $\Psi(x_1, x_2, x_3) = \check{\Phi}(x_1, 2x_2, x_3)$ evaluated at $(\xi_1, 2\xi_2, \xi_3)$, where we use similar coordinates for ξ. With the obvious (abusive) notation, from formal properties of Fourier transforms we have

$$\check{\Phi}(x_1, 2x_2, x_3)\hat{}(\xi_1, 2\xi_2, \xi_3) = \int_{W(\mathbb{R})} \bar{\tau}(x_1\xi_1 + 2x_2\xi_2 + x_3\xi_3)\check{\Phi}(x_1, 2x_2, x_3)\,d^+x =$$

$$= \tfrac{1}{2}\int_{W(\mathbb{R})} \bar{\tau}(x_1\xi_1 + x_2\xi_2 + x_3\xi_3)\check{\Phi}(x_1, x_2, x_3)\,d^+x =$$

$$= \tfrac{1}{2}\Phi^{\check{}\,\hat{}}(\xi_1, \xi_2, \xi_3) = \tfrac{1}{2}\Phi(\xi),$$

by changing variables in the Fourier integral and invoking Fourier inversion on $\mathbb{R}^3$.

Thus, putting this all together, we have found that the desired local integral is

$$(\pi^{-\kappa}\pi^{-(\kappa-1/2)}\Gamma(\kappa)\Gamma(\kappa - \tfrac{1}{2})) \times i^{-\kappa} \times (\tfrac{1}{2}\Gamma_2(\kappa)^{-1}2^{\kappa+1}\Phi(\xi)) =$$

$$= 2^{\kappa}i^{-\kappa}\Phi(\xi) =$$

$$= \begin{cases} 2^{\kappa}i^{-\kappa}|\det\xi|^{\kappa-3/2}\exp(-2\pi\,\mathrm{Tr}\,\xi M) & \text{if } \xi \in C_2 \\ 0 & \text{otherwise.}\end{cases}$$

In particular, this shows that we only get non-zero contributions for ξ *totally positive definite.*

Now consider a finite prime $\mathfrak{p}$ dividing $\mathfrak{n}$, using the fact that ξ is totally positive definite. We claim that the local integral can be evaluated in the form

$$\int_{W(F_\mathfrak{p})}\int_{GL(2,F_\mathfrak{p})} \|\det t\|^{\kappa}\chi(\det t)\varphi^*(t(1_2 x)g)\bar{\tau}(\mathrm{Tr}(\xi x))\,dt\,d^+x =$$

$$= \mathrm{measure}(\mathbb{K}(\mathfrak{n})_\mathfrak{p})[\mathfrak{o}_\mathfrak{p}^* : \mathfrak{o}_\mathfrak{p}]^{-3/2}\sum_{x\in X} \|\det t_x\|^{\kappa}\chi(\det t_x)\varphi^*(t_x(1_2 x)g)\bar{\tau}(\mathrm{Tr}(\xi x)),$$

where t_x depends on x, and where

$$\mathbb{K}(\mathfrak{n})_\mathfrak{p} = \{k \in GL(2, \mathfrak{o}_\mathfrak{p}) \colon h \equiv 1 \text{ modulo } \mathfrak{n}\mathfrak{o}_\mathfrak{p}\}.$$

If we can prove this, then the value of the integral clearly lies in $\mathbb{Q}_{ab}$, and we can apply elements of $\mathrm{Gal}(\mathbb{Q}_{ab}/\mathbb{Q})$ termwise. We first claim that the integrand is uniformly locally constant in x (uniformly in t). Let

$$(c \quad d) = \begin{pmatrix} 1 & 1 & 0 & 0 \\ 0 & 0 & -1 & 1 \end{pmatrix}$$

and let Λ be the module of two-by-four matrices over $\mathfrak{o}_\mathfrak{p}$. Then the congruence condition

$$(ty \quad txy^{T-1}) \in (c \quad d) + \mathfrak{n}\Lambda$$

is equivalent to

$$(t \quad tx) \in (cy^{-1} \quad dy^{T}) + \mathfrak{n}\Lambda\ell'(y)^{-1}.$$

Since y is fixed, it is clear that $\varphi^*(t_x(1_2 x)g)$ is uniformly locally constant in x (uniformly in t). Certainly the $\bar{\tau}(\mathrm{Tr}(x\xi))$ is uniformly locally constant.

Similarly, it is not hard to see that there is a proper ideal $\mathfrak{m} \subset \mathfrak{n}$ (depending on y) in $\mathfrak{o}$ so that for h in

$$H = \{h \in GL(2, \mathfrak{o}_{\mathfrak{p}}) \colon h \equiv 1_2 \text{ modulo } \mathfrak{m}\}$$

we have (for every x and t)

$$\varphi^*(t_x(1_2 x)g) = \varphi^*(t_x(h^{-1}xh^{T})g).$$

Therefore,

$$\int_{GL(2,F_{\mathfrak{p}})} \|\det t\|^{\kappa}\chi(\det t)\varphi^*(t(1_2 x)g)\,dt =$$

$$= \int_{GL(2,F_{\mathfrak{p}})} \|\det t\|^{\kappa}\chi(\det t)\varphi^*(t(h^{-1}xh^{T})g)\,dt =$$

$$= \int_{GL(2,F_{\mathfrak{p}})} \|\det t\|^{\kappa}\chi(\det t)\varphi^*(t(1_2 hxh^{T})g)\,dt$$

by replacing t by th. That is, as a function of x this integral over t is constant on cosets $H(x) = \{hxh^{T} \colon h \in H\}$ (uniformly in t). Now take $h = 1 + \eta$ in H; we have

$$hxh^{T} = x + \eta x + x\eta^{T} + \eta x \eta^{T}.$$

Note that

$$\mathrm{Tr}(\xi hxh^{T}) = \mathrm{Tr}(\xi x) + \mathrm{Tr}(\xi\eta x) + \mathrm{Tr}(\xi x\eta^{T}) + \mathrm{Tr}(\xi\eta x\eta^{T}) =$$

$$= \mathrm{Tr}(\xi x) + 2\,\mathrm{Tr}(\xi\eta x) + \mathrm{Tr}(\xi\eta x\eta^{T}).$$

For a positive integer n, define

$$S_n = \{x \in W(F_{\mathfrak{p}}) \colon \text{the sup of the norms of the entries of } x \text{ is } N\mathfrak{p}^{n}\}.$$

Clearly S_n is stabilized by H, for $n > 0$. Let k be such that $\xi \in S_k$. Let

$$Y_n = \{\eta \text{ two-by-two: } \eta \equiv 0 \bmod \mathfrak{p}^N,\ N \geq \tfrac{1}{2}(n+k)\} \subset Y$$

$$H_n = \{1 + \eta : \eta \in Y_n\}.$$

Then for fixed x in S_n

$$\int_{H_n(x)} \int_{GL(2,F_\mathfrak{p})} \|\det t\|^\kappa \chi(\det t)\varphi^*(t(1_2 x)g)\bar{\tau}(\mathrm{Tr}(\xi x))\,dt\,d^+x =$$

$$= \text{constant} \times \int_{Y_n} \bar{\tau}(2\,\mathrm{Tr}\,\xi x\eta^T + \mathrm{Tr}\,\xi\eta x\eta^T)\,d^+\eta =$$

$$= \text{constant} \times \int_{Y_n} \bar{\tau}(2\,\mathrm{Tr}\,\xi x\eta^T)\,d'\eta$$

since $\bar{\tau}(\mathrm{Tr}\,\xi\eta x\eta^T) = 1$, by the construction of Y_n. We claim that for n sufficiently large (depending on ξ), for any $x \in S_n$ the latter integral is 0. To this end, it suffices to find (for given x and ξ) one η_0 in Y_n so that $\bar{\tau}(2\,\mathrm{Tr}\,\xi x\eta^T) \neq 1$; this is easy. Since S_n is a disjoint union of H_n-orbits, the integral over x in $\bigcup_{n \geq n_0} S_n$ is 0 for n_0 sufficiently large. That is, we need only integrate over x in some large compact set (uniformly in t).

Further, because of the specific nature of the congruence condition, it is clear that for a given x there is at most one t left modulo $\mathbb{K}(\mathfrak{n})_\mathfrak{p}$ so that the congruence condition is fulfilled. Therefore, combining the compact support in x and the uniform local constancy in x, up to the measure of $\mathbb{K}(\mathfrak{n})_\mathfrak{p}$ (which is a rational number) the whole integral at $\mathfrak{p}$ has the same value as a *finite* sum

$$[\mathfrak{o}_\mathfrak{p}^* : \mathfrak{o}_\mathfrak{p}]^{-3/2} \sum_{x \in X} \|\det t_x\|^\kappa \chi(\det t_x)\varphi^*(t_x(1_2 x)g)\bar{\tau}(\mathrm{Tr}(\xi x))$$

where the leading factor occurs because of the normalization of additive Haar measure. It is clear that the value of the sum is in $\mathbb{Q}_{ab}$, in fact, in the field Ω generated over $\mathbb{Q}$ by some $N\mathfrak{p}$-power root of unity.

Let $\sigma = \sigma_\eta$ be in $\mathrm{Gal}(\Omega/\mathbb{Q})$ sending a root of unity ζ to $\zeta^{\eta^{-1}}$, with $\eta \in \mathbb{Z}_p^x$ where p is the rational prime lying under $\mathfrak{p}$; for present purposes we may take η of finite order. Then we have

$$\sigma\left(\sum_{x \in X} \|\det t_x\|^\kappa \chi(\det t_x)\varphi^*(t_x(1_2 x)g)\bar{\tau}(\mathrm{Tr}(\xi x))\right) =$$

$$= \sum_{x \in X} \|\det t_x\|^\kappa \chi(\det t_x)\varphi^*(t_x(1_2 x)g)\bar{\tau}(\mathrm{Tr}(\xi\eta^{-1}x)) =$$

$$= \sum_{x \in X} \|\det t_{\eta x}\|^\kappa \chi(\det t_{\eta x})\varphi^*(t_{\eta x}(1_2 \eta x)g)\bar{\tau}(\mathrm{Tr}(\xi x))$$

since we can certainly arrange that X be stable under multiplication by η. It is easy to check that if t_x is such $\varphi^*(t_x(1_2x)g) \neq 0$, then we can take

$$t_{\eta x} = \begin{pmatrix} 1 & 0 \\ 0 & \eta^{-1} \end{pmatrix} t_x.$$

Therefore,

$$\sigma\left(\sum_{x \in X} \|\det t_x\|^\kappa \chi(\det t_x)\varphi^*(t_x(1_2x)g)\bar{\tau}(\mathrm{Tr}(\xi x)) \right) =$$

$$= \sum_{x \in X} \|\det t_{\eta x}\|^\kappa \chi(\det t_{\eta x})\varphi^*(t_{\eta x}(1_2\eta x)g)\bar{\tau}(\mathrm{Tr}(\xi x)) =$$

$$= \chi(\eta^{-1}) \sum_{x \in X} \|\det t_x\|^\kappa \chi(\det t_x)\varphi^*(t_{\eta x}(1_2\eta x)g)\bar{\tau}(\mathrm{Tr}(\xi x)) =$$

$$= \chi(\eta) \sum_{x \in X} \|\det t_x\|^\kappa \chi(\det t_x)\bar{\tau}(\mathrm{Tr}(\xi x)) =$$

$$= \chi(\eta) \sum_{x \in X} \|\det t_x\|^\kappa \chi(\det t_x)\varphi^*(t_x(1_2x)g)\bar{\tau}(\mathrm{Tr}(\xi x)).$$

From this we see that application of σ_η multiplies this sum by $\chi(\eta)$ for $\eta \in \mathbb{Z}_p^\times$.

For a finite prime $\mathfrak{p}$ not dividing $\mathfrak{n}$ we use Siegel's original method. There are two cases to treat. First, there is the case of the all-but-finitely-many $\mathfrak{p}$ so that $\mathfrak{p}$ does not divide 2, $\mathfrak{p}$ is absolutely unramified, and Ay and ξ are in $GL(2, \mathfrak{o}_p)$; this applies to almost all primes $\mathfrak{p}$. Call such primes *good*. Second, we consider the finitely many other primes, which we call *bad*. Still we have g of the special form $g = \ell'(y)$.

LEMMA Let δ_1 and δ_2 be the elementary divisors of $y^{-1}xy^{T-1}$; let $\tilde{\delta}_i$ be the greatest common divisor of δ_i and 1. Then we have

$$\int_{GL(2,F_\mathfrak{p})} \|\det t\|^\kappa \chi(\det t)\varphi^*(t(1_2x)g)\, dt =$$

$$= \|\tilde{\delta}_1\|^{-\kappa}\chi(\tilde{\delta}_1)\|\tilde{\delta}_2\|^{-\kappa}\chi(\tilde{\delta}_2)/(1 - \chi(\mathfrak{p})N\mathfrak{p}^{-\kappa})(1 - \chi(\mathfrak{p})N\mathfrak{p}^{1-\kappa}).$$

Further, if $\mathfrak{p}$ does not divide 2 and is absolutely unramified,

$$\|\tilde{\delta}_1\|^{-\kappa}\chi(\tilde{\delta}_1)\,\|\tilde{\delta}_2\|^{-\kappa}\chi(\tilde{\delta}_2) = \int_M \tau(\mathrm{Tr}\, y^{-1}xy^{T-1}vv^T)\, dv$$

where M is the set of $2 \times 2\kappa$ matrices over $\mathfrak{o}_\mathfrak{p}$, with the standard additive measure.

Proof In our integral, replace t by ty^{-1} to transform the integral to

$$\chi(\det y)\int_{GL(2,F_\mathfrak{p})} \|\det t\|^\kappa \chi(\det t)\varphi^*(t(1_2\ y^{-1}xy^{T-1}))\,dt.$$

Let $C \in SL(2, \mathfrak{o}_\mathfrak{p})$, $D \in GL(2, \mathfrak{o}_\mathfrak{p})$ be so that

$$C(y^{-1}xy^{T-1})D = \delta = \begin{pmatrix} \delta_1 & 0 \\ 0 & \delta_2 \end{pmatrix}$$

is in elementary divisor form. Now φ^* is right $GL(4, \mathfrak{o}_\mathfrak{p})$-invariant, so by changing variables the latter integral becomes

$$\int_{GL(2,F_\mathfrak{p})} \|\det t\|^\kappa \chi(\det t)\varphi^*(t(1_2\quad \delta))\,dt.$$

Since the integrand is left $GL(2, \mathfrak{o}_\mathfrak{p})$-invariant, we may integrate over $P(F_\mathfrak{p})$ by invoking the Iwasawa decomposition. With coordinates

$$t = \begin{pmatrix} t_1 & t_2t_3 \\ 0 & t_3 \end{pmatrix}$$

the integral becomes

$$\int \|t_1t_3\|^\kappa \chi(t_1t_3)\,\mathrm{ch}(\tilde{\delta}_1 t_1)\,\mathrm{ch}(\tilde{\delta}_2 t_2 t_3)\,\mathrm{ch}(\tilde{\delta}_2 t_3)\,d^\times t_1\, d^+ t_2\, d^\times t_3$$

where ch denotes the characteristic function of $\mathfrak{o}_\mathfrak{p}$ and $\tilde{\delta}_i$ is the greatest common divisor of δ_i and 1. Integrating first over t_2 gives

$$\int \|t_1t_3\|^\kappa \chi(t_1t_3)\,\mathrm{ch}(\tilde{\delta}_1 t_1)\,\|\tilde{\delta}_2 t_3\|^{-1}\,\mathrm{ch}(\tilde{\delta}_2 t_3)\,d^\times t_1\, d^\times t_3.$$

Replacing t_1 by $t_1/\tilde{\delta}_1$ and t_3 by $t_3/\tilde{\delta}_2$ turns this into

$$\|\tilde{\delta}_1\|^{-\kappa}\chi(\tilde{\delta}_1)\,\|\tilde{\delta}_2\|^{-\kappa}\chi(\tilde{\delta}_2)\int \|t_1t_3\|^\kappa \chi(t_1t_3)\,\mathrm{ch}(t_1)\,\|t_3\|^{-1}\,\mathrm{ch}(t_3)\,d^\times t_1\, d^\times t_3.$$

The remaining integral is a product of two local zeta-type integrals, and the whole is

$$\|\tilde{\delta}_1\|^{-\kappa}\chi(\tilde{\delta}_1)\,\|\tilde{\delta}_2\|^{-\kappa}\chi(\tilde{\delta}_2)/(1-\chi(\mathfrak{p})N\mathfrak{p}^{-\kappa})(1-\chi(\mathfrak{p})N\mathfrak{p}^{1-\kappa}).$$

Now we evaluate the other integral expression when $\mathfrak{p}$ does not divide 2 and is absolutely unramified. Since $\mathfrak{p}$ is odd, there is $C \in GL(2, \mathfrak{o}_\mathfrak{p})$ so that $Cy^{-1}xy^{T-1}C^T =$

δ. Therefore, by changing variables, the second integral in the assertion of the theorem becomes

$$\int_{v \in M} \tau(\operatorname{Tr} \delta v v^T)\, dv = \left(\int_{\mathfrak{o}_\mathfrak{p}} \tau(\delta_1 v)\, dv\right)^{2\kappa} \times \left(\int_{\mathfrak{o}_\mathfrak{p}} \tau(\delta_2 v)\, dv\right)^{2\kappa}.$$

Since $\mathfrak{p}$ is absolutely unramified, it is not difficult to compute that

$$\left(\int_{\mathfrak{o}_\mathfrak{p}} \tau(\delta_i v)\, dv\right)^2 = (-1/\mathfrak{a})_2 \|\tilde{\delta}_i\|^{-1}$$

where $\mathfrak{a}$ is the power of $\mathfrak{p}$ so that $\tilde{\delta}_i \mathfrak{o}_\mathfrak{p} = \mathfrak{a}\mathfrak{o}_\mathfrak{p}$ (e.g., by converting these into classical exponential sums). Since $\mathfrak{p}$ is absolutely unramified and does not divide 2, $(-1/\mathfrak{a})_2^\kappa = \chi(\tilde{\delta}_i)$. Combining these facts yields the assertion of the lemma. ■

LEMMA Let f be the function defined by

$$f(x) = \|\tilde{\delta}_1\|^{-\kappa}\chi(\tilde{\delta}_1)\,\|\tilde{\delta}_2\|^{-\kappa}\chi(\tilde{\delta}_2)$$

(in the notation of the previous lemma). For each prime $\mathfrak{p}$, there is a non-zero ideal $\mathfrak{a}$ in $\mathfrak{o}_\mathfrak{p}$ so that f satisfies $f(x + u) = f(x)$ for $u \in W(\mathfrak{a}\mathfrak{o}_\mathfrak{p})$. Further, for good $\mathfrak{p}$, we may take $\mathfrak{a} = \mathfrak{o}_\mathfrak{p}$.

Proof The first assertion is clear from the first integral expression in the previous lemma; the second is clear from the second. ■

Now consider a good (finite) prime $\mathfrak{p}$. We want to compute the Fourier transform of the function $f(y^{-1}xy^{T-1})$ (with f as in the previous lemma), evaluated at $\xi \in GL(2, \mathfrak{o}_\mathfrak{p})$. First, since $y \in GL(2, \mathfrak{o}_\mathfrak{p})$, the elementary divisors of $y^{-1}xy^{T-1}$ are the same as those of x, so we can merely compute the Fourier transform of f. We use the integral expression from the first lemma:

$$f(x) = \int_{v \in M} \tau(\operatorname{Tr} x v v^T)\, dv.$$

Thus, the Fourier transform along $W(F_\mathfrak{p})$ is

$$\hat{f}(\xi) = \int\int_M \tau(\operatorname{Tr} x(vv^T - \xi))\, dv\, d^+x.$$

We claim that this integral is equal to the same integral but merely over $x \in W(\mathfrak{p}^{-1}\mathfrak{o}_\mathfrak{p})$ and $v \in M$. Let $H = \{1_2 + \eta \colon \eta \in W(\mathfrak{p}\mathfrak{o}_\mathfrak{p})\}$. For a positive integer n, define

$$S_n = \{x \in W(F_\mathfrak{p})\text{: the sup of the norms of the entries of } x \text{ is } N\mathfrak{p}^n\}.$$

Clearly S_n is stabilized by H, for $n > 0$. Let

$$Y_n = \{\eta \in W(\mathfrak{p}^N \mathfrak{o}_\mathfrak{p}) \colon N \geq \tfrac{1}{2}n\}$$

$$H_n = \{1 + \eta \colon \eta \in Y_n\} \subset GL(2, \mathfrak{o}_\mathfrak{p}).$$

Then for fixed x in S_n

$$\int_{H_n(x)} f(x)\bar{\tau}(\operatorname{Tr} \xi x) d^+ x = \int_{Y_n} f(hxh^T)\bar{\tau}(\operatorname{Tr} \xi hxh^T)\, d^+\eta =$$

$$= \int_{Y_n} f(x)\bar{\tau}(\operatorname{Tr} \xi hxh^T)\, d^+\eta =$$

$$= f(x)\bar{\tau}(\operatorname{Tr} \xi x) \times \int_{Y_n} \bar{\tau}(\operatorname{Tr} \xi(2x\eta^T + \eta x \eta^T))\, d^+\eta =$$

$$= f(x)\bar{\tau}(\operatorname{Tr} \xi x) \times \int_{Y_n} \bar{\tau}(2 \operatorname{Tr} \xi x \eta^T)\, d^+\eta$$

since $\bar{\tau}(\operatorname{Tr} \xi \eta x \eta^T) = 1$, by the construction of Y_n. We claim that for $n > 1$, for any $x \in S_n$ the latter integral is 0. To this end, it suffices to find (for given x and ξ) one η_0 in Y_n so that $\bar{\tau}(2 \operatorname{Tr} \xi x \eta^T) \neq 1$; this is easy. Since S_n is a disjoint union of H_n-orbits, the integral over x in $\bigcup_{n \geq 2} S_n$ is 0. This proves the claim. Also, $f(x)$ is invariant under translations by $W(\mathfrak{o}_\mathfrak{p})$. Therefore,

$$\int_{H_n(x)} f(x)\bar{\tau}(\operatorname{Tr} \xi x)\, d^+ x = \sum_x f(x)\bar{\tau}(\operatorname{Tr} \xi x) \qquad (x \text{ in } W(\mathfrak{p}^{-1}\mathfrak{o}_\mathfrak{p})/W(\mathfrak{o}_\mathfrak{p})).$$

Using the integral expression for f, this is

$$\sum_x \int_M \tau(\operatorname{Tr} x(vv^T - \xi))\, dv = \sum_x N\mathfrak{p}^{-4\kappa} \sum_v \tau(\operatorname{Tr} x(vv^T - \xi)) \quad (v \text{ in } W(\mathfrak{o}_\mathfrak{p}) \bmod \mathfrak{p}\mathfrak{o}_\mathfrak{p})$$

since x lies in $W(\mathfrak{p}\mathfrak{o}_\mathfrak{p})$. Interchanging summations, this is

$$N\mathfrak{p}^{-4\kappa} \sum_v \sum_x \tau(\operatorname{Tr} x(vv^T - \xi)) =$$

$$= N\mathfrak{p}^{3-4\kappa} \times \text{number of } v \text{ in } M \bmod \mathfrak{p} \text{ so that } vv^T \equiv \xi \text{ modulo } \mathfrak{p}$$

since the inner sum over x is zero unless $vv^T \equiv \xi$, and in that case is $N\mathfrak{p}^3$. Since $\mathfrak{p}$ does not divide 2, by a change of variables we may suppose that ξ is diagonal, with entries ξ_1 and ξ_2 (neither 0 mod $\mathfrak{p}$).

LEMMA Let $k = \mathfrak{o}_\mathfrak{p}/\mathfrak{p}\mathfrak{o}_\mathfrak{p}$, and let q be the cardinality of k. Let Q be a symmetric matrix in $GL(n, k)$, and let $\zeta \in k^\times$. Then

$$\#\{v \in k^n : vQv^T = \zeta\} = q^{n-1} - q^{-1}(\det Q/\mathfrak{p})_2((-1/\mathfrak{p})_2 q)^{n/2} \qquad (n \text{ even})$$

$$= q^{n-1} + q^{-1}(-\zeta \det Q/\mathfrak{p})_2((-1/\mathfrak{p})_2 q)^{(n+1)/2} \qquad (n \text{ odd}).$$

Proof We consider elements of k^n as row vectors. Since $\mathfrak{p}$ does not divide 2, we can change coordinates to suppose that Q is diagonal with diagonal entries Q_i. Then

$$vQv^T = \sum_i Q_i v_i^2$$

where v_i is the i-th coordinate of v. Let τ be a non-trivial additive character on k. Then

$$\#\{v : vQv^T = \zeta\} = \sum_v q^{-1} \sum_{x \in k} \tau(x(\zeta - vQv^T)) =$$

$$= q^{-1} \sum_{x \in k} \sum_{v_i \in k} \prod_i \tau(x(\zeta - Q_i v_i^2)) = q^{-1} \sum_{x \in k} \tau(x\zeta) \prod_i \mathfrak{G}(xQ_i)$$

where

$$\mathfrak{G}(y) = \sum_{w \in k} \tau(yw^2).$$

We know that

$$\mathfrak{G}(xQ_i) = (xQ_i/\mathfrak{p})_2 \mathfrak{G}(1)$$

for x and Q_i not 0 in k; otherwise it is q. Also, $\mathfrak{G}(1)^2 = (-1/\mathfrak{p})_2 q$. Therefore,

$$\#\{v : vQv^T = \zeta\} = q^{-1} \sum_{x \in k^\times} \tau(x\zeta) \prod_i (xQ_i/\mathfrak{p})_2 \mathfrak{G}(1) + q^{-1} \prod_i q =$$

$$= q^{n-1} + q^{-1}(\det Q/\mathfrak{p})_2 \mathfrak{G}(1)^n \sum_{x \in k^\times} \tau(x\zeta)(x/\mathfrak{p})_2^n.$$

If n is even, this is

$$\#\{v : vQv^T = \zeta\} = q^{n-1} - q^{-1}(\det Q/\mathfrak{p})_2((-1/\mathfrak{p})_2 q)^{n/2}.$$

If n is odd, this is

$$\#\{v : vQv^T = \zeta\} = q^{n-1} + q^{-1}(\det Q/\mathfrak{p})_2 \mathfrak{G}(1)^n \sum_{x \in k^\times} \tau(x\zeta)(x/\mathfrak{p})_2 =$$

$$= q^{n-1} + q^{-1}(\det Q/\mathfrak{p})_2 \mathfrak{G}(1)^n (-\zeta/\mathfrak{p})_2 \sum_{x \in k^\times} \tau(-x)(x/\mathfrak{p})_2 =$$

$$= q^{n-1} + q^{-1}(\det Q/\mathfrak{p})_2 \mathfrak{G}(1)^n (-\zeta/\mathfrak{p})_2 \mathfrak{G}(1) =$$

$$= q^{n-1} + q^{-1}(-\zeta \det Q/\mathfrak{p})_2((-1/\mathfrak{p})_2 q)^{(n+1)/2}.$$

This gives the assertion of the lemma. ∎

Therefore, the number of $w \in k^{2\kappa}$ so that $ww^T = \xi_1$ is

$$q^{2\kappa-1} - q^{-1}((-1/\mathfrak{p})_2 q)^\kappa$$

for any ξ_1. For each such w, the space $w^\perp$ of $v \in k^{2\kappa}$ so that $wv^T = 0$ is $2\kappa - 1$ dimensional, and on $w^\perp$ the restriction of the quadratic form $v \to vv^T$ is of the form $v \to vQv^T$ with $\det Q = \xi_1^{-1}$. Therefore, for each w so that $ww^T = \xi_1$, the number of v so that $wv^T = 0$ and $vv^T = \xi_2$ is

$$q^{2\kappa-2} + q^{-1}(-\xi_2\xi_1/\mathfrak{p})_2((-1/\mathfrak{p})_2 q)^\kappa = q^{2\kappa-2} + q^{\kappa-1}(-\det \xi/\mathfrak{p})_2(-1/\mathfrak{p})_2^\kappa.$$

Applying this to the problem at hand, we have

$$\int_{W(F\mathfrak{p})} f(x)\bar{\tau}(\xi x)\, d^+x = \sum_x N\mathfrak{p}^{-4\kappa} \sum_v \tau(\operatorname{Tr} x(vv^T - \xi)) =$$

$$= q^{3-4\kappa} \times (q^{2\kappa-1} - q^{\kappa-1}(-1/\mathfrak{p})_2^\kappa)(q^{2\kappa-2} + q^{\kappa-1}(-\det \xi/\mathfrak{p})_2(-1/\mathfrak{p})_2^\kappa) =$$

$$= (1 - (-1/\mathfrak{p})_2^\kappa q^{-\kappa})(1 + (-\det \xi/\mathfrak{p})_2(-1/\mathfrak{p})_2^\kappa q^{1-\kappa}) =$$

$$= (1 - (-1/\mathfrak{p})_2^\kappa q^{-\kappa})(1 - q^{2-2\kappa})/(1 - (-\det \xi/\mathfrak{p})_2(-1/\mathfrak{p})_2^\kappa q^{1-\kappa}) =$$

$$= (1 - \chi(\mathfrak{p})q^{-\kappa})(1 - q^{2-2\kappa})/(1 - (-\det \xi/\mathfrak{p})_2\chi(\mathfrak{p})q^{1-\kappa}).$$

This times

$$[\mathfrak{o}_\mathfrak{p}^* : \mathfrak{o}_\mathfrak{p}]^{-3/2}/(1 - \chi(\mathfrak{p})N\mathfrak{p}^{-\kappa})(1 - \chi(\mathfrak{p})N\mathfrak{p}^{1-\kappa})$$

is the $\mathfrak{p}$-integral for $\mathfrak{p}$ a good prime; it is

$$[\mathfrak{o}_\mathfrak{p}^* : \mathfrak{o}_\mathfrak{p}]^{-3/2}(1 - q^{2-2\kappa})/(1 - \chi(\mathfrak{p})N\mathfrak{p}^{1-\kappa})(1 - (-\det \xi/\mathfrak{p})_2\chi(\mathfrak{p})q^{1-\kappa}).$$

For $\mathfrak{p}$ a bad prime (not dividing $\mathfrak{n}$), the integral is

$$1/(1 - \chi(\mathfrak{p})N\mathfrak{p}^{-\kappa})(1 - \chi(\mathfrak{p})N\mathfrak{p}^{1-\kappa})$$

times

$$\int_{W(F_\mathfrak{p})} f(y^{-1}xy^{T-1})\bar{\tau}(\xi x)\, d^+x.$$

We have already noted that the integrand is uniformly locally constant in x. We claim that this integral is equal to

$$[\mathfrak{o}_\mathfrak{p}^* : \mathfrak{o}_\mathfrak{p}]^{-3/2} \sum_{x \in X} f(y^{-1}xy^{T-1})\bar{\tau}(\xi x)$$

for some finite set X. Let

$$H = \{h = 1_2 + \eta \colon \eta \in W(\mathfrak{p}\mathfrak{o}_\mathfrak{p}) \text{ so that } f(y^{-1}xy^{T-1}) = f(y^{-1}hxh^T y^{T-1})\}.$$

For a positive integer n, define

$$S_n = \{x \in W(F_\mathfrak{p})\colon \text{the sup of the norms of the entries of } x \text{ is } N\mathfrak{p}^n\}.$$

Clearly S_n is stabilized by H, for $n > 0$. Let $\xi \in S_k$.

$$Y_n = \{\eta \in W(\mathfrak{p}^N\mathfrak{o}_\mathfrak{p})\colon N \geq \tfrac{1}{2}(n + k)\}$$

$$H_n = \{1 + \eta \colon \eta \in Y_n\} \subset GL(2, \mathfrak{o}_\mathfrak{p}).$$

Then for fixed x in S_n

$$\int_{H_n(x)} f(y^{-1}xy^{T-1})\bar{\tau}(\operatorname{Tr} \xi x)\, d^+x =$$

$$= \int_{Y_n} f(y^{-1}hxh^T y^{T-1})\bar{\tau}(\operatorname{Tr} \xi hxh^T)\, d^+\eta =$$

$$= \int_{Y_n} f(y^{-1}xy^{T-1})\bar{\tau}(\operatorname{Tr} \xi hxh^T)\, d^+\eta =$$

$$= f(y^{-1}xy^{T-1})\bar{\tau}(\operatorname{Tr} \xi x) \times \int_{Y_n} \bar{\tau}(\operatorname{Tr} \xi(2x\eta^T + \eta x\eta^T))\, d^+\eta =$$

$$= f(y^{-1}xy^{T-1})\bar{\tau}(\operatorname{Tr} \xi x) \times \int_{Y_n} \bar{\tau}(2 \operatorname{Tr} \xi x\eta^T)\, d^+\eta$$

since $\bar{\tau}(\operatorname{Tr} \xi\eta x\eta^T) = 1$, by the construction of Y_n. We claim that for n sufficiently large, for any $x \in S_n$ the latter integral is 0. To this end, it suffices to find (for given x and ξ) one η_0 in Y_n so that $\bar{\tau}(2 \operatorname{Tr} \xi x\eta^T) \neq 1$; this is easy. Since S_n is a disjoint union of H_n-orbits, the integral over x in $\bigcup_{n \geq n_0} S_n$ is 0 for n_0 sufficiently large. Therefore, we need only integrate x over a compact set. By the local constancy, such an integral is indeed equal to a finite sum

$$[\mathfrak{o}_\mathfrak{p}^* : \mathfrak{o}_\mathfrak{p}]^{-3/2} \sum_{x \in X} f(y^{-1}xy^{T-1})\bar{\tau}(\xi x).$$

The value of the sum is certainly in a field Ω generated by some $N\mathfrak{p}$-power root of unity ζ. Take $\sigma = \sigma_\eta$ with $\eta \in \mathbb{Z}_p^\times$ giving an element $\zeta \to \zeta^{\eta^{-1}}$ of $\operatorname{Gal}(\Omega/\mathbb{Q})$; we may take η of finite order. Then

$$\sigma\left(\sum_{x\in X} f(y^{-1}xy^{T-1})\bar{\tau}(\xi x)\right) = \sum_{x\in X} f(y^{-1}xy^{T-1})\bar{\tau}(\xi x\eta^{-1}) =$$

$$= \sum_{x\in X} f(y^{-1}\eta xy^{T-1})\bar{\tau}(\xi x)$$

since we may choose X to be stable under multiplication by η. But the elementary divisors of $y^{-1}\eta xy^{T-1}$ are exactly those of $y^{-1}xy^{T-1}$, so $f(y^{-1}\eta xy^{T-1}) = f(y^{-1}xy^{T-1})$. Therefore,

$$\sum_{x\in X} f(y^{-1}xy^{T-1})\bar{\tau}(\xi x)$$

is a *rational number*. Therefore, for bad $\mathfrak{p}$ (not dividing $\mathfrak{n}$), the $\mathfrak{p}$-integral is a *rational multiple* of

$$[\mathfrak{o}_\mathfrak{p}^* : \mathfrak{o}_\mathfrak{p}]^{-3/2}/(1 - \chi(\mathfrak{p})N\mathfrak{p}^{-\kappa})(1 - \chi(\mathfrak{p})N\mathfrak{p}^{1-\kappa}).$$

7.8 Summary of Fourier Coefficient Computations

Let D be the absolute discriminant of F. In the following, when we write $L_F(s, \omega)$ for an L-function over F (with character ω of finite order), we mean to include only those Euler factors at primes not dividing the conductor of ω and not dividing the *proper* ideal $\mathfrak{n}$. Collecting the calculations of the previous three sections, we have the following, *omitting the exponential functions*: The Fourier coefficients of the special Siegel's Eisenstein series E of interest to us are

i. From the small cell, there is no contribution because $\mathfrak{n}$ is proper.
ii. From the middle cell, at totally positive semi-definite indices ξ of rank one, we have a *rational multiple of*

$$\pi^{m-m\kappa}D^{1/2}L_F(\kappa - 1, \chi).$$

iii. From the big cell, only for ξ totally positive definite, we have a *rational multiple of*

$$(N_{F/\mathbb{Q}} \det \xi)^{1/2} D^{1/2} \zeta_F(\kappa - 1) L_F(\kappa - 1, \chi_\xi) \zeta_F(2\kappa - 2)^{-1} \qquad \text{(for } \kappa \text{ even)}$$

$$\beta i^{-\kappa m}(N_{F/\mathbb{Q}} \det \xi)^{1/2} D^{1/2} L_F(\kappa - 1, \chi) L_F(\kappa - 1, \chi_\xi\chi) \zeta_F(2\kappa - 2)^{-1} \quad \text{(for } \kappa \text{ odd)}$$

where χ_ξ is the quadratic character $\mathfrak{p} \to (-\det \xi/\mathfrak{p})_2$ defined for $\mathfrak{p}$ prime to the fractional ideal $\xi\mathfrak{o}$, and where $\beta \in \mathbb{Q}_{ab}$ is as follows. There is a rational integer N, whose prime factors lie among those rational primes dividing $\mathfrak{n} \cap \mathbb{Z}$, so that β lies in $\mathbb{Q}(\zeta)$, where ζ is an N-th root of unity. Further, for $\eta \in (\mathbb{Z}/N\mathbb{Z})^\times$, let $\sigma = \sigma_\eta$ be the automorphism of $\mathbb{Q}(\zeta)$ given by $\sigma\colon \zeta \to \zeta^\eta$. Then

$$\beta^\sigma = \chi(\eta)\beta.$$

LEMMA The number $i^{\kappa m}\beta$ lies in $\mathbb{Q}$.

Proof Note that κ is odd. Represent $\eta \in \hat{\mathbb{Z}}^{\times}$ by $\eta \in (\mathbb{Z}/N\mathbb{Z})^{\times}$. By Dirichlet's theorem on primes in an arithmetic progression, it suffices to suppose that η is represented modulo N by a prime p. Further, we may certainly suppose that p is not 2, and does not ramify in F. If $p \equiv 1 \bmod 4$, then -1 is a square $\bmod\, p$, so -1 is a square modulo any prime $\mathfrak{p}$ lying over p in F. Therefore, for $p \equiv 1 \bmod 4$, $\chi(\eta) = 1$. If $p \equiv 3 \bmod 4$, then for a prime $\mathfrak{p}$ lying over p we have

$$(-1/\mathfrak{p})_2 = (-1)^{f(\mathfrak{p})}$$

where $f(\mathfrak{p})$ is the residue class field extension degree. As no prime over p ramifies in F,

$$\chi(p) = \prod_{\mathfrak{p} \text{ over } p} \chi(\mathfrak{p}) = \prod_{\mathfrak{p} \text{ over } p} (-1)^{f(\mathfrak{p})} = (-1)^m.$$

Therefore, if $m = [F:\mathbb{Q}]$ is even, $\chi \equiv 1$ on $\hat{\mathbb{Z}}^{\times}$, and $\beta \in \mathbb{Q}$; also, $i^{\kappa m} \in \mathbb{Q}$. If m is odd, then apparently χ coincides with the character $p \to (-1/p)_2$ over $\mathbb{Q}$. This is exactly the same character by which the number i is equivariant under $\hat{\mathbb{Z}}^{\times} \approx \mathrm{Gal}(\mathbb{Q}_{ab}/\mathbb{Q})$. Since $m\kappa$ is odd, we have

$$\sigma_p(i^{\kappa m}\beta) = (-1/p)_2^{\kappa m} i^{\kappa m}(-1/p)_2\beta = i^{\kappa m}\beta \times (-1/p)_2^{\kappa m+1} = i^{\kappa m}\beta.$$

That is, $i^{\kappa m}\beta$ is in $\mathbb{Q}$, as claimed. ∎

Therefore, in any case the big-cell contribution occurs only for ξ totally positive definite, and is a *rational multiple of*

$$(N_{F/\mathbb{Q}} \det \xi)^{1/2} D^{1/2} L_F(\kappa - 1, \chi) L_F(\kappa - 1, \chi_\xi \chi) \zeta_F(2\kappa - 2)^{-1}.$$

Summarizing these results, we have the following theorem.

THEOREM The Fourier coefficients of our special Siegel's Eisenstein series E are *rational multiples* of $\pi^{m-m\kappa} D^{1/2} L_F(\kappa - 1, \chi)$ times

i. 0 (small-cell contribution)

ii. 1 (middle-cell contribution)

iii. $(N_{F/\mathbb{Q}} \det \xi)^{1/2} \pi^{m(\kappa-1)} L_F(\kappa - 1, \chi_\xi \chi) \zeta_F(2\kappa - 2)^{-1}$ (big-cell contribution),

where in the big-cell case ξ is totally positive definite. ∎

Remarks Note that in each case if the argument of one of the L-functions is odd (respectively, even), then the character is totally odd (respectively, totally even).

PROPOSITION The expression (iii) in the previous theorem lies in $\mathbb{Q}$.

Proof First consider $F = \mathbb{Q}$. From Appendix A.5, when $k \geq 2$ is of the same parity as a Dirichlet character ω, we have

$$L_{\mathbb{Q}}(k, \omega)/i^k \pi^k \gamma(\omega) \in \mathbb{Q}(\omega)$$

where

$$\gamma(\omega) = \sum_{x \bmod N} \bar{\omega}(x) \exp(2\pi i x/N),$$

where N is finite-prime part of the conductor of ω, and where $\mathbb{Q}(\omega)$ is the field generated over $\mathbb{Q}$ by the values of ω. Now suppose that ω is $\mathbb{Q}$-valued. Take $\sigma = \sigma_\eta$ in $\mathrm{Gal}(\mathbb{Q}_{ab}/\mathbb{Q})$ sending a root of unity ζ to $\zeta\eta$; we may take $\eta \in (\mathbb{Z}/N\mathbb{Z})^x$. We have

$$\gamma(\omega)^\sigma = \sum_{x \bmod N} \bar{\omega}(x) \exp(2\pi i \eta x/N) = \omega(\eta) \sum_{x \bmod N} \omega(x) \exp(2\pi i x/N) =$$

$$= \omega(\eta)\gamma(\omega).$$

Now take $\omega = (M/*)_2$ for an integer M. Then, taking η to be an arbitrary prime not dividing N, we see that $M^{1/2}\gamma(\omega)$ is rational. By taking $\eta = -1$, we find that $M > 0$ for ω even, and $M < 0$ for ω odd.

Regarding the big-cell contribution (iii) for $F = \mathbb{Q}$, first,

$$\zeta_{\mathbb{Q}}(2\kappa - 2)^{-1} \in \pi^{-2\kappa}\mathbb{Q}.$$

For κ odd—i.e., for $\kappa - 1$ even—$\omega = \chi_\xi \chi = (\det \xi/*)_2$ is an even character, and

$$(\det \xi)^{1/2} L_F(\kappa - 1, \chi_\xi \chi)/\pi^{(\kappa-1)} \in L_F(\kappa - 1, \chi_\xi \chi)/\pi^{(\kappa-1)}\gamma(\omega)\mathbb{Q} \subset \mathbb{Q}.$$

by our prior discussion. For κ even—i.e., for $\kappa - 1$ odd—$\omega = \chi_\xi \chi = \chi_\xi = (-\det \xi/*)_2$ is an odd character, and

$$(\det \xi)^{1/2} L_F(\kappa - 1, \chi_\xi \chi)/\pi^{(\kappa-1)} = \pm(-\det \xi)^{1/2} L_F(\kappa - 1, \chi_\xi \chi)/i\pi^{(\kappa-1)} \in$$

$$\in L_F(\kappa - 1, \chi_\xi \chi)/i\pi^{(\kappa-1)}\gamma(\omega)\mathbb{Q} \in \mathbb{Q}$$

For $F \neq \mathbb{Q}$, this is a direct translation of the corollary to the theorem of section 6.2. ■

Remarks Only the results of section 6.1 *in the case* $F = \mathbb{Q}$ are used to obtain the results of section 6.2. For $F = \mathbb{Q}$, the results of Appendix A.5 are sufficient to yield the rationality of the (normalized) Fourier coefficients of the Siegel's Eisenstein series. That is, we do not use section 6.2 in the present discussion when $F = \mathbb{Q}$. This is noted to make clear that there is no circularity in our proof of the results of section

6.1 (i.e., the present discussion) and those of section 6.2 (regarding L-functions over totally real number fields other than $\mathbb{Q}$).

7.9 The Main Formula

Still $\mathfrak{n}$ is a fixed proper ideal of $\mathfrak{o}$, as in sections 7.1–7.8. We fix $0 < \kappa \in \mathbb{Z}$ large enough to ensure the convergence of the Siegel's Eisenstein series E considered in sections 7.1–7.8. Consider the imbedding

$$\iota\colon SL(2, F) \times SL(2, F) \to Sp(2, F)$$

given by

$$\begin{pmatrix} a & b \\ c & d \end{pmatrix} \times \begin{pmatrix} a' & b' \\ c' & d' \end{pmatrix} \to \begin{pmatrix} a & 0 & b & 0 \\ 0 & a' & 0 & b' \\ c & 0 & d & 0 \\ 0 & c' & 0 & d' \end{pmatrix}.$$

Let

$$\mathbb{K} = \{g \in SL(2, \hat{\mathfrak{o}})\colon g \equiv 1 \text{ modulo } \mathfrak{n}\}.$$

Define a $\mathbb{C}$-conjugate linear map $f \to f^{\natural}$ by

$$f^{\natural}(g) = \bar{f}(\varepsilon g \varepsilon^{-1})$$

where $\varepsilon = \left(\begin{smallmatrix} -1 & 0 \\ 0 & 1 \end{smallmatrix}\right)$. For each prime $\mathfrak{p}$ not dividing $\mathfrak{n}$, as in section 6.1, define

$$H_1(\mathfrak{p}) = \{\text{left and right } SL(2, \mathfrak{o}_\mathfrak{p})\text{-invariant continuous compactly supported functions on } SL(2, F_\mathfrak{p}), \text{ with convolution}\}.$$

As in section 6.1, put

$$\mathscr{H}_1(\mathfrak{n}) = \bigotimes_{\mathfrak{p} \text{ not dividing } \mathfrak{n}} (\text{spherical Hecke algebra at } \mathfrak{p} \text{ for } SL(2)).$$

Let Λ be the collection of algebra homomorphisms λ of $\mathscr{H}_1(\mathfrak{n})$ to $\mathbb{C}$. Let

$$\mathrm{Cfm}_1(\mathfrak{n}, \kappa) = \{f\colon f \text{ is a right } \mathbb{K}\text{-invariant weight } \kappa \text{ holomorphic cuspform}\}$$

$$\mathrm{Cfm}_1(\mathfrak{n}, \kappa, \lambda) = \{f\colon f \text{ is a right } \mathbb{K}\text{-invariant weight } \kappa \text{ holomorphic cuspform and (by right convolution) is a } \lambda\text{-eigenfunction for } \mathscr{H}_1(\mathfrak{n})\}.$$

Just as for $GL(2)$, it is easy to prove that $\mathscr{H}_1(\mathfrak{n})$ is commutative, and that the collection of operators given by them on $\mathrm{Cfm}_1(\mathfrak{n}, \kappa)$ is closed under adjoints with respect to the inner product

$$\langle f_1, f_2 \rangle_1 = \int_{SL(2,F)\backslash SL(2,\mathbb{A})} f_1(g)\bar{f}_2(g)\, dg.$$

Also, this is an immediate corollary of the lemma of section 6.1. Therefore,

$$\mathrm{Cfm}_1(\mathfrak{n}, \kappa) = \bigoplus_{\lambda} \mathrm{Cfm}_1(\mathfrak{n}, \kappa, \lambda) \qquad \text{(orthogonal sum).}$$

It is not difficult to check that the map $f \to f^{\natural}$ sends $\mathrm{Cfm}_1(\mathfrak{n}, \kappa)$ to itself, and sends $\mathrm{Cfm}_1(\mathfrak{n}, \kappa, \lambda)$ to $\mathrm{Cfm}_1(\mathfrak{n}, \kappa, \bar{\lambda})$: This is formal. Then we have the "Main Formula":

THEOREM Suppose that $\mathfrak{n} \neq \mathfrak{o}$. Then there are *non-zero* complex numbers $S(\lambda)$ so that for all g and g' in $SL(2, \mathbb{A})$ we have

$$E(\iota(g, g')) = \sum_{\lambda \in \Lambda} S(\lambda) \sum_{f} f(g) f^{\natural}(g')$$

where for each β and λ the sum over f is over an orthonormal basis for $\mathrm{Cfm}_1(\mathfrak{n}, \kappa, \lambda)$ with respect to the inner product above.

Proof Recall from section 7.1 that the Eisenstein series E is defined by a sum of the form

$$E(g) = \sum_{\gamma} q(\gamma g) \qquad (\gamma \text{ summed over } \tilde{P}(F)\backslash Sp(2, F)).$$

This induces consideration of the double coset space of the following lemma:

LEMMA The double coset space

$$\tilde{P}(F)\backslash Sp(2, F)/\iota(SL(2, F) \times SL(2, F))$$

has representatives 1_4 and

$$\theta = \begin{pmatrix} 0 & 0 & -1 & 0 \\ 0 & 1 & 0 & 0 \\ 1 & 1 & 0 & 0 \\ 0 & 0 & -1 & 1 \end{pmatrix}.$$

Proof of Lemma In section 7.2 we saw that

$$\tilde{P}(F)\backslash Sp(2, F) \approx GL(2, F)\backslash SCP(F)$$

(as right $Sp(2, F)$-spaces), where the map is given by taking the lower half of a matrix in $Sp(2, F)$. Thus, it suffices to consider

$$GL(2, F)\backslash SCP(F)/\iota(SL(2, F) \times SL(2, F)).$$

Given $(c \quad d)$ in $SCP(F)$, it is easy to show that there is

$$u \in U(F) \times U(F) \subset SL(2, F) \times SL(2, F)$$

so that the second entry d' of $(c' \quad d') = (c \quad d)\iota(u)$ has rank two. Therefore, by a further left multiplication by $GL(2, F)$ we may suppose that $(c \quad d)$ is of the form $(c \quad 1_2)$ where necessarily $c = c^T$. Suitable right multiplication by the image under ι of elements in $JU(F)J^{-1} \times JU(F)J^{-1}$ obviously can be used to make the diagonal entries of such c become 0. That is, in any double coset we have a representative

$$\begin{pmatrix} 0 & c_1 & 1 & 0 \\ c_1 & 0 & 0 & 1 \end{pmatrix}.$$

If $c_1 = 0$, we have the obvious representative 1_4. If $c_1 \neq 0$, then right multiplication by

$$\iota\left(\begin{pmatrix} c_1^{-1} & 0 \\ 0 & c_1 \end{pmatrix} \times 1_2\right)$$

and left multiplication by

$$\begin{pmatrix} 1 & 0 \\ 0 & c_1^{-1} \end{pmatrix}$$

put this into the form

$$\begin{pmatrix} 0 & 1 & 1 & 0 \\ 1 & 0 & 0 & 1 \end{pmatrix}.$$

This matrix in $SCP(F)$ has the obvious representative in $Sp(2, F)$ given by

$$\begin{pmatrix} 1 & 0 & 0 & 0 \\ 0 & 1 & 0 & 0 \\ 0 & 1 & 1 & 0 \\ 1 & 0 & 0 & 1 \end{pmatrix}.$$

A further right multiplication by $\iota(J \times 1_2)$ gives the indicated representative. ■

LEMMA The coset $\tilde{P}(F)$ in $\tilde{P}(F)\backslash Sp(2,F)$ has isotropy group

$$P_1(F) \times P_1(F)$$

in $SL(2,F) \times SL(2,F)$, where P_1 is the group of upper triangular matrices. The coset $\tilde{P}(F)\theta$ has isotropy group

$$\Theta(F) = \{g \times \varepsilon^{-1}g\varepsilon : g \in SL(2,F)\} \subset SL(2,F) \times SL(2,F).$$

In particular, from this the irredundancy of these two representatives is clear.

Proof The first case is even easier than the second, so we treat only the second. We wish to determine the set of $g \times g'$ in $SL(2,F) \times SL(2,F)$ so that

$$\theta\iota(g \times g')\theta^{-1} = \begin{pmatrix} * & * \\ 0 & * \end{pmatrix}$$

in a two-by-two block decomposition. Letting $g = \left(\begin{smallmatrix} a & b \\ c & d \end{smallmatrix}\right)$ and $g' = \left(\begin{smallmatrix} a' & b' \\ c' & d' \end{smallmatrix}\right)$ and multiplying out, we have exactly the four requirements

$$-b - b' = -a + a' = d - d' = c + c' = 0.$$

This gives the asserted result. ∎

COROLLARY

$$E(\iota(g,g')) = \sum_{\gamma,\gamma' \in P_1(F)\backslash SL(2,F)} q(\iota(\gamma g, \gamma' g)) + \sum_{\gamma \in SL(2,F)} q(\theta\iota(\gamma g, g')).$$

Proof The only observation needed here is that $\Theta(F)$ is a "diagonally imbedded" copy of $SL(2,F)$ in $SL(2,F) \times SL(2,F)$, i.e., by $g \to (g, \varepsilon^{-1}g\varepsilon)$, so that

$$\Theta(F)\backslash SL(2,F) \times SL(2,F)$$

has representatives $SL(2,F) \times \{1_2\}$. ∎

LEMMA When $\mathfrak{n}$ is a proper ideal of $\mathfrak{o}$, for all (g,g') in $SL(2,\mathbb{A}) \times SL(2,\mathbb{A})$ we have

$$q(\iota(g,g')) = 0.$$

Proof For non-vanishing of the kernel q for the Siegel's Eisenstein series E, it is required of $\tilde{g} = \left(\begin{smallmatrix} a & b \\ c & d \end{smallmatrix}\right)$ in $Sp(2,\mathbb{A})$ that

$$(c \quad d) \equiv \begin{pmatrix} 1 & 1 & 0 & 0 \\ 0 & 0 & -1 & 1 \end{pmatrix} \text{ modulo } \mathfrak{n}.$$

In particular, the two-by-two matrix formed by the first and third columns (or second and fourth) is of rank two. However, it is visible that no element of the form

$$\begin{pmatrix} 0 & 0 & 1 & 0 \\ 0 & 0 & 0 & 1 \end{pmatrix} \iota(g, g') = \begin{pmatrix} * & 0 & * & 0 \\ 0 & * & 0 & * \end{pmatrix}$$

can have this property. That is, q is identically zero on

$$\tilde{P}(F) 1_4 \iota(SL(2, \mathbb{A}) \times SL(2, \mathbb{A})). \qquad \blacksquare$$

COROLLARY By the previous lemma, when $\mathfrak{n}$ is a *proper* ideal, we have

$$E(\iota(g, g')) = \sum_{\gamma \in SL(2, F)} q(\iota(\gamma g, g')). \qquad \blacksquare$$

LEMMA For $h \in SL(2, \mathbb{A})$ we have

$$q(\theta\iota(hg, g')) = q(\theta\iota(g, \varepsilon^{-1} h^{-1} \varepsilon g')).$$

Proof This is formal, from the integral definition of q in section 7.1, and from the nature of θ, as discussed above regarding the F-*rational* isotropy group of the coset $\tilde{P}(F)\theta$. ■

PROPOSITION When $\mathfrak{n}$ is a proper ideal, $E(\iota(g, g'))$ is a holomorphic cuspform on $SL(2, \mathbb{A}) \times SL(2, \mathbb{A})$. Precisely, as a function of $g \in SL(2, \mathbb{A})$, for fixed g', $E(\iota(g, g'))$ is a holomorphic cuspform of weight κ and is right $\mathbb{K}$-invariant; as a function of $g' \in SL(2, \mathbb{A})$, for fixed g, $E(\iota(g, g'))$ is a holomorphic cuspform of weight κ and is right $\mathbb{K}$-invariant.

Proof Since E is left $Sp(2, F)$-invariant and $\iota(SL(2, F) \times SL(2, F)) \subset Sp(2, F)$, the left invariance required of a cuspform follows. By its definition, E is right-invariant under

$$\mathbb{K}^{\#} = \{g \in Sp(2, \hat{\mathfrak{o}}) \colon g \equiv 1_4 \text{ modulo } \mathfrak{n}\}.$$

Since

$$\iota(\mathbb{K} \times \mathbb{K}) \subset \mathbb{K}^{\#},$$

the assertion about right $\mathbb{K} \times \mathbb{K}$-invariance follows. At an infinite prime, from section 7.1 we have a right equivariance by $U^*(2)$ by the character

$$\begin{pmatrix} A & B \\ -B & A \end{pmatrix} \to \det(A + iB)^{\kappa}.$$

Therefore, on $SO(2) \times SO(2)$ we have

$$\iota\left(\begin{pmatrix} a & b \\ c & d \end{pmatrix} \times \begin{pmatrix} a' & b' \\ c' & d' \end{pmatrix}\right) \to (ic+d)^\kappa \times (ic'+d')^\kappa.$$

This shows that $E(\iota(g,g'))$ is of weight κ in both g and g'.

Next, we verify that $E(\iota(g,g'))$ is holomorphic. For the first of these, it suffices to check that the kernel is "holomorphic," in the sense that it is annihilated by the lowering operators. For this, it suffices to take

$$g = u(x)\delta(y) = \begin{pmatrix} 1 & x \\ 0 & 1 \end{pmatrix}\begin{pmatrix} y^{1/2} & 0 \\ 0 & y^{-1/2} \end{pmatrix} \in SL(2, \mathbb{R})$$

$$g' = u(x')\delta(y')$$

with $x, x' \in \mathbb{R}$ and $y, y' > 0$. From section 7.4, $q(\theta\iota(g,g'))$ is (up to a factor independent of g and g')

$$\det(iC + D)^{\ \kappa}$$

where

$$(C \quad D) = \theta\iota(g,g') = \begin{pmatrix} y^{1/2} & y'^{1/2} & x/y^{1/2} & x'/y'^{1/2} \\ 0 & 0 & -1/y^{1/2} & 1/y'^{1/2} \end{pmatrix}.$$

Therefore,

$$\det(iC + D)^{-\kappa} = y^{\kappa/2} y'^{\kappa/2}(x + iy + x' + iy')^{-\kappa}.$$

As this is $y^{\kappa/2}y'^{\kappa/2}$ times a holomorphic function of $x + iy \in \mathfrak{H}$ and $x' + iy' \in \mathfrak{H}$, it is holomorphic of weight κ. (See section 3.3 for similar calculations.)

All that remains is to see that $E(\iota(g,g'))$ is a cuspform, for $\mathfrak{n}$ proper. This will follow from the nature of the infinite part. Since the treatment of $E(\iota(g,g'))$ as a function of g' is essentially identical to that as a function of g, we consider only the latter. The 0-th Fourier coefficient in g is

$$\int_{U(F)\backslash U(\mathbb{A})} \sum_{\gamma \in SL(2,F)} q(\theta\iota(\gamma u g, g'))\, du = \int_{U(\mathbb{A})} \sum_{\gamma \in SL(2,F)/U(F)} q(\theta\iota(\gamma u g, g'))\, du =$$

$$= \int_{U(\mathbb{A})} \sum_{\gamma \in U(F)\backslash SL(2,F)} q(\theta\iota(ug, \gamma g'))\, du$$

by the nature of the isotropy group of $\tilde{P}(F)\theta$. That is, the 0-th Fourier coefficient is a sum of translates in g' of integrals over $U(\mathbb{A})$ in g. Since q factors over primes, to prove vanishing it suffices to prove vanishing of

$$\int_{U(F_\mathfrak{p})} q(\theta\iota(ug, g'))\, du$$

for an infinite prime $\mathfrak{p}$ and for arbitrary $g, g' \in SL(2, F_\mathfrak{p})$. Since the $\mathfrak{p}$-part of q is right $SO(2) \times SO(2)$-equivariant, we can suppose (by an Iwasawa decomposition) that g and g' are upper triangular; indeed, we may take $g = \delta(y)$ as above, and $g' = u(x')\delta(y')$, with $y, y' > 0$. Then, noting the explicit expression for q obtained in the discussion of holomorphy just above, it suffices to show that

$$\int_{\mathbb{R}} \det(x + iy + x' + iy')^{-\kappa}\, dx = 0.$$

This is clear from residues. Thus, we have shown that $E(\imath(g, g'))$ is a holomorphic cuspform of weight κ in both g and g', etc. ■

From the last proposition, and from general facts about holomorphic cuspforms, it follows that for a proper ideal $\mathfrak{n}$ there are constants $A(f, f')$ so that

$$E(\imath(g, g')) = \sum_{\lambda, \lambda'} \sum_{f, f'} A(f, f') f(g) f'(g'),$$

where λ and λ' range over Λ, and f (respectively, f') ranges over an orthonormal basis for $\mathrm{Cfm}_1(\mathfrak{n}, \kappa, \lambda)$ (respectively, $\mathrm{Cfm}_1(\mathfrak{n}, \kappa, \lambda')$). We note that holomorphic cuspforms with distinct Hecke eigenvalues must be orthogonal: The commutative ring of operators $\mathscr{H}_1(\mathfrak{n})$ is closed under adjoints. Therefore,

$$\langle E(\imath(*, g')), f \rangle_1 = \sum_{\lambda'} \sum_{f'} A(f, f') f'(g'),$$

where the inner product is the integral over g in $SL(2, F)\backslash SL(2, \mathbb{A})$. We compute directly:

$$\int_{SL(2,F)\backslash SL(2,\mathbb{A})} E(\imath(g, g'))\bar{f}(g)\, dg =$$

$$= \int_{SL(2,F)\backslash SL(2,\mathrm{A})} \sum_{\gamma \in SL(2,F)} q(\theta\imath(\gamma g, g'))\bar{f}(g)\, dg =$$

$$= \int_{SL(2,\mathbb{A})} q(\theta\imath(g, g')\bar{f}(g)\, dg.$$

From the invariance of $q(\theta\imath(g, g'))$ under

$$(g, g') \to (hg, \varepsilon^{-1}h\varepsilon g'),$$

we can rewrite this as

$$\int_{SL(2,\mathbb{A})} q(\theta\imath(\varepsilon g'^{-1}\varepsilon^{-1} g, 1))\bar{f}(g)\, dg.$$

Replacing g by $\varepsilon g'\varepsilon^{-1}g^{-1}\varepsilon^{-1}$ makes this

$$\int_{SL(2,\mathbb{A})} q(\theta\iota(g^{-1},1)\bar{f}(\varepsilon g'\varepsilon^{-1}g^{-1})\,dg = \left(\int_{SL(2,\mathbb{A})} f(g'g^{-1})q(\theta\iota(g^{-1},1))\,dg\right)^{\natural}.$$

The expression inside the last set of parentheses is the effect upon f of application of a product (over primes) of convolution operators. *We claim that a holomorphic cuspform which is an eigenfunction for Hecke operators in $\mathscr{H}_1(\mathfrak{n})$ at primes not dividing $\mathfrak{n}$ is an eigenfunction for this operator, with non-zero eigenvalue.* Given the previous discussion, this would prove the theorem. To verify this claim, we consider three cases: primes dividing $\mathfrak{n}$, infinite primes, and finite primes not dividing $\mathfrak{n}$.

At a finite prime $\mathfrak{p}$ dividing $\mathfrak{n}$, from the definition of q (in section 7.1) we have

$$q(\theta\iota(g^{-1},1)) \neq 0$$

if and only if $g \equiv 1_2$ modulo $\mathfrak{n}$, i.e., if and only if $g \in \mathbb{K} \cap SL(2,\mathfrak{o}_\mathfrak{p})$. Therefore, at such $\mathfrak{p}$, the convolution operator

$$f(g') \to \int_{SL(2,F_\mathfrak{p})} f(g'g^{-1})q(\theta\iota(g^{-1},1))\,dg$$

is simply a multiplication by a *non-zero* (rational) constant $S_\mathfrak{p}$ depending only on $\mathfrak{n}$.

Regarding infinite primes $\mathfrak{p}$, we claim that *holomorphic cuspforms are eigenfunctions for these convolutions appearing in the infinite-prime part, with non-zero eigenvalues.* The kernel $\psi(g) = q(\theta\iota(g^{-1},1))$ is readily verified to be left and right $SO(2)^m$-equivariant:

$$\psi\left(\begin{pmatrix} \cos\theta & \sin\theta \\ -\sin\theta & \cos\theta \end{pmatrix} g\right) = \psi(g)e^{-\kappa i\theta}$$

$$\psi\left(g\begin{pmatrix} \cos\theta & \sin\theta \\ -\sin\theta & \cos\theta \end{pmatrix}\right) = \psi(g)e^{\kappa i\theta}.$$

Therefore, using the right equivariance of f, in these infinite-prime integrals we may integrate over the subgroup $P^+(F_\mathfrak{p})$ consisting of upper triangular matrices with positive diagonal entries, and also assume that $g' \in P^+(F_\mathfrak{p})SL(2,\mathbb{A}_0)$. Put

$$g = u(x)\delta(y) \qquad g' = u(x')\delta(y')g_0$$

with $g_0 \in SL(2,\mathbb{A}_0)$, where

$$u(x) = \begin{pmatrix} 1 & x \\ 0 & 1 \end{pmatrix} \qquad \delta(y) = \begin{pmatrix} y & 0 \\ 0 & y^{-1} \end{pmatrix}$$

with $x,\ x' \in \mathbb{R}$ and $y,\ y' > 0$. Recall that $f(u(x)\delta(y)g_0)$ has a Fourier expansion (depending on g_0)

$$f(u(x)\delta(y)g_0) = \|y\|^\kappa \sum_{\xi \in F^\times} c(\xi)\tau_\infty(\xi x)\exp(-\pi \operatorname{Tr} \xi y^2)$$

where Tr denotes the $\mathbb{C}$-linear extension of the trace from F to $\mathbb{Q}$. Therefore, we have

$$f(g'g_0g^{-1}) = f(u(x')\delta(y')g^{-1}g_0) = f(u(x' + y'^2x)\delta(yy')g_0) =$$

$$= \|yy'\|^\kappa \sum_{\xi \in F^\times} c(\xi)\tau_\infty(\xi(x' + y'^2x))\exp(-2\pi \operatorname{Tr} \xi y^2y'^2).$$

Combining the second lemma in section 7.4 and calculations as just above, we have

$$q(\theta\iota(g^{-1}, 1)) = \pi^{-\kappa}\pi^{-(\kappa-1/2)}\Gamma(\kappa)\Gamma(\kappa - \tfrac{1}{2})y^\kappa \det(i(1 + y^2) + x)^{-\kappa}.$$

Therefore, for the ξ-th term, at an infinite prime $\mathfrak{p}$ the integral is

$$\int \tau(-\xi(x' + y'^2x))(yy')^\kappa \exp(-2\pi \operatorname{Tr} \xi y^2y'^2)y^\kappa \det(i(1 + y^2) + x)^{-\kappa}\, d^+x'y^{-2}\, d^\times y =$$

$$= y'^\kappa\tau(\xi x') \int \tau(-\xi y'^2x)y^{2\kappa-2} \exp(-2\pi \operatorname{Tr} \xi y^2y'^2) \det(i(1 + y^2) + x)^{-\kappa}\, d^+x'\, d^\times y.$$

By residues, integrating over x gives 0 unless $\xi > 0$ (at $\mathfrak{p}$), in which case it yields a non-zero multiple (independent of ξ, y, x') of

$$y'^\kappa\tau(\xi x')(\xi y'^2)^{\kappa-1} \int y^{2\kappa-2} \exp(-2\pi \operatorname{Tr} \xi y^2y'^2)\exp(-2\pi \operatorname{Tr} \xi y'^2(1 + y^2))\, d^\times y =$$

$$= y'^\kappa\tau(\xi x')\exp(-2\pi \operatorname{Tr} \xi y'^2)(\xi y'^2)^{\kappa-1} \int y^{2\kappa-2} \exp(-4\pi \operatorname{Tr} \xi y^2y'^2)\, d^\times y =$$

$$= y'^\kappa\tau(\xi x')\exp(-2\pi \operatorname{Tr} \xi y'^2) \times \int y^{2\kappa-2} \exp(-4\pi \operatorname{Tr} y^2)\, d^\times y =$$

$$= y'^\kappa\tau(\xi x')\exp(-2\pi \operatorname{Tr} \xi y'^2) \times \tfrac{1}{2}(4\pi)^{1-\kappa}\Gamma(\kappa - 1).$$

Thus, we have shown that holomorphic cuspforms are eigenfunctions for these convolutions appearing in the infinite-prime part, with *non-zero eigenvalues.*

For finite $\mathfrak{p}$ not dividing $\mathfrak{n}$ our previous discussion makes apparent the following:

LEMMA The function $g \to q(\theta\iota(g^{-1}, 1))$ is left and right $SL(2, \mathfrak{o}_\mathfrak{p})$-invariant. ∎

Therefore, the operator

$$f(g') \to \int_{SL(2,F_\mathfrak{p})} f(g'g^{-1})q(\theta\iota(g^{-1}, 1))\, dg$$

is an infinite sum of (spherical) Hecke operators in $SL(2, F_{\mathfrak{p}})$; f was assumed to be an eigenfunction for this algebra of operators, so this convolution operator multiplies f by some constant $S(\lambda)_{\mathfrak{p}}$ depending only on the Hecke eigenvalue. We need to determine the form of $S(\lambda)_{\mathfrak{p}}$.

Put $K = SL(2, \mathfrak{o}_{\mathfrak{p}})$. Let η_i be the characteristic function of

$$K\begin{pmatrix} \pi^i & 0 \\ 0 & \pi^{-i} \end{pmatrix}K$$

where π is a local parameter and $0 \leq i \in \mathbb{Z}$. Clearly these functions are a $\mathbb{C}$-basis for the spherical Hecke algebra in $SL(2, F_{\mathfrak{p}})$. Further, it is not hard to check that the operators induced on cuspforms by these functions η_i are *self-adjoint*. We first will express $q(\theta\iota(g^{-1}, 1))$ in terms of these functons η_i. Since $q(\theta\iota(g^{-1}, 1))$ is left and right K-invariant, it suffices to take g of the form

$$g = \delta(y) = \begin{pmatrix} y & 0 \\ 0 & y^{-1} \end{pmatrix}$$

with $\|y\| \leq 1$. Then

$$(0_2 1_2)\theta\iota(g^{-1}, 1) = \begin{pmatrix} y^{-1} & 1 & 0 & 0 \\ 0 & 0 & -y & 1 \end{pmatrix}.$$

In the integral expression

$$q(\theta\iota(g^{-1}, 1)) = \int_{GL(2, F_{\mathfrak{p}})} \|\det t\|^{\kappa}\chi(\det t)\varphi^*(t(0, 1)\theta\iota(g^{-1}, 1))\,dt$$

replace t by

$$t\begin{pmatrix} y & 0 \\ 0 & 1 \end{pmatrix}$$

to obtain

$$q(\theta\iota(g^{-1}, 1)) = \|y\|^{\kappa}\chi(y)\int_{GL(2, F_{\mathfrak{p}})} \|\det t\|^{\kappa}\chi(\det t)\varphi^{**}(t)\,dt$$

where φ^{**} is the characteristic function of the set of integral two-by-two matrices. As we have computed before, this integral is

$$1/(1 - \chi(\pi)N\mathfrak{p}^{-\kappa})(1 - \chi(\pi)N\mathfrak{p}^{1-\kappa}).$$

Therefore,

$$q(\theta\iota(g^{-1}, 1)) = \left(\sum_{i \geq 0} \chi(\pi^i) N\mathfrak{p}^{-\kappa} \eta_i(g)\right) \Big/ (1 - \chi(\pi) N\mathfrak{p}^{-\kappa})(1 - \chi(\pi) N\mathfrak{p}^{1-\kappa}).$$

Next, we express all the η_i's in terms of η_1. To this end, we compute $\eta_i * \eta_1$. Let $q = N\mathfrak{p}$. Certainly $\eta_0 * \eta_1 = \eta_1$. By the left and right K-invariance, we may as well evaluate this convolution at $g = \delta(\pi^k)$. Then

$$(\eta_i * \eta_1)(g) = \int_{K\delta(\pi)K} \eta_i(gh^{-1})\, dh = \sum_{h \in K \backslash K\delta(\pi)K} \eta_i(gh^{-1}).$$

We may take representatives

$$\begin{pmatrix} \pi & 0 \\ 0 & \pi^{-1} \end{pmatrix}$$

$$\begin{pmatrix} 1 & \beta/\pi \\ 0 & 1 \end{pmatrix} \qquad \beta \text{ in } \mathfrak{o}_\mathfrak{p}/\mathfrak{p}\mathfrak{o}_\mathfrak{p} - \{0\}$$

$$\begin{pmatrix} \pi^{-1} & \beta/\pi \\ 0 & \pi \end{pmatrix} \qquad \beta \text{ in } \mathfrak{o}_\mathfrak{p}/\mathfrak{p}^2\mathfrak{o}_\mathfrak{p}$$

for $K \backslash K\delta(\pi)K$. Therefore, with $g = \delta(\pi^k)$, the convolution is the sum of the values of η_i at

$$\begin{pmatrix} \pi^{k-1} & 0 \\ 0 & \pi^{1-k} \end{pmatrix}$$

$$\begin{pmatrix} \pi^k & \pi^{k-1}\beta \\ 0 & \pi^{-k} \end{pmatrix} \qquad \beta \text{ in } \mathfrak{o}_\mathfrak{p}/\mathfrak{p}\mathfrak{o}_\mathfrak{p} - \{0\}$$

$$\begin{pmatrix} \pi^{k+1} & \pi^{k-1}\beta \\ 0 & \pi^{-1-k} \end{pmatrix} \qquad \beta \text{ in } \mathfrak{o}_\mathfrak{p}/\mathfrak{p}^2\mathfrak{o}_\mathfrak{p}.$$

In the first family, the elementary divisors are π^{k-1} and π^{1-k}. In the second family, the elementary divisors are π^k and π^{-k} if $k > 0$, but are π and π^{-1} if $k = 0$. In the third family, the elementary divisors are π^{k+1} and π^{-1-k}. Therefore,

$$\eta_i * \eta_1 = \eta_{i+1} + (q - 1)\eta_i + q^2 \eta_{i-1} \qquad (\text{for } i \geq 2)$$

$$\eta_1 * \eta_1 = (q + q^2)\eta_0 + (q - 1)\eta_1 + (q^2 + 1)\eta_2.$$

This gives a recursion for η_i in terms of η_1 and these convolutions: Formally, one obtains

$$(1 + (q - 1 - \eta_1)X + q^2X^2)\left(\sum_{n \geq 0} \eta_n X^n\right) = (1 - X)(1 + qX).$$

Therefore,

$$\sum_{i \geq 0} \chi(\pi^i) q^{-\kappa} \eta_i =$$

$$= (1 - \chi(\pi)q^{-\kappa})(1 + \chi(\pi)q^{1-\kappa})/(1 + (q - 1 - \eta_1)\chi(\pi)q^{-\kappa} + q^{2-2\kappa}).$$

Altogether we have

$$q(\theta\iota(g^{-1}, 1)) =$$

$$= (1 + \chi(\pi)q^{1-\kappa})/(1 + (q - 1 - \eta_1)\chi(\pi)q^{-\kappa} + q^{2-2\kappa})(1 - \chi(\pi)q^{1-\kappa})$$

(noting that onc factor from the denominator and one from the numerator cancel). From this discussion, if $f * \eta_1 = \lambda f$, then

$$f * q(\theta\iota(g^{-1}, 1)) =$$

$$= ((1 + \chi(\pi)q^{1-\kappa})/(1 + (q - 1 - \lambda)\chi(\pi)q^{-\kappa} + q^{2-2\kappa})(1 - \chi(\pi)q^{1-\kappa})) \times f.$$

Due to the fact that the operator given by η_1 is self-adjoint, the eigenvalue λ must be *real*. Therefore, the eigenvalue of f under convolution with $q(\theta\iota(g^{-1}, 1))$ is *real*.

In summary, at a finite prime $\mathfrak{p}$ dividing $\mathfrak{n}$, the $\mathfrak{p}$-part of the convolution operator on f is multiplication by a non-zero rational number (coming from a volume constant); at an infinite prime, this convolution operator multiplies f by a non-zero number (explicitly determined above); at a finite prime $\mathfrak{p}$ not dividing $\mathfrak{n}$, the effect is multiplication by

$$S_\mathfrak{p} = (1 + \chi(\pi)q^{1-\kappa})/(1 + (q - 1 - \lambda)\chi(\pi)q^{-\kappa} + q^{2-2\kappa})(1 - \chi(\pi)q^{1-\kappa})$$

(where $q = N\mathfrak{p}$, $\lambda \in \mathbb{R}$ depends on $\mathfrak{p}$, and π is a local parameter at $\mathfrak{p}$). Since the Siegel's Eisenstein series is absolutely (etc.) convergent, the infinite product

$$\prod_{\mathfrak{p} \text{ finite not dividing } \mathfrak{n}} S_\mathfrak{p}$$

must be absolutely convergent. Therefore, for this product to vanish some numerator $(1 + \chi(\pi)q^{1-\kappa})$ must be 0. Since $\kappa > 1$ this is not possible.

Therefore, we have shown that

$$\langle E(\iota(*, g')), f \rangle_1 = S(\lambda) f^\natural(g'),$$

where $S(\lambda) \neq 0$ depends only upon the homomorphism $\lambda\colon \mathscr{H}_1(\mathfrak{n}) \to \mathbb{C}$ so that

$$f * \eta = \lambda(\eta) f$$

for all $\eta \in \mathscr{H}_1(\mathfrak{n})$. Therefore, from earlier remarks

$$E(\iota(g, g')) = \sum_{\lambda} S(\lambda) \sum_{f} f(g) f^{\natural}(g')$$

where for each $\lambda\colon \mathscr{H}_1(\mathfrak{n}) \to \mathbb{C}$ the sum over f is over an orthonormal basis for λ-eigenfunctions (among holomorphic cuspforms). ∎

7.10 Proof of the Arithmetic Structure Theorem

At last we can prove the Arithmetic Structure Theorem stated in section 6.1. There are two parts to the proof. First, we combine the Main Formula of section 7.9 with the results of section 7.8 on the Fourier coefficients of the Siegel's Eisenstein series to obtain a result on the Fourier coefficients and inner products of cuspforms on $SL(2, \mathbb{A})$. Then we must compare the cuspforms, spherical Hecke algebras, and inner products on $SL(2, \mathbb{A})$ to those on $GL(2, \mathbb{A})$ to obtain the Arithmetic Structure Theorem for cuspforms on $GL(2, \mathbb{A})$.

Again, (as remarked in section 7.8) we note that this discussion for $F = \mathbb{Q}$ depends only upon the special-value results for Dirichlet L-functions over $\mathbb{Q}$, given in Appendix A.5. For $[F : \mathbb{Q}] > 1$, in section 7.8 we used Klingen's result from section 6.2, which depends only upon the Arithmetic Structure Theorem for $\mathbb{Q}$; i.e., there is no circularity in our argument.

Fix a weight $(\kappa, \ldots, \kappa)$, abbreviated simply as κ. Fix a non-zero ideal $\mathfrak{n}$ in $\mathfrak{o}$, and define

$$\tilde{\mathbb{K}} = \left\{ g \in GL(2, \hat{\mathfrak{o}}) \colon g \equiv \begin{pmatrix} * & 0 \\ 0 & 1 \end{pmatrix} \text{ modulo } \mathfrak{n} \right\}$$

$$\mathbb{K} = \{ g \in SL(2, \hat{\mathfrak{o}}) \colon g \equiv 1_2 \text{ modulo } \mathfrak{n} \}.$$

For a finite prime $\mathfrak{p}$ not dividing $\mathfrak{n}$, let

$$H(\mathfrak{p}) = \{\text{left and right } GL(2, \mathfrak{o}_{\mathfrak{p}})\text{-invariant compactly supported functions on } GL(2, F_{\mathfrak{p}})\}$$

$$H_1(\mathfrak{p}) = \{\text{left and right } SL(2, \mathfrak{o}_{\mathfrak{p}})\text{-invariant compactly supported functions on } SL(2, F_{\mathfrak{p}})\}.$$

Put

$$\mathscr{H}(\mathfrak{n}) = \bigotimes H(\mathfrak{p}) \qquad \mathscr{H}_1(\mathfrak{n}) = \bigotimes H_1(\mathfrak{p}).$$

We use notation

$$\delta(y) = \begin{pmatrix} y & 0 \\ 0 & y^{-1} \end{pmatrix}$$

for $y \in F_{\mathfrak{p}}^{x}$ or $y \in \mathbb{J}$, and

$$\ell(y) = \begin{pmatrix} y & 0 \\ 0 & 1 \end{pmatrix}.$$

PROPOSITION Let $K = GL(2, \mathfrak{o}_{\mathfrak{p}})$ and $K_1 = SL(2, \mathfrak{o}_{\mathfrak{p}})$. The map α: $H_1(\mathfrak{p}) \to H(\mathfrak{p})$ given by

α: characteristic function of $K_1 \delta(y) K_1 \to$ characteristic function of $K\delta(y)K$

is an injective convolution algebra homomorphism (where we give maximal compact subgroups measure one in either case). (Proof is easy and formal.) ■

Define

$$\mathrm{Cfm}(\mathfrak{n}, \kappa) = \{\text{right } \tilde{\mathbb{K}}\text{-invariant holomorphic cuspforms of weight } \kappa \text{ on } GL(2, \mathbb{A})\}$$

$$\mathrm{Cfm}(\mathfrak{n}, \kappa, \omega) = \{\text{right } \tilde{\mathbb{K}}\text{-invariant holomorphic cuspforms of weight } \kappa \text{ with central character } \omega \text{ on } GL(2, \mathbb{A})\}$$

$$\mathrm{Cfm}_1(\mathfrak{n}, \kappa) = \{\text{right } \mathbb{K}\text{-invariant holomorphic cuspforms of weight } \kappa \text{ on } SL(2, \mathbb{A})\}.$$

Though we have not overtly discussed this before, a cuspform on $SL(2, \mathbb{A})$ (as opposed to $GL(2, \mathbb{A})$) is required to have its 0-th Fourier coefficient (in an adelic sense) vanish; the condition of holomorphy is the same. Let Z^+ be the connected component of the identity in the archimedean part of $Z(\mathbb{A})$. Since $\mathrm{Cfm}(\mathfrak{n}, \kappa)$ decomposes as a finite sum

$$\bigoplus_{\omega} \mathrm{Cfm}(\mathfrak{n}, \kappa, \omega)$$

where the characters ω are of finite order, it is clear that elements of $\mathrm{Cfm}(\mathfrak{n}, \kappa)$ are Z^+-invariant. We have another inner product $\langle\ ,\ \rangle^{\sim}$ on $\mathrm{Cfm}(\mathfrak{n}, \kappa)$ given by

$$\langle f_1, f_2 \rangle^{\sim} = \int_{Z^+ GL(2,F) \backslash GL(2,\mathbb{A})} f_1(g) \bar{f}_2(g)\, dg$$

and an inner product $\langle\ ,\ \rangle_1$ on $\mathrm{Cfm}_1(\mathfrak{n}, \kappa)$ given by

$$\langle f_1, f_2 \rangle_1 = \int_{SL(2,F) \backslash SL(2,\mathbb{A})} f_1(g) \bar{f}_2(g)\, dg.$$

It is easy to check that the inner product $\langle\ ,\ \rangle^{\sim}$ and the standard inner product $\langle\ ,\ \rangle$ differ by a uniform rational constant. Let λ: $\mathscr{H}_1(\mathfrak{n}) \to \mathbb{C}$ be an algebra homomorphism; define

$$\mathrm{Cfm}(\mathfrak{n},\kappa)_\lambda = \{f \in \mathrm{Cfm}(\mathfrak{n},\kappa): f * \alpha(\eta) = \lambda(\eta) f \text{ for all } \eta \in \mathscr{H}_1(\mathfrak{n})\}$$

$$\mathrm{Cfm}(\mathfrak{n},\kappa,\omega)_\lambda = \{f \in \mathrm{Cfm}(\mathfrak{n},\kappa,\omega): f * \alpha(\eta) = \lambda(\eta) f \text{ for all } \eta \in \mathscr{H}_1(\mathfrak{n})\}$$

$$\mathrm{Cfm}_1(\mathfrak{n},\kappa)_\lambda = \mathrm{Cfm}_1(\mathfrak{n},\kappa,\lambda) = \{f: \mathrm{Cfm}_1(\mathfrak{n},\kappa): f * \eta = \lambda(\eta) f \text{ for all } \eta \in \mathscr{H}_1(\mathfrak{n})\}.$$

Let $\lambda: \mathscr{H}(\mathfrak{n}) \to \mathbb{C}$ be an algebra homomorphism; define

$$\mathrm{Cfm}(\mathfrak{n},\kappa,\omega,\lambda) = \{f \in \mathrm{Cfm}(\mathfrak{n},\kappa,\omega): \mathrm{f} * \eta = \lambda(\eta) f \text{ for all } \eta \in \mathscr{H}(\mathfrak{n})\}.$$

Definition For either sort of algebra homomorphism λ, for a fixed weight κ, let $\mathbb{Q}(\lambda)$ be the field generated over $\mathbb{Q}$ by the set of values $\lambda(\eta_\mathfrak{p})$, where for $GL(2)$

$$\eta_\mathfrak{p} = N\mathfrak{p}^{\kappa/2} \times \text{characteristic function of } GL(2,\mathfrak{o}_\mathfrak{p})\ell(\pi^{-1})GL(2,\mathfrak{o}_\mathfrak{p})$$

and for $SL(2)$

$$\eta_\mathfrak{p} = \text{characteristic function of } GL(2,\mathfrak{o}_\mathfrak{p})\delta(\pi)GL(2,\mathfrak{o}_\mathfrak{p})$$

where π is a local parameter at $\mathfrak{p}$ in either case. *In the former case, the normalizing constant (which depends on the weight) is necessary for preservation of rationality properties.*

For a character ω on $\mathbb{J}/F_\infty^+ F^\times$, put

$$\mathbb{Q}(\omega) = \text{field generated over } \mathbb{Q} \text{ by the values of } \omega.$$

Also, let $\mathbb{Q}(\omega,\lambda) = \mathbb{Q}(\omega)\mathbb{Q}(\lambda)$, etc. For a $\mathbb{C}$-valued function ψ and an element σ of $\mathrm{Aut}(\mathbb{C}/\mathbb{Q})$, let ψ^σ be the function whose value $\psi^\sigma(x)$ at a point x in the domain of ψ is $\psi(x)^\sigma$.

Let $\mathbb{J}_\infty^+$ be the connected component of the identity in $\mathbb{J}_\infty$, and put $\mathbb{J}^+ = \mathbb{J}_\infty^+\mathbb{J}_0$. On elements $g \in GL(2,\mathbb{A})$ of the form

$$g = u(x)\ell(y) = \begin{pmatrix} 1 & x \\ 0 & 1 \end{pmatrix}\begin{pmatrix} y & 0 \\ 0 & 1 \end{pmatrix}$$

with $x \in \mathbb{A}$ and $y \in \mathbb{J}^+$, a holomorphic automorphic form f on $GL(2,\mathbb{A})$ has a *Fourier expansion*

$$f(g) = \|y\|^{\kappa/2} \sum_{\xi \in F} \tau(\xi x) V_\xi(y_0) \exp(-2\pi \operatorname{Tr} \xi_\infty y_\infty)$$

where $V_\xi(y_0) = V((\xi y)_0)$ is a function of only the finite-prime parts y_0 and ξ_0 of y and ξ. On the other hand, a holomorphic cuspform f on $SL(2,\mathbb{A})$ has a different sort of *Fourier expansion*: Taking

$$g = u(x)\delta(y) = \begin{pmatrix} 1 & x \\ 0 & 1 \end{pmatrix}\begin{pmatrix} y & 0 \\ 0 & y^{-1} \end{pmatrix}$$

with $x \in \mathbb{A}$ and $y \in \mathbb{J}$, we have

$$f(g) = \|y\|^\kappa \sum_{\xi \in F} \tau(\xi x) V_\xi(y_0)_1 \exp(-2\pi \operatorname{Tr} \xi_\infty y_\infty^2)$$

where $V_\xi(y_0)_1$ is a function of only the finite-prime parts y_0 and ξ_0 of y and ξ.

Definition In the previous notation:

i. A holomorphic modular form f of weight κ on $GL(2, \mathbb{A})$ is said to have *Fourier coefficients in a field* $\Omega \subset \mathbb{C}$ *in a* $GL(2)$*-sense* if $V_\xi(y_0) \in \Omega$ for all $\xi \in F$ and $y_0 \in \mathbb{J}_0$.
ii. A holomorphic modular form f of weight κ on $SL(2, \mathbb{A})$ is said to have *Fourier coefficients in a field* $\Omega \subset \mathbb{C}$ *in an* $SL(2)$*-sense* if $V_\xi(y_0)_1 \in \Omega$ for all $\xi \in F$ and $y_0 \in \mathbb{J}_0$.

Remarks It should be noted that these definitions of rationality of Fourier coefficients depend only on the Fourier expansions at special elements of the groups $SL(2)$ and $GL(2)$. In particular, for $GL(2)$, we consider only elements $u(x)\ell(y)$ with $y \in \mathbb{J}^+$.

As above, χ is the idele class character associated to the κ-th power $(-1/*)_2^\kappa$ of the quadratic symbol $(-1/*)_2$ over F. Regarding the special Siegel's Eisenstein series $E(g)$ on $Sp(2, \mathbb{A})$ defined in section 7.1, in section 7.8 we found that there are *rational* numbers $c(\xi; y)$ and a uniform non-zero constant c (whose nature is essentially *irrelevant*) so that

$$E(u'(x)\ell'(y)) = c\|\det y\|^\kappa \chi(\det y) \sum_{\xi} c(\xi_0; y_0)\tau(\operatorname{Tr} \xi x) \exp(-2\pi \operatorname{Tr} \xi_\infty y_\infty y_\infty^T)$$

where (by abuse of notation) in the first instance Tr denotes the matrix trace, and in the second denotes the composition of the matrix trace and the $\mathbb{C}$-linear extension of the trace from F to $\mathbb{Q}$. Here

$$x \in W(\mathbb{A}) = \{\text{symmetric two-by-two matrices over } \mathbb{A}\}$$

$$y \in GL(2, \mathbb{A})$$

where ξ runs over *totally positive semi-definite* elements of $W(F)$, using the notation

$$u'(x) = \begin{pmatrix} 1 & x \\ 0 & 1 \end{pmatrix} \qquad (\text{for } x \in W(\mathbb{A}))$$

$$\ell'(A) = \begin{pmatrix} A & 0 \\ 0 & A^{T-1} \end{pmatrix} \qquad (\text{for } A \in GL(2, \mathbb{A})).$$

In particular, $c(\xi_0; y_0)$ only depends on the finite-prime parts y_0 and ξ_0 of y and ξ. We need a little more information.

PROPOSITION For fixed $y_0 \in GL(2, \mathbb{A}_0)$, the collection of totally positive semi-definite ξ in $W(F)$ so that $c(\xi_0; y_0) \neq 0$ is contained in a set of the form

$$\Lambda^+(N) = \{\text{totally positive semi-definite elements } \xi \text{ of } W(F) \text{ so that } N\xi \in W(\mathfrak{o})\}$$

for some positive rational integer N.

Proof Recall the right invariance of $E(g)$ under

$$\mathbb{K}^\# = \{g \in Sp(2, \hat{\mathfrak{o}})\colon g \equiv 1_4 \text{ modulo } \mathfrak{n}\}.$$

From this, for any $v \in \mathfrak{n}W(\hat{\mathfrak{o}})$ we have

$$E(u'(x)\ell'(y)) = E(u'(x)\ell'(y)u'(v)) = E(u'(x + y_0 v y_0^{-1})\ell'(y)).$$

From the uniqueness of the Fourier expansion in x, if $c(\xi_0; y_0) \neq 0$ this implies that

$$1 = \tau(\operatorname{Tr} \xi y_0 v y_0^{-1}).$$

It is easy to see that this imposes a condition of the sort asserted. ■

PROPOSITION Using notation

$$u(x) = \begin{pmatrix} 1 & x \\ 0 & 1 \end{pmatrix} \qquad \delta(y) = \begin{pmatrix} y & 0 \\ 0 & y^{-1} \end{pmatrix},$$

take two elements g and g' of $SL(2, \mathbb{A})$ of the form $g = u(x)\delta(y)$ and $g' = u(x')\delta(y')$ with $x, x' \in \mathbb{A}$ and $y, y' \in \mathbb{J}$. Then we have a Fourier expansion

$$E(\imath(g, g')) =$$

$$= c\,\|yy'\|^\kappa \chi(yy') \sum_{\xi, \xi'} c(\xi, \xi'; y_0, y_0')\tau(\xi x + \xi' x')\exp(-2\pi \operatorname{Tr}(\xi_\infty y_\infty^2 + \xi_{\infty i} y_\infty'^2))$$

where Tr is the $\mathbb{C}$-linear extension of the trace from F to $\mathbb{Q}$, and

$$c(\xi, \xi'; y_0, y_0') = \sum_x c\left(X; \begin{pmatrix} y_0 & 0 \\ 0 & y_0' \end{pmatrix}\right)$$

where X is summed over all totally positive semi-definite elements of $W(F)$ of the form

$$X = \begin{pmatrix} \xi & * \\ * & \xi' \end{pmatrix}.$$

For each fixed y_0 and y_0', this sum over X is *finite*.

Proof From the nature of the imbedding

$$\imath\colon SL(2,\mathbb{A}) \times SL(2,\mathbb{A}) \to Sp(2,\mathbb{A}),$$

we have

$$\imath(u(x), u(x')) = u'\left(\begin{pmatrix} x & 0 \\ 0 & x' \end{pmatrix}\right)$$

and

$$\imath(\delta(y), \delta(y')) = \ell'\left(\begin{pmatrix} y & 0 \\ 0 & y' \end{pmatrix}\right).$$

It follows immediately that the Fourier coefficient $c(\xi, \xi'; y_0, y_0')$ is the sum over X as indicated, since

$$\operatorname{Tr} X \begin{pmatrix} x & 0 \\ 0 & x' \end{pmatrix} = \xi x + \xi' x'$$

exactly for X of the indicated form. Regarding the finiteness of the sum: For fixed y_0 and y_0' the index X must lie in some set $\Lambda^+(N)$ as above. That is, X is totally positive semi-definite and has fixed diagonal entries ξ and ξ', and the entries lie in $N^{-1}\mathfrak{o}$ for some positive integer N. If z/N is the off-diagonal entry, with $z \in \mathfrak{o}$, then

$$\xi\xi' \to z^2/N^2$$

must either be 0 or be a totally positive element of F. It is elementary to see that there are only finitely many such $z \in \mathfrak{o}$. Thus, the sum is finite. ∎

The point of the previous proposition is that the coefficients $c(\xi, \xi; y_0, y_0')$ are *rational numbers*, for all y, y', since they are *finite* sums of the coefficients $c(X; *)$ of the Siegel's Eisenstein series, which are rational numbers. Note that we need not take great care with the uniform constant c, other than that it is non-zero. Combining this with the Main Formula of section 7.9, for a *proper* ideal $\mathfrak{n}$ we obtain the following corollary.

COROLLARY

$$\sum_{\lambda} c^{-1} S(\lambda) \sum_{f} f(u(x)\delta(y)) f^{\natural}(u(x')\delta(y'))$$

has *rational Fourier coefficients* (for all $y, y' \in \mathbb{J}$) in the $SL(2)$-sense of section 6.1. ∎

As before, λ runs over algebra homomorphisms of the $SL(2)$ spherical Hecke algebra $\mathscr{H}_1(\mathfrak{n})$ (away from primes dividing $\mathfrak{n}$), and for each λ the sum over f is over an orthonormal basis for

$$\mathrm{Cfm}_1(\mathfrak{n}, \kappa, \lambda) = \{\text{weight } \kappa \text{ right } \mathbb{K}\text{-invariant holomorphic cuspforms on } SL(2, \mathbb{A}) \text{ which are } \lambda\text{-eigenfunctions}\}$$

where still

$$\mathbb{K} = \{g \in SL(2, \hat{\mathfrak{o}}): g \equiv 1_2 \text{ modulo } \mathfrak{n}\}.$$

From this corollary, we can prove that the space of cuspforms

$$\mathrm{Cfm}_1(\mathfrak{n}, \kappa) = \bigoplus_{\lambda} \mathrm{Cfm}_1(\mathfrak{n}, \kappa, \lambda)$$

has a $\mathbb{C}$*-basis consisting of cuspforms* φ *with Fourier coefficients in* $\mathbb{Q}$, in the sense that

$$\varphi(u(x)\delta(y)) = \|y\|^{\kappa}\chi(y) \sum_{\xi} c(\xi; y_0; \varphi)\tau(\xi x)\exp(-2\pi \operatorname{Tr} \xi_\infty y_\infty)$$

with $c(\xi; y_0; \varphi) \in \mathbb{Q}$ for all $y \in \mathbb{J}^+$. We use the *finite dimensionality* of this space of cuspforms; this was proven in detail in section 1.7 in a classical setting. The comparisons of adelic and classical cuspforms in section 3.1 (via the Strong Approximation Theorem for $SL(2)$) then yield the result for $SL(2, \mathbb{A})$.

Let $\{f_i\}$ be an orthonormal basis for $\mathrm{Cfm}_1(\mathfrak{n}, \kappa)$ consisting of eigenfunctions for $\mathscr{H}_1(\mathfrak{n})$, and write

$$\lambda_i = c^{-1}S(\lambda)$$

if f_i is a λ-eigenvector. From sections 7.8 and 7.9 we know that

$$\sum_i \lambda_i f_i(g) f_i^{\natural}(g')$$

has Fourier coefficients in $\mathbb{Q}$, in the $SL(2)$-sense. Let X be a *finite* set of indices so that, for any index ξ' and $y_0 \in \mathbb{J}_0$,

$$c(\xi'; y_0; \varphi) = \sum_{\xi \in X} \psi(\xi, \xi') c(\xi'; y_0; \varphi)$$

for some constants $\psi(\xi, \xi')$ for any φ in $\mathrm{Cfm}_1(\mathfrak{n}, \kappa)$; such a set X exists because of the finite dimensionality of $\mathrm{Cfm}_1(\mathfrak{n}, \kappa)$, and X has cardinality $n = \dim_{\mathbb{C}} \mathrm{Cfm}_1(\mathfrak{n}, \kappa)$. By choice of a bijection $\{1, \ldots, n\} \to X$, we may index the f_i by elements of X. Let Ψ be the n-by-n matrix with (ξ, ξ')-th entry

$$\Psi(\xi, \xi') = c(\xi'; y_0'; f_\xi^{\natural}) \qquad (\text{with } \xi \text{ and } \xi' \text{ in } X).$$

Then Ψ is invertible for y_0' in a dense set of $\mathbb{J}_0$, by previous remarks. Still

$$\|y'\|^{\kappa}\chi(y') \sum_{\xi,\xi' \in X} f_\xi(g)\lambda_\xi c(\xi'; y_0; f_\xi^\natural)\tau(\xi' x')\exp(-2\pi \operatorname{Tr} \xi'_\infty y'_\infty)$$

has rational Fourier coefficients, in the present sense. Therefore, fixing y'_0 at a point where Ψ is invertible, every entry in the row vector

$$(\ldots, \varphi_\xi, \ldots) = (\ldots, \lambda_\xi f_\xi, \ldots)\Psi \qquad \text{(indicating matrix multiplication)}$$

of cuspforms has Fourier coefficients in $\mathbb{Q}$. Therefore, as Ψ is invertible, $\{\varphi_\xi\}$ is a $\mathbb{C}$-basis for $\mathrm{Cfm}_1(\mathfrak{n}, \kappa)$. *This proves that there is a* $\mathbb{C}$*-basis for* $\mathrm{Cfm}_1(\mathfrak{n}, \kappa)$ *consisting of cuspforms with rational Fourier coefficients, in the* $SL(2)$*-sense, for a proper ideal* $\mathfrak{n}$.

Now we can obtain the result for $\mathrm{Cfm}_1(\mathfrak{o}, \kappa)$. First, for $\mathfrak{n}_1$ and $\mathfrak{n}_2$ relatively prime, it is immediate that

$$\mathrm{Cfm}_1(\mathfrak{o}, \kappa) = \mathrm{Cfm}_1(\mathfrak{n}_1, \kappa) \cap \mathrm{Cfm}_1(\mathfrak{n}_2, \kappa).$$

Therefore, any f in $\mathrm{Cfm}_1(\mathfrak{o}, \kappa)$ can certainly be written as a linear combination $\sum_i c_i \varphi_i$ of elements φ_i of $\mathrm{Cfm}(\mathfrak{n}_1, \kappa)$ with Fourier coefficients in $\mathbb{Q}$. Then for $\sigma \in \mathrm{Aut}(\mathbb{C}/\mathbb{Q})$

$$f^\sigma = \sum_i c_i^\sigma \varphi_i \in \mathrm{Cfm}_1(\mathfrak{n}_1, \kappa),$$

by the result for *proper* $\mathfrak{n}_1$. But the same argument shows that

$$f^\sigma \in \mathrm{Cfm}_1(\mathfrak{n}_2, \kappa).$$

The following lemma will allow us to finish the proof.

LEMMA* Let V be a $\mathbb{C}$-subspace of $\mathbb{C}^n$ so that V is stable under the componentwise action of $\mathrm{Aut}(\mathbb{C}/\mathbb{Q})$ on $\mathbb{C}^n$. Then V has a $\mathbb{C}$-basis consisting of vectors in $\mathbb{Q}^n \subset \mathbb{C}^n$.

*Proof of Lemma** This is by induction on n and on $\dim_{\mathbb{C}} V$. If $\dim_{\mathbb{C}} V = 1$, then the assertion is easy to verify: Take a vector v in V with one component equal to 1; then $v - v^\sigma = 0$, since $\dim_{\mathbb{C}} V = 1$. Suppose that $\dim_{\mathbb{C}} V > 1$. Identify $\mathbb{C}^{n-1}$ with the subspace of $\mathbb{C}^n$ consisting of vectors with first component 0. If V does not contain any vector with first component non-zero, then $V \subset \mathbb{C}^{n-1}$, and we are done, by induction. So suppose that $v = (1, \ldots)$ is in V. Then for any $\sigma \in \mathrm{Aut}(\mathbb{C}/\mathbb{Q})$, $v - v^\sigma \in V \cap \mathbb{C}^{n-1}$. By induction, $V \cap \mathbb{C}^{n-1}$ has a basis $v_2, \ldots, v_k$ of vectors in $\mathbb{Q}^{n-1} \subset \mathbb{C}^{n-1}$. By a $\mathbb{Q}$-linear change of coordinates, we may suppose that

$$v_j = (0, \ldots, 0, 1, 0, \ldots, 0) \qquad \text{(with the 1 at the } j\text{-th component).}$$

Then there is a linear combination $w = \sum_j c_j v_j$ of the v_j's so that $v - w$ has 2nd through k-th component 0. We have

$$(v - w) - (v - w)^\sigma = v - v^\sigma + w^\sigma - w \in V \cap \mathbb{C}^{n-1}.$$

That is, the components of $v - w$ other than the 2nd through k-th are in $\mathbb{Q}$; but the 2nd through k-th are 0, so are surely in $\mathbb{Q}$. Therefore, $v - w$ is in $\mathbb{Q}^n$. Obviously

$$V = \mathbb{C}(v - w) \oplus V \cap \mathbb{C}^{n-1},$$

so that lemma is proven. ∎

Let $\{f_i\}$ be a $\mathbb{C}$-basis for $\mathrm{Cfm}_1(\mathfrak{o}, \kappa)$, and let $\{\varphi_j\}$ be a collection of cuspforms on $SL(2, \mathbb{A})$ with Fourier coefficients in $\mathbb{Q}$ so that each f_i is expressible as a linear combination of the φ_j's. Then the lemma, together with the remarks preceding it, yield the claim. ∎

Next, we consider the *nature of the eigenvalues of the spherical Hecke algebra for SL(2)*. For each finite prime $\mathfrak{p}$ not dividing $\mathfrak{n}$, let $\eta_1 = \eta_{1\mathfrak{p}}$ be the element of the spherical Hecke algebra of $SL(2, F_\mathfrak{p})$ introduced in the proof of the Main Formula in section 7.9:

$$\eta_{1\mathfrak{p}} = \text{characteristic function of } SL(2, \mathfrak{o}_\mathfrak{p})\delta(\pi_\mathfrak{p})SL(2, \mathfrak{o}_\mathfrak{p})$$

where $\pi_\mathfrak{p}$ is a local parameter at $\mathfrak{p}$.

CLAIM The eigenvalues of the maps $f \to f * \eta_{1\mathfrak{p}}$ on $\mathrm{Cfm}_1(\mathfrak{n}, \kappa)$ are totally real algebraic numbers in $\mathbb{R}$. The field generated over $\mathbb{Q}$ by

$$\{\text{eigenvalues for } f \to f * \eta_{1\mathfrak{p}} \text{ on } \mathrm{Cfm}(\mathfrak{n}, \kappa), \mathfrak{p} \text{ not dividing } \mathfrak{n}\}$$

is finitely generated over $\mathbb{Q}$. For a fixed homomorphism λ of $\mathscr{H}_1(\mathfrak{n})$ to $\mathbb{C}$, there is a basis for $\mathrm{Cfm}_1(\mathfrak{n}, \kappa, \lambda)$ consisting of cuspforms with Fourier coefficients in

$$\mathbb{Q}(\lambda) = \text{field generated over } \mathbb{Q} \text{ by } \{\lambda(\eta_{1\mathfrak{p}}): \mathfrak{p} \text{ not dividing } \mathfrak{n}\}.$$

Also,

$$\mathrm{Cfm}_1(\mathfrak{n}, \kappa, \lambda) = \mathrm{Cfm}_1(\mathfrak{n}, \kappa, \bar{\lambda}).$$

Finally, for $\sigma \in \mathrm{Aut}(\mathbb{C}/\mathbb{Q})$, writing $\lambda^\sigma(\eta) = \lambda(\eta)^\sigma$, we have

$$\mathrm{Cfm}_1(\mathfrak{n}, \kappa, \lambda)^\sigma = \mathrm{Cfm}_1(\mathfrak{n}, \kappa, \lambda^\sigma),$$

where on the lefthand side we apply σ to the Fourier coefficients.

Proof of Claim As in the proof of the Main Formula, with $K = SL(2, F_\mathfrak{p})$, we find that

$$K \backslash K\delta(\pi)K$$

has representatives

$$\delta(\pi)$$

$$u(\beta/\pi) \qquad (\text{with } \beta \in \mathfrak{o}_\mathfrak{p}/\mathfrak{p}\mathfrak{o}_\mathfrak{p} - \{0\})$$

$$\delta(\pi^{-1})u(\beta) \qquad (\text{with } \beta \in \mathfrak{o}_\mathfrak{p}/\mathfrak{p}^2\mathfrak{o}_\mathfrak{p}).$$

Therefore, the convolution of a cuspform f with $\eta_{1\mathfrak{p}}$ is just a sum

$$\sum_h f(gh^{-1})$$

as h ranges over these representatives. Writing a Fourier expansion

$$f(u(x)\delta(y)) = \sum_{\xi \in F^\times} \tau(\xi x) W_f(y; \xi),$$

we find that, if $f' = f * \eta_{1\mathfrak{p}}$, then

$$W_{f'}(y; \xi) = W_f(y/\pi; \xi) + W_f(y; \xi) \sum_\beta \tau(\xi\beta y^2/\pi) + W_f(\pi y; \xi) \sum_\beta \tau(\xi\beta\eta^2/\pi)$$

where in the first sum β ranges over $\mathfrak{o}_\mathfrak{p}/\mathfrak{p}\mathfrak{o}_\mathfrak{p} - \{0\}$, and in the second over $\mathfrak{o}_\mathfrak{p}/\mathfrak{p}^2\mathfrak{o}_\mathfrak{p}$. Note that the sums are *rational numbers*. Therefore, (values of) the Fourier coefficients of f' are finite linear combinations over $\mathbb{Q}$ of (the values of) the Fourier coefficients of f; in particular, if f has Fourier coefficients in $\mathbb{Q}$, then so does f'. Since there is a basis $\{f_i\}$ for $\mathrm{Cfm}_1(\mathfrak{n}, \kappa)$ consisting of cuspforms with rational Fourier coefficients, the (finite) matrix expressing the effect of each $\eta_{1\mathfrak{p}}$ with respect to the basis $\{f_i\}$ has rational entries. From this, the eigenvalues are algebraic. As remarked before, it is easy to check that the $\eta_{1\mathfrak{p}}$'s give self-adjoint operators; therefore, their eigenvalues must be totally real.

From the expression just above for the effect on Fourier coefficients of these Hecke operators, from the existence of a basis of cuspforms with rational (values of) Fourier coefficients, and from the finite dimensionality of spaces of cuspforms, it is immediate that the condition $f * \eta_{1\mathfrak{p}} = \lambda(\eta_{1\mathfrak{p}})f$ imposes a (family of) linear relation(s) *over the field* $\mathbb{Q}(\lambda(\eta_{1\mathfrak{p}}))$ on the Fourier coefficients of f. From this it follows that $\mathrm{Cfm}_1(\mathfrak{n}, \kappa, \lambda)$ has a basis consisting of cuspforms with Fourier coefficients in $\mathbb{Q}(\lambda)$. The very last assertion of the claim follows in a similar way.

The fact that $\mathbb{Q}(\lambda)$ is finitely generated over $\mathbb{Q}$ follows from the fact that $\mathrm{Cfm}_1(\mathfrak{n}, \kappa)$ is finite-dimensional; necessarily there is a finite collection $\{\eta_{1\mathfrak{p}}: \mathfrak{p} \in S\}$ so that eigenfunctions for this collection of operators are eigenfunctions for all of $\mathscr{H}_1(\mathfrak{n})$. Also, since $\mathscr{H}_1(\mathfrak{n})$ is generated by the operators $\eta_{1\mathfrak{p}}$ (which are all self-adjoint), it follows that $\mathrm{Cfm}_1(\mathfrak{n}, \kappa, \lambda) = \mathrm{Cfm}_1(\mathfrak{n}, \kappa, \bar{\lambda})$. This finishes the proof of this claim. ∎

Now we can compare the inner products of elements of $\mathrm{Cfm}_1(\mathfrak{n}, \kappa, \lambda)$. Even though we have only the Main Formula of section 7.9 for *proper* ideals $\mathfrak{n}$, for this argument we can always shrink $\mathfrak{n}$ without harm; therefore, the present discussion also includes the level one case $\mathfrak{n} = \mathfrak{o}$.

We note that the map $f \to f^\natural$ on $\mathrm{Cfm}_1(\mathfrak{n}, \kappa, \lambda)$ merely has the effect of taking the Fourier coefficients of f to their complex conjugates. Also, it is formal that

$$\langle f^\natural, \varphi^\natural \rangle_1 = \langle \varphi, f \rangle_1.$$

We have already seen that there is a basis of $\mathrm{Cfm}_1(\mathfrak{n}, \kappa, \lambda)$ consisting of cuspforms with Fourier coefficients (values) in $\mathbb{Q}(\lambda) \subset \mathbb{R}$. We claim that there is an orthonormal basis for $\mathrm{Cfm}_1(\mathfrak{n}, \kappa, \lambda)$ with Fourier coefficients in $\mathbb{R}$. Take any f_1 in $\mathrm{Cfm}_1(\mathfrak{n}, \kappa, \lambda)$ with real Fourier coefficients; by dividing f_1 by $\langle f_1, f_1 \rangle_1^{1/2}$ we can suppose that $\langle f_1, f_1 \rangle_1 = 1$. Given f_2 in $\mathrm{Cfm}_1(\mathfrak{n}, \kappa, \lambda)$ orthogonal to f_1, we have

$$\langle f_1, f_2^\natural \rangle_1 = \langle f_2^{\natural\natural}, f_1^\natural \rangle_1 = \langle f_2, f_1 \rangle_1 = 0.$$

Therefore, $f_2^\natural$ is also orthogonal to f_1. Certainly either $f_2 + f_2^\natural$ or $i(f_2 - f_2^\natural)$ is not identically zero, both are orthogonal to f_1, and both have real Fourier coefficients. An induction finishes the proof of this claim.

Therefore, there is an orthonormal basis for each $\mathrm{Cfm}_1(\mathfrak{n}, \kappa, \lambda)$ consisting of cuspforms f so that $f = f^\natural$. With such basis, we have

$$\sum_\lambda c^{-1} S(\lambda) \sum_i f_{\lambda i}(g) f_{\lambda i}(g')$$

which has rational Fourier coefficients, where each inner sum is over an orthonormal basis $\{f_{\lambda i}\}$ for $\mathrm{Cfm}_1(\mathfrak{n}, \kappa, \lambda)$ consisting of cuspforms with real Fourier coefficients. For each λ, let $\{\varphi_{\lambda i}: i = 1, \ldots, d(\lambda)\}$ be a basis for $\mathrm{Cfm}_1(\mathfrak{n}, \kappa, \lambda)$ consisting of cuspforms with Fourier coefficients in $\mathbb{Q}(\lambda) \subset \mathbb{R}$. Then, because the Fourier coefficients are real,

$$\langle \varphi_{\lambda i}, \varphi_{\lambda j} \rangle_1 = \langle \varphi_{\lambda j}^\natural, \varphi_{\lambda i}^\natural \rangle_1 = \langle \varphi_{\lambda j}, \varphi_{\lambda i} \rangle_1.$$

That is, *the inner products of the* $\varphi_{\lambda i}$ *are real.* We certainly may take

$$f_{\lambda 1} = \varphi_{\lambda 1} / \langle \varphi_{\lambda 1}, \varphi_{\lambda 1} \rangle_1^{1/2}.$$

Letting

$$\Phi_{\lambda 2} = \varphi_{\lambda 2} \langle \varphi_{\lambda 2}, \varphi_{\lambda 2} \rangle_1^{-1/2} - \langle \varphi_{\lambda 2}, \varphi_{\lambda 1} \rangle_1 \langle \varphi_{\lambda 1}, \varphi_{\lambda 1} \rangle_1^{-1} \langle \varphi_{\lambda 2}, \varphi_{\lambda 2} \rangle_1^{-1/2} \varphi_{\lambda 1}$$

we can take

$$f_2 = \Phi_{\lambda 2} / \langle \Phi_{\lambda 1}, \Phi_{\lambda 1} \rangle_1^{1/2}.$$

Generally, by this (Gram-Schmidt) process there are *real numbers* C_{ij} $(i \geq j)$ so that

$$f_{\lambda n} = C_{\lambda nn} \varphi_{\lambda n} + \cdots + C_{\lambda n1} \varphi_{\lambda 1}.$$

Substituting these expressions for $f_{\lambda i}$, and rearranging, we find that

$$\sum_{\lambda} c^{-1}S(\lambda) \sum_{ij} \varphi_{\lambda i}(g)\varphi_{\lambda j}(g')\left\{\sum_{k\geq i,j} C_{\lambda ki}C_{\lambda kj}\right\}$$

has Fourier coefficients in $\mathbb{Q}$. From the linear independence of the $\varphi_{\lambda i}$'s, this implies that, for all i and j,

$$c^{-1}S(\lambda) \sum_{k\geq i,j} C_{\lambda ki}C_{\lambda kj} \in \mathbb{Q}(\lambda)$$

since the $\varphi_{\lambda i}$'s have Fourier coefficients in $\mathbb{Q}(\lambda)$, using the last assertion of the claim just above. Now the (necessarily non-singular) matrix $C = C_\lambda$ with (k, i)-th entry $C_{ij} = C_{\lambda ij}$ is lower triangular, and we have found that the matrix

$$c^{-1}S(\lambda)C^TC$$

has entries in $\mathbb{Q}(\lambda)$. From this, an induction shows that C_{kk}^2/C_{ii}^2 and C_{ij}/C_{ii} are all in $\mathbb{Q}(\lambda)$: First, take $i = d(\lambda)$; all the numbers

$$c^{-1}S(\lambda)C_{ii}C_{ii} \qquad c^{-1}S(\lambda)C_{ii}C_{ii-1} \quad \ldots \quad c^{-1}S(\lambda)C_{ii}C_{i1}$$

are in $\mathbb{Q}(\lambda)$. Therefore, C_{ij}/C_{ii} is in $\mathbb{Q}(\lambda)$ for all j. Assume that we have shown that C_{kk}^2/C_{ii}^2 and C_{ij}/C_{ii} are all in $\mathbb{Q}(\lambda)$ for $i, j, k > n$. Let $d = d(\lambda)$. We know that

$$c^{-1}S(\lambda)C_{dn}C_{dn} + \cdots + c^{-1}S(\lambda)C_{nn}C_{nn}$$

$$\cdots\cdots\cdots\cdots$$

$$c^{-1}S(\lambda)C_{dn}C_{di} + \cdots + c^{-1}S(\lambda)C_{nn}C_{ni}$$

are all in $\mathbb{Q}(\lambda)$. Dividing the top line by $c^{-1}S(\lambda)C_{dd}^2 \in \mathbb{Q}(\lambda)$, we find that

$$C_{dn}C_{dn}/C_{dd}^2 + \cdots + C_{nn}C_{nn}/C_{dd}^2 \in \mathbb{Q}(\lambda).$$

For $k > n$, we have

$$(C_{kn}C_{kn}/C_{dd}^2)\mathbb{Q}(\lambda) = (C_{kn}C_{kn}/C_{kk}^2)\mathbb{Q}(\lambda) \subset \mathbb{Q}(\lambda),$$

by induction. Therefore,

$$C_{nn}^2/C_{dd}^2 \in \mathbb{Q}(\lambda).$$

Similarly,

$$C_{dn}C_{di}/C_{dd}^2 + \cdots + C_{nn}C_{ni}/C_{dd}^2 \in \mathbb{Q}(\lambda).$$

Once again, for $k > n$ we have

$$(C_{kn}C_{ki}/C_{dd}^2)\mathbb{Q}(\lambda) = (C_{kn}C_{ki}/C_{kk}^2)\mathbb{Q}(\lambda) \subset \mathbb{Q}(\lambda),$$

by induction. Therefore,

$$C_{nn}C_{ni}/C_{dd}^2 \in \mathbb{Q}(\lambda).$$

Taking ratios gives $C_{ni}/C_{nn} \in \mathbb{Q}(\lambda)$.

From $\langle f_{\lambda 1}, f_{\lambda 2}\rangle_1 = 0$, we have

$$0 = C_{11}C_{22}\langle \varphi_{\lambda 1}, \varphi_{\lambda 2}\rangle_1 + C_{11}C_{21}\langle \varphi_{\lambda 1}, \varphi_{\lambda 1}\rangle_1$$

or

$$\langle \varphi_{\lambda 1}, \varphi_{\lambda 2}\rangle_1 / \langle \varphi_{\lambda 1}, \varphi_{\lambda 1}\rangle_1 = C_{21}/C_{22} \in \mathbb{Q}(\lambda).$$

But $\varphi_{\lambda 1}$ and $\varphi_{\lambda 2}$ were completely arbitrary (with Fourier coefficients in $\mathbb{Q}(\lambda)$); already this makes clear that (via a Gram-Schmidt process) we can construct an orthogonal basis for $\mathrm{Cfm}_1(\mathfrak{n}, \kappa, \lambda)$ consisting of cuspforms with Fourier coefficients in $\mathbb{Q}(\lambda)$. In particular, the same rationality property holds upon interchanging $\varphi_{\lambda 1}$ and $\varphi_{\lambda 2}$, so that if $\langle \varphi_{\lambda 1}, \varphi_{\lambda 2}\rangle_1$ is non-zero we have

$$\langle \varphi_{\lambda 2}, \varphi_{\lambda 2}\rangle_1 / \langle \varphi_{\lambda 1}, \varphi_{\lambda 1}\rangle_1 \in \mathbb{Q}(\lambda).$$

If $\langle \varphi_{\lambda 1}, \varphi_{\lambda 2}\rangle_1 = 0$, then let $\varphi = \varphi_{\lambda 1} + \varphi_{\lambda 2}$;

$$\langle \varphi_{\lambda 1}, \varphi\rangle_1 = \langle \varphi_{\lambda 1}, \varphi_{\lambda 1}\rangle_1 \neq 0.$$

Therefore,

$$\langle \varphi, \varphi\rangle_1 / \langle \varphi_{\lambda 1}, \varphi_{\lambda 1}\rangle_1 \in \mathbb{Q}(\lambda).$$

From this it follows once again that

$$\langle \varphi_{\lambda 2}, \varphi_{\lambda 2}\rangle_1 / \langle \varphi_{\lambda 1}, \varphi_{\lambda 1}\rangle_1 \in \mathbb{Q}(\lambda).$$

This gives the desired comparison result for $SL(2)$.

Last, we wish to show that *if f and φ are both in* $\mathrm{Cfm}_1(\mathfrak{n}, \kappa, \lambda)$ *and* $\langle f, \varphi\rangle = 0$, *then for any* $\sigma \in \mathrm{Aut}(\mathbb{C}/\mathbb{Q})$ *we still have* $\langle f^\sigma, \varphi^\sigma\rangle = 0$. The action of σ is on Fourier coefficients (in an $SL(2)$-sense). By previous results, we already know that f^σ and φ^σ lie in $\mathrm{Cfm}_1(\mathfrak{n}, \kappa, \lambda^\sigma)$. What we will show is that, if $\{f_{\lambda i}\}$ is an orthogonal basis for $\mathrm{Cfm}_1(\mathfrak{n}, \kappa, \lambda)$ then $\{f_{\lambda i}^\sigma\}$ is an orthogonal basis for $\mathrm{Cfm}_1(\mathfrak{n}, \kappa, \lambda^\sigma)$. From above, we know that there is such a basis $\{f_{\lambda i}\}$ so that $f_i^\natural = f_i$ (i.e., having real Fourier coefficients), and each f_i has Fourier coefficients in $\mathbb{Q}(\lambda)$. And we know that

$$\sum_\lambda c^{-1} S(\lambda) \sum_i f_{\lambda i} \otimes f_{\lambda i} / \langle f_{\lambda i}, f_{\lambda i}\rangle_1$$

has rational Fourier coefficients. Put $c_{\lambda i} = c^{-1}S(\lambda)/\langle f_{\lambda i} f_{\lambda i}\rangle$. Then we have

$$\sum_{\lambda}\sum_{i} c_{\lambda i} f_{\lambda i} \otimes f_{\lambda i} = \sum_{\lambda}\sum_{i} c_{\lambda i}^{\sigma} f_{\lambda i}^{\sigma} \otimes f_{\lambda i}^{\sigma}.$$

Since $f_{\lambda i}^{\sigma}$ is in $\mathrm{Cfm}_1(\mathfrak{n}, \kappa, \lambda^{\sigma})$, this implies that (for all λ)

$$\sum_{i} c_{\lambda^{\sigma} i} f_{\lambda^{\sigma} i} \otimes f_{\lambda^{\sigma} i} = \sum_{i} c_{\lambda i}^{\sigma} f_{\lambda i}^{\sigma} \otimes f_{\lambda i}^{\sigma}.$$

From this point it is elementary to see that, since the cuspforms $f_{\lambda^{\sigma} i}$ are mutually orthogonal, the cuspforms $f_{\lambda i}^{\sigma}$ must also be mutually orthogonal. This is the desired result.

This finishes the proofs for SL(2).

Now we prove the theorem for GL(2) *by reduction to the case of SL*(2).

Remarks In fact, this reduction is quite elementary. The only burdens are some relatively minor ones necessitated by the fact that quotients of $GL(2, \mathbb{A})$ are finite disjoint unions of quotients of $SL(2, \mathbb{A})$. After this annoyance is overcome, the remaining arguments are identical in form to those for $SL(2)$; therefore, after the appropriate point we will leave the imitation of proof to the reader.

Let $\mathfrak{m}$ be a non-zero ideal relatively prime to $\mathfrak{n}$. We use notation

$$\mathbb{K}(\mathfrak{m}\mathfrak{n}) = \{g \in SL(2, \hat{\mathfrak{o}}) \colon g \equiv 1_2 \text{ modulo } \mathfrak{m}\mathfrak{n}\}$$

$$\tilde{\mathbb{K}}(\mathfrak{m}, \mathfrak{n}) = \left\{g \in GL(2, \hat{\mathfrak{o}}) \colon g \equiv 1_2 \text{ modulo } \mathfrak{m},\ g \equiv \begin{pmatrix} * & 0 \\ 0 & 1 \end{pmatrix} \text{ modulo } \mathfrak{n}\right\}$$

$$\tilde{\mathbb{K}}(\mathfrak{n}) = \tilde{\mathbb{K}}(\mathfrak{o}, \mathfrak{n}).$$

Let $\mathfrak{p}_0$ be a prime not dividing $\mathfrak{n}$. It is elementary that for some power $\mathfrak{m}$ of $\mathfrak{p}_0$ the natural map

$$SL(2, F)\backslash SL(2, \mathbb{A})/\mathbb{K}(\mathfrak{m}, \mathfrak{n}) \to Z^{+}GL(2, F)\backslash GL(2, \mathbb{A})/\tilde{\mathbb{K}}(\mathfrak{m}, \mathfrak{n})$$

is injective *for all* $\mathfrak{n}$ *prime to* $\mathfrak{m}$. Further, by the finiteness of generalized ideal class groups, there is a finite set $\{\xi_i \colon i = 1, \ldots, h\}$ *depending only upon* $\mathfrak{m}$ of finite ideles so that with

$$\tilde{\xi}_i = \ell(\xi_i)$$

we have

$$GL(2, \mathbb{A}) = \bigcup_{i} Z^{+}GL(2, F)SL(2, \mathbb{A})\tilde{\xi}_i\tilde{\mathbb{K}}(\mathfrak{m}, \mathfrak{n}) \qquad \text{(disjoint union).}$$

Indeed, by the general version of Dirichlet's theorem on primes in an arithmetic progression, we may take the ξ_i's to be local parameters at distinct primes $\mathfrak{p}_i$ not dividing $\mathfrak{m}\mathfrak{n}$. Therefore,

$$Z^+ GL(2, F)\backslash GL(2, \mathbb{A})/\tilde{\mathbb{K}}(\mathfrak{m}, \mathfrak{n}) \approx \bigcup_i SL(2, F)\backslash SL(2, \mathbb{A})/\mathbb{K}(\mathfrak{m}\mathfrak{n})_i$$

where

$$\mathbb{K}(\mathfrak{m}\mathfrak{n})_i = SL(2, \mathbb{A}) \cap \tilde{\xi}_i \mathbb{K}(\mathfrak{m}, \mathfrak{n}) \tilde{\xi}_i^{-1}.$$

Taking the ξ_i to be local parameters as mentioned just above, let $\mathfrak{a} = \prod_i \mathfrak{p}_i$. Then for all i

$$\mathbb{K}(\mathfrak{m}\mathfrak{n}\mathfrak{a}) \subset \mathbb{K}_i.$$

Given a right $\tilde{\mathbb{K}}(\mathfrak{m}, \mathfrak{n})$-invariant, Z^+-invariant, left $GL(2, F)$-invariant function f on $GL(2, \mathbb{A})$, the isomorphism

$$Z^+ GL(2, F)\backslash GL(2, \mathbb{A})/\tilde{\mathbb{K}}(\mathfrak{m}, \mathfrak{n}) \approx \bigcup_i SL(2, F)\backslash SL(2, \mathbb{A})/\mathbb{K}(\mathfrak{m}\mathfrak{n})_i$$

yields an h-tuple

$$\beta(f) = \{\beta_i(f) \colon i = 1, \ldots, h\}$$

of right $\mathbb{K}(\mathfrak{m}\mathfrak{n})_i$-invariant, left $SL(2, F)$-invariant functions on $SL(2, \mathbb{A})$ by

$$\beta_i(f)(g) = f(g\tilde{\xi}_i).$$

Conversely, every such h-tuple of functions on $SL(2, \mathbb{A})$ gives such a function on $GL(2, \mathbb{A})$. Let $V_\xi(y_0)$ be the ξ-th Fourier coefficient of f in the $GL(2)$-sense, and let $V_\xi(y_0; i)_1$ be the ξ-th Fourier coefficient of $\beta_i(f)$ in the $SL(2)$-sense. For a given choice of $\{\xi_i\}$, let $\mathfrak{a}$ be the ideal in $\mathfrak{o}$ so that

$$\mathfrak{a}\hat{\mathfrak{o}} = \left(\prod_i \xi_i\right)\hat{\mathfrak{o}}.$$

CLAIM $\mathrm{Cfm}(\mathfrak{n}, \kappa)$ has a basis consisting of cuspforms with Fourier coefficients in $\mathbb{Q}$ (in the $GL(2)$ sense).

Proof Let notation be as just above. By the choice of the collection of representatives $\tilde{\xi}_i$, the values of the $SL(2)$-sense Fourier coefficients of all the $\beta_i(f)$ determine the $GL(2)$-sense Fourier coefficients of f completely: Given $y \in \mathbb{J}^+$, we can write

$$y = t y_\infty^2 \xi_i \varepsilon$$

with $t \in F^x$, $y_\infty \in \mathbb{J}_\infty$, with some ξ_i, and with $\varepsilon \in \hat{\mathfrak{o}}^x$ so that $\varepsilon \equiv 1$ modulo $\mathfrak{m}$. (Since the infinite parts of both y and y_∞^2 are totally positive, $t \in F^\times$ is also totally positive.) Then it is formal, from the definition of these Fourier coefficients, that

$$V_\xi(y_0) = V_\xi(1; i)_1.$$

That is, the (values of) $GL(2)$ Fourier coefficients are a subset of the (values of) $SL(2)$ Fourier coefficients. From this, it follows that for $\sigma \in \mathrm{Aut}(\mathbb{C}/\mathbb{Q})$

$$\beta^{-1}(\{\beta_i(f)^\sigma\}) = f^\sigma$$

where the lefthand side action of σ is in the $SL(2)$-sense and the righthand side action is in the $GL(2)$ sense.

From the $SL(2)$ result on Fourier coefficients, each $\beta_i(f)^\sigma$ is still a holomorphic cuspform of weight κ on $SL(2, \mathbb{A})$ with respect to $\mathbb{K}(\mathfrak{mna})$, which certainly contains $\mathbb{K}(\mathfrak{mna})_i$. We also have

$$Z^+(F_\infty)GL(2, F)\backslash GL(2, \mathbb{A})/\tilde{\mathbb{K}}(\mathfrak{m}, \mathfrak{na}) \approx \bigcup_i SL(2, F)\backslash SL(2, \mathbb{A})/\mathbb{K}(\mathfrak{mna})_i$$

for the same reasons as before, *by the same map*. Therefore, f^σ is a holomorphic cuspform which is right-invariant by $\tilde{\mathbb{K}}(\mathfrak{m}, \mathfrak{na})$. Now $\mathfrak{a}$ was arbitrary (prime to $\mathfrak{mn}$), but the map β *does not depend upon* $\mathfrak{a}$; therefore, by varying the choice of $\{\xi_i\}$ (via Dirichlet's theorem on primes in arithmetic progressions) we conclude that f^σ is a holomorphic cuspform which is right-invariant by $\tilde{\mathbb{K}}(\mathfrak{m}, \mathfrak{n})$. Further, by varying the choice of $\mathfrak{p}_0$ (of which $\mathfrak{m}$ is a power) we find that f^σ is right $\tilde{\mathbb{K}}(\mathfrak{n})$-invariant. That is, $\mathrm{Cfm}(\mathfrak{n}, \kappa)$ is *stable* under the action of $\mathrm{Aut}(\mathbb{C}/\mathbb{Q})$.

Since the $GL(2)$ Fourier coefficients are a subset of the $SL(2)$ Fourier coefficients, each f in $\mathrm{Cfm}(\mathfrak{n}, \kappa)$ can be written as a linear combination $\sum_k c_k \varphi_k$ of cuspforms φ_k on $GL(2, \mathbb{A})$ with $GL(2)$ Fourier coefficients in $\mathbb{Q}$, although these cuspforms may not be right $\tilde{\mathbb{K}}(\mathfrak{n})$-invariant. This fact, together with the finite dimensionality of $\mathrm{Cfm}(\mathfrak{n}, \kappa)$, and the fact that $\mathrm{Cfm}(\mathfrak{n}, \kappa)$ is stable under the action of $\mathrm{Aut}(\mathbb{C}/\mathbb{Q})$ will yield the assertion of the claim. Let $\{f_i\}$ be a $\mathbb{C}$-basis for $\mathrm{Cfm}(\mathfrak{n}, \kappa)$, and let $\{\varphi_j\}$ be a collection of cuspforms on $GL(2, \mathbb{A})$ with $GL(2)$ Fourier coefficients in $\mathbb{Q}$ so that each f_i is expressible as a linear combination of the φ_j's. Then the lemma* above, together with the preceding remarks, yield the claim. ■

CLAIM The space $\mathrm{Cfm}(\mathfrak{n}, \kappa)$ decomposes as a direct sum (with finitely many summands)

$$\bigoplus_\omega \mathrm{Cfm}(\mathfrak{n}, \kappa, \omega)$$

over central characters, and each space $\mathrm{Cfm}(\mathfrak{n}, \kappa, \omega)$ has a basis consisting of cuspforms with Fourier coefficients in $\mathbb{Q}(\omega)$, in the $GL(2)$-sense. Further, for $\sigma \in \mathrm{Aut}(\mathbb{C}/\mathbb{Q})$

$$\mathrm{Cfm}(\mathfrak{n}, \kappa, \omega)^\sigma = \mathrm{Cfm}(\mathfrak{n}, \kappa, \omega^\sigma).$$

Proof The first assertion is immediate from the fact that there are only finitely many central characters which are trivial on Z^+ and trivial on $\{\eta \in \hat{\mathfrak{o}}^\times : \eta \equiv 1 \text{ modulo } \mathfrak{n}\}$. Using notation as above, we have another formal relation between $SL(2)$ and $GL(2)$ Fourier coefficients: For $y_0 \in \mathbb{J}_0$

$$V_\xi(y_0; i)_1 = V_\xi(y_0^2 \xi_i)\bar{\omega}(y_0).$$

From this, the result of previous claim, together with the same type of argument as in that claim, yields the second assertion. The third assertion on conjugation also follows similarly. ■

CLAIM $\mathrm{Cfm}(\mathfrak{n}, \kappa, \omega)$ admits a finite direct sum decomposition

$$\mathrm{Cfm}(\mathfrak{n}, \kappa, \omega) = \bigoplus_\lambda \mathrm{Cfm}(\mathfrak{n}, \kappa, \omega)_\lambda$$

over homomorphisms $\lambda: \mathscr{H}_1(\mathfrak{n}) \to \mathbb{C}$, where each $\mathrm{Cfm}(\mathfrak{n}, \kappa, \omega)_\lambda$ is the λ-eigenspace. Each space $\mathrm{Cfm}(\mathfrak{n}, \kappa, \omega)_\lambda$ has a basis consisting of cuspforms with Fourier coefficients in $\mathbb{Q}(\omega, \lambda)$, in a $GL(2)$-sense. Further, for $\sigma \in \mathrm{Aut}(\mathbb{C}/\mathbb{Q})$

$$\mathrm{Cfm}(\mathfrak{n}, \kappa, \omega)_\lambda^\sigma = \mathrm{Cfm}(\mathfrak{n}, \kappa, \omega^\sigma)_{\lambda^\sigma}.$$

Proof First, note that the action of $\mathscr{H}_1(\mathfrak{n})$ certainly commutes with translation by the center. Then this claim follows by an argument which is but a minor variant of the argument used for $\mathrm{Cfm}_1(\mathfrak{n}, \kappa, \lambda)$. ■

Remarks The effect of $\mathscr{H}_1(\mathfrak{n})$ on $GL(2)$ Fourier coefficients on the whole of $\mathrm{Cfm}(\mathfrak{n}, \kappa)$ cannot be discussed sensibly from this viewpoint; rather, the effect is best understood on the summands $\mathrm{Cfm}(\mathfrak{n}, \kappa, \omega)$.

CLAIM Let $f \in \mathrm{Cfm}(\mathfrak{n}, \kappa, \omega_1)_\lambda$ with $GL(2)$ Fourier coefficients in $\mathbb{Q}(\omega_1, \lambda)$ and let $\varphi \in \mathrm{Cfm}(\mathfrak{n}, \kappa, \omega_2)_\lambda$ with $GL(2)$ Fourier coefficients in $\mathbb{Q}(\omega_2, \lambda)$. Then

$$\langle \varphi, \varphi \rangle / \langle f, f \rangle \qquad \langle f, \varphi \rangle / \langle f, f \rangle$$

are in $\mathbb{Q}(\omega_1, \omega_2, \lambda)$.

Proof Again, we use notation as above. Let $\alpha: \mathscr{H}_1(\mathfrak{mna}) \to \mathscr{H}(\mathfrak{mna})$ be the imbedding of spherical Hecke algebras as in section 6.1. Identifying $\mathscr{H}_1(\mathfrak{mna})$ with its image, we have an action of $\mathscr{H}_1(\mathfrak{mna})$ on right $\tilde{\mathbb{K}}(\mathfrak{mna})$-invariant functions on $GL(2, \mathbb{A})$. It is clear that each of the maps β_i is an $\mathscr{H}_1(\mathfrak{mna})$-homomorphism. In particular, as f is an $\mathscr{H}_1(\mathfrak{mna})$-eigenfunction, each $\beta_i(f)$ is such, with the same eigenvalues as those of f. It is clear that

$$\langle f, \varphi \rangle^\sim = \sum_i \langle \beta_i(f), \beta_i(\varphi) \rangle_1$$

for any two cuspforms f and φ, with obvious normalizations of measures. Therefore, from the $SL(2)$ result on comparability of inner products, as f and φ are both λ-eigenfunctions for $\mathscr{H}_1(\mathfrak{mna})$ with Fourier coefficients in $\mathbb{Q}(\omega_1, \omega_2, \lambda)$, the inner products $\langle \beta_i(f), \beta_i(\varphi) \rangle_1$, $\langle \beta_i(f), \beta_i(f) \rangle_1$, and $\langle \beta_i(\varphi), \beta_i(\varphi) \rangle_1$ are $\mathbb{Q}(\omega_1, \omega_2, \lambda)$-multiples of $\langle \beta_j(f), \beta_j(f) \rangle_1$ for all i and j so that $\langle \beta_j(f), \beta_j(f) \rangle_1$ is non-zero. Therefore, $\langle f, \varphi \rangle^{\sim}$, $\langle f, f \rangle^{\sim}$, and $\langle \varphi, \varphi \rangle^{\sim}$ are $\mathbb{Q}(\omega_1, \omega_2, \lambda)$-multiples of $\langle \beta_j(f), \beta_j(f) \rangle_1$ for all j so that $\langle \beta_j(f), \beta_j(f) \rangle_1$ is non-zero. Noting that $\langle\ ,\ \rangle^{\sim}$ differs from $\langle\ ,\ \rangle$ by a uniform rational constant, this gives the $GL(2)$ version of the comparability of inner products. ■

Remarks In the previous assertion, if $\omega_1 \neq \omega_2$ then the inner product $\langle f, \varphi \rangle$ is necessarily zero. Nevertheless, even then we have a result comparing $\langle f, f \rangle$ and $\langle \varphi, \varphi \rangle$.

CLAIM Let f and φ be in $\mathrm{Cfm}(\mathfrak{n}, \kappa)_\lambda$ so that $\langle f, \varphi \rangle = 0$. Then for any $\sigma \in \mathrm{Aut}(\mathbb{C}/\mathbb{Q})$ we have $\langle f^\sigma, \varphi^\sigma \rangle = 0$, where the action of σ is on $GL(2)$-sense Fourier coefficients.

Proof Via the relations

$$\beta^{-1}(\{\beta_i(f)^\sigma\}) = f^\sigma \qquad \beta^{-1}(\{\beta_i(\varphi)^\sigma\}) = \varphi^\sigma$$

and from the comparison of $\langle\ ,\ \rangle^{\sim}$ and $\langle\ ,\ \rangle_1$ as in the proof of the previous claim, the $SL(2)$ version of this assertion yields the $GL(2)$ version. ■

CLAIM Let f be an eigenfunction for $\mathscr{H}_1(\mathfrak{n})$ in $\mathrm{Cfm}(\mathfrak{n}, \kappa)$, and let φ be in $\mathrm{Cfm}(\mathfrak{n}, \kappa)$. Then for any $\sigma \in \mathrm{Aut}(\mathbb{C}/\mathbb{Q})$

$$\{\langle \varphi, f \rangle / \langle f, f \rangle\}^\sigma = \langle \varphi^\sigma, f^\sigma \rangle / \langle f^\sigma, f^\sigma \rangle.$$

Proof By previous results, we can write

$$\varphi = cf + \sum_i \psi_i$$

where the ψ_i's are now mutually orthogonal cuspforms each lying in some space $\mathrm{Cfm}(\mathfrak{n}, \kappa)_\lambda$ and all are orthogonal to f. Then

$$\varphi^\sigma = c^\sigma f^\sigma + \sum_i \psi_i^\sigma.$$

Because each ψ_i^σ is still orthogonal to f^σ, we have

$$\langle \varphi^\sigma, f^\sigma \rangle = c^\sigma \langle f^\sigma, f^\sigma \rangle.$$

This, combined with the obvious relation

$$c = \langle \varphi, f \rangle / \langle f, f \rangle,$$

yields the result. ■

For a finite prime $\mathfrak{p}$ not dividing $\mathfrak{n}$, for fixed weight κ, let $\eta_\mathfrak{p}$ be the element of the spherical Hecke algebra at $\mathfrak{p}$ given by

$$\eta_\mathfrak{p} = N\mathfrak{p}^{\kappa/2} \times \text{characteristic function of } GL(2, \mathfrak{o}_\mathfrak{p})\ell(\pi^{-1})GL(2, \mathfrak{o}_\mathfrak{p})$$

(where the constant arises for elementary reasons to preserve rationality properties of Fourier coefficients).

CLAIM The space $\mathrm{Cfm}(\mathfrak{n}, \kappa, \omega)$ decomposes as a finite direct sum

$$\mathrm{Cfm}(\mathfrak{n}, \kappa, \omega) = \bigoplus_\lambda \mathrm{Cfm}(\mathfrak{n}, \kappa, \omega, \lambda)$$

where λ ranges over homomorphisms λ: $\mathscr{H}(\mathfrak{n}) \to \mathbb{C}$. Each space $\mathrm{Cfm}(\mathfrak{n}, \kappa, \omega, \lambda)$ has a basis consisting of cuspforms with Fourier coefficients in $\mathbb{Q}(\omega, \lambda)$, in a $GL(2)$-sense.

Proof The decomposition assertion is formal. That each space $\mathrm{Cfm}(\mathfrak{n}, \kappa, \omega, \lambda)$ has a basis as indicated follows from an explicit calculation of the effect of the Hecke operators $\eta_\mathfrak{p}$ on Fourier coefficients, very much as did the $SL(2)$ version of this result, noting that the constant normalizing $\eta_\mathfrak{p}$ is necessary to make this work out exactly right. It is not hard to verify that the algebra of operators given by $\mathscr{H}(\mathfrak{n})$ on $\mathrm{Cfm}(\mathfrak{n}, \kappa, \omega)$ is generated by the operators $\eta_\mathfrak{p}$, although this is not true on all of $\mathrm{Cfm}(\mathfrak{n}, \kappa)$: Specification of a central character obviously imposes some relations. The comparison of inner products is a weakening of the assertion of a previous claim, as $\mathscr{H}(\mathfrak{n})$ gives a finer decomposition of spaces $\mathrm{Cfm}(\mathfrak{n}, \kappa, \omega)$ than does the algebra $\mathscr{H}_1(\mathfrak{n})$. ■

This concludes the proof of the Arithmetic Structure Theorem of section 6.1.

Appendix A.1 Integration on Homogeneous Spaces

Let G be a locally compact Hausdorff topological group. It is well-known that G is completely regular (one can give an easy version of the proof of Urysohn's lemma). This fact is important in itself, and is also used in the proof of the following very standard result.

THEOREM On a locally compact Hausdorff topological group G there is a positive Borel measure μ so that for every measurable set $E \subset G$ and $g \in G$ one has $\mu(gE) = \mu(E)$. Further, any two such measures are positive real multiples of one another. Such a measure is a *left Haar measure* on the group G. ∎

COROLLARY Let μ be a left Haar measure on a locally compact Hausdorff topological group G. There is a continuous group homomorphism $\Delta: G \to \{\text{positive real numbers}\}$ so that for $g_1 \in G$ and f a continuous compactly supported function on G,

$$\int f(gg_1)\,d\mu(g) = \Delta(g_1) \int f(g)\,d\mu(g).$$

This Δ is the *modular function* of G.

Proof It is clear that

$$f \to \int f(gg_1)\,d\mu(g)$$

gives a left G-invariant positive additive functional on continuous compactly supported functions. Therefore, there is a left Haar measure ν so that

$$\int f(gg_1)\,d\mu(g) = \int f(g)\,d\nu(g).$$

By the uniqueness of Haar measure, there is a positive real number $\Delta(g_1)$ so that $\nu = \Delta(g_1)\mu$:

$$\int f(gg_1)\,d\mu(g) = \Delta(g_1)\int f(g)\,d\mu(g).$$

With g_1 and g_2 in G, we have

$$\int f(gg_1g_2)\,d\mu(g) = \Delta(g_1)\int f(gg_2)\,d\mu(g) = \Delta(g_1)\Delta(g_2)\int f(g)\,d\mu(g),$$

where to obtain the first equality we apply the previous to the function $F(g) = f(gg_1)$. But also

$$\int f(gg_1g_2)\,d\mu(g) = \Delta(g_1g_2)\int f(g)\,d\mu(g).$$

Therefore, Δ is a group homomorphism to the positive real numbers.

Since f is uniformly continuous, the function

$$\Delta(g_1) = \int f(gg_1)\,d\mu(g)\Big/\int f(g)\,d\mu(g)$$

is a continuous function of $g_1 \in G$. Note that we can compute Δ by consideration of just one non-negative not identically vanishing compactly supported continuous function f. ■

Definition We say that the group G is *unimodular* if $\Delta \equiv 1$. In that case, a left Haar measure is also a right Haar measure, and vice-versa.

Note Abelian, discrete, or compact groups G are unimodular. Or, let H be the topological closure in G of $\{ghg^{-1}h^{-1}: g, h \in G\}$. If $H = G$, then G is unimodular.

THEOREM Let G be a topological group with left Haar measure μ and modular function Δ, and let H be a closed subgroup of G with left Haar measure ν and modular function δ. There is a (left) G-invariant (positive Borel) measure σ on G/H if and only if

$$\Delta|_H = \delta.$$

If σ exists, then it is uniquely determined by the condition that for a compactly supported continuous function f on G we have

$$\int_G f\,d\mu = \int_{G/H} d\sigma \int_H f(gh)\,d\nu(h)$$

LEMMA The map $f \to Af$ given by

$$(Af)(gH) = \int_H f(gh)\,d\nu(h)$$

is a surjection from compactly supported continuous functions on G to compactly supported continuous functions on H.

Proof Clearly the map A sends compactly supported continuous functions on G to compactly supported continuous functions on H. Given a compactly supported continuous function ψ on H, let C be a compact subset of G/H containing the support of ψ. Let $\{U_\alpha\}$ be a family of open subsets of G with compact closures so that $\bigcup_\alpha U_\alpha = G$; since $\bigcup_\alpha U_\alpha H$ covers C, there is a finite subset $\{U_i\}$ so that $\{U_i H\}$ covers C. Let f be a compactly supported continuous function on G which is non-negative, and identically 1 on $\bigcup_i U_i$. Let $F(g) = \int_H f(gh)\,d\nu(h)$. We claim that

$$\varphi(g) = \psi(gH)f(g)/F(g)$$

defines a compactly supported continuous function on G so that $A\varphi = \psi$. Indeed, for $F(g) = 0$ it must be that g is outside the closure of $\bigcup_i U_i H$, so that already g is in the interior of the set where ψ vanishes. Therefore, φ is continuous. It is clearly of compact support, since f is. Then

$$\int_H \varphi(gh)\,d\nu(h) = \int_H \psi(ghH)f(gh)/F(gh)\,d\nu(h) =$$

$$= \psi(gH)F(gH)^{-1} \int_H f(gh)\,d\nu(h),$$

since ψ and F are right H-invariant. But the latter integral is just $F(gH)$. ■

Proof of Theorem Suppose that there were a left G-invariant measure σ on G/H. Then for f a compactly supported continuous function on G,

$$f \to \int_{G/H} d\sigma(gH) \int_H f(gh)\,d\nu(h)$$

gives a left-invariant positive linear functional, so must be given by integration against a multiple of the left Haar measure on G. This shows *uniqueness* up to multiples. In particular, for $h_1 \in G$, we must have

$$\int_{G/H} d\sigma(gH) \int_H f(ghh_1)\,d\nu(h) = \Delta(h_1) \int_{G/H} d\sigma \int_H f(gh)\,d\nu(h).$$

But already

$$\int_H f(ghh_1)\,d\nu(h) = \delta(h_1)\int_H f(gh)\,d\nu(h).$$

Therefore, it is *necessary* that Δ restricted to H is δ.

On the other hand, if Δ restricted to H is δ, then the lemma allows us to *define* a positive linear functional on compactly supported continuous functions on G/H by

$$\int_{G/H} (Af)(gH)\,d\sigma(gH) = \int_G f(g)\,d\mu(g).$$

It is immediate that this is left G-invariant, etc. ■

COROLLARY There is certainly a G-invariant measure on G/H (with H a closed subgroup) in the following cases:

i. G unimodular and H unimodular, e.g.,
   a. G unimodular and H compact
   b. G unimodular and H discrete
   c. G abelian

ii. G arbitrary and H has no non-trivial continuous homomorphisms to the positive real numbers, e.g.,
   a. G arbitrary and H compact. ■

Appendix A.2
Harmonic Analysis on the Adeles

We develop the standard facts about Fourier transforms, Fourier series, and the Poisson Summation Formula on the adeles, giving direct proofs of all assertions, rather than referring to abstract harmonic analysis. Though abstract harmonic analysis (for locally compact abelian groups) can be used to establish the fundamentals, it is of interest to see simple direct proofs which follow the lines of the classical archimedean versions. Also, we give the standard normalizations of the additive and multiplicative Haar measures on local fields, adeles, and ideles.

Though for convenience we consider only the "one-dimensional" case here, the results (e.g., the adelic Poisson Summation Formula) in the n-dimensional case ($1 \leq n \in \mathbb{Z}$) are exactly analogous, as are the proofs.

Let F be a number field, $\mathfrak{o}$ its integers, $\mathbb{A}$ its adeles, $\mathbb{A}_0$ its finite adele, $\mathbb{A}_\infty$ its infinite adeles, $F_\mathfrak{p}$ the completion of F at a prime $\mathfrak{p}$, $\mathfrak{o}_\mathfrak{p}$ the $\mathfrak{p}$-adic integers if $\mathfrak{p}$ is finite. Recall that the set

$$\left(\prod_{\mathfrak{p}\text{ infinite}} F_\mathfrak{p}\right)\Big/\mathfrak{o} \times \prod_{\mathfrak{p}\text{ finite}} \mathfrak{o}_\mathfrak{p}$$

provides a set of irredundant representatives for $\mathbb{A}/F$, by the Strong Approximation Theorem. Let $\| \ \|_\mathfrak{p}$ be the $\mathfrak{p}$-adic norms so that the product formula holds.

Regarding *normalization of additive Haar measures*, use the usual Lebesgue measure on $\mathbb{R}$. For $\mathfrak{p}$ finite, take the translation-invariant measure on $F_\mathfrak{p}$ so that $\mathfrak{p}^n\mathfrak{o}_\mathfrak{p}$ has measure $N\mathfrak{p}^{-n}[\mathfrak{o}_\mathfrak{p} : \mathfrak{f}_\mathfrak{p}]^{-1/2}$, where $\mathfrak{f}_\mathfrak{p} \subset \mathfrak{o}_\mathfrak{p}$ is the *local different* of $F_\mathfrak{p}$ over $\mathbb{Q}_p$. This normalization is necessary so that

$$\text{measure}(\mathfrak{o}_\mathfrak{p}) \times \text{measure}(\mathfrak{f}_\mathfrak{p}^{-1}) = 1,$$

which in turn will be seen to be the condition for Fourier inversion to hold. For a similar reason (i.e., the choice of normalization of the exponential functions giving the characters below) for $\mathbb{C}$ we take *twice* the usual Lebesgue measure. Take the product measure on $\mathbb{A}$.

Regarding *normalization of multiplicative Haar measures*, on $\mathbb{R}^\times$ use $dy/|y|$. On $\mathbb{C}^\times$ use $(2\pi)^{-1}\, dm(z)/\|z\|$ when m is the usual additive Haar measure, and $\|z\| = |z|^2$

is the product-formula norm. For $\mathfrak{p}$ finite, give $F_{\mathfrak{p}}^{\times}$ the multiplicative Haar measure so that the local units have measure 1. (The latter choice is consistent with the dictum that a maximal compact open subgroup of a topological group, if it exists, should always have measure 1.) Give the ideles the product measure.

Now we define the *standard additive characters*. When $\mathfrak{p}$ is a real infinite prime, define

$$\tau_{\mathfrak{p}}(x) = \exp(2\pi i x) \qquad (x \in F_{\mathfrak{p}}).$$

For a finite rational prime p, one has an obvious

$$\lambda_p\colon \mathbb{Q}_p \to \mathbb{Q}_p/\mathbb{Z}_p \to \mathbb{Q}/\mathbb{Z},$$

where the second arrow approximates an element x of $\mathbb{Q}_p$ by an element x_0 of $\mathbb{Q}$ so that $\|x - x_0\|_p \le 1$. Let $\mathrm{Tr}_{\mathfrak{p}}$ be the trace from $F_{\mathfrak{p}}$ to $\mathbb{Q}_p$, where $\mathfrak{p}$ lies over p. Then we have a composite map

$$\lambda_{\mathfrak{p}} \circ \mathrm{Tr}_{\mathfrak{p}}\colon F_{\mathfrak{p}} \to \mathbb{Q}_p \to \mathbb{Q}_p/\mathbb{Z}_p \to \mathbb{Q}/\mathbb{Z}.$$

For any prime $\mathfrak{p}$ of F lying over a rational prime p, define

$$\tau_{\mathfrak{p}}(x) = \exp(-2\pi i \lambda_{\mathfrak{p}} \circ \mathrm{Tr}_{\mathfrak{p}}(x)) \qquad (\mathfrak{p} \text{ finite})$$

$$\tau_{\mathfrak{p}}(x) = \exp(2\pi i \lambda_{\mathfrak{p}} \circ \mathrm{Tr}_{\mathfrak{p}}(x)) \qquad (\mathfrak{p} \text{ infinite}).$$

Note that the sign in the non-archimedean case is opposite that of the archimedean case. For $x \in \mathbb{A}$, put

$$\tau_0(x) = \prod_{\mathfrak{p} \text{ finite}} \tau_{\mathfrak{p}}(x) \qquad \tau_\infty(x) = \prod_{\mathfrak{p} \text{ infinite}} \tau_{\mathfrak{p}}(x)$$

$$\tau(x) = \tau_\infty(x)\tau_0(x).$$

For an integrable $\mathbb{C}$-valued function f on $F_{\mathfrak{p}}$, define the $\mathfrak{p}$-adic Fourier transform $\hat{f}$ (and inverse transform $\check{f}$) by

$$\hat{f}(\xi) = \int_{F_{\mathfrak{p}}} f(x)\bar{\tau}_{\mathfrak{p}}(x\xi)\,dx$$

$$\check{f}(\xi) = \int_{F_{\mathfrak{p}}} f(x)\tau_{\mathfrak{p}}(x\xi)\,dx.$$

The adelic Fourier transform (and inverse transform) of an integrable function on $\mathbb{A}$ are defined by

$$\hat{f}(\xi) = \int_{\mathbb{A}} f(x)\bar{\tau}_{\mathfrak{p}}(x\xi)\,dx$$

$$\check{f}(\xi) = \int_{\mathbb{A}} f(x)\tau_{\mathfrak{p}}(x\xi)\,dx.$$

(We define the Fourier transform on the finite or infinite adeles analogously.)

Note We have the usual formal properties. For $x_0 \in F_{\mathfrak{p}}$, $\lambda \in F_{\mathfrak{p}}^{x}$ the following identities follow immediately from the definition:

$$f(x + x_0)\hat{\ }(\xi) = \tau(\xi x_0)\hat{f}(\xi)$$

$$f(\lambda x)\hat{\ }(\xi) = \|\lambda\|_{\mathfrak{p}}^{-1}\hat{f}(\lambda^{-1}\xi)$$

$$\{\tau(xx_0)f(x)\}\hat{\ }(\xi) = \hat{f}(\xi - x_0).$$

Of course, similar identities hold for the inverse transform. For an integrable function f on $\mathbb{A}/F$, also write (abusing notation)

$$\hat{f}(\xi) = \int_{\mathbb{A}/F} f(x)\bar{\tau}(x\xi)\,dx$$

$$\check{f}(\xi) = \int_{\mathbb{A}/F} f(x)\tau(x\xi)\,dx.$$

LEMMA The characters on $\mathbb{A}$ given by

$$x \to \tau(x\xi) \qquad (\text{with } \xi \in F)$$

are invariant under $x \to x + \alpha$, for $\alpha \in F$.

Proof It suffices to show this for $\xi = 1$. Now

$$\sum_{\mathfrak{p}\text{ infinite}} \lambda_{\mathfrak{p}}(\mathrm{Tr}_{\mathfrak{p}}(\alpha)) = \sum_{\mathfrak{p}\text{ infinite}} \{\mathrm{Tr}_{\mathfrak{p}}(\alpha) \bmod \mathbb{Z}\} = [F:\mathbb{Q}]\,\mathrm{Trace}_{F/\mathbb{Q}}(\alpha),$$

$$\sum_{\mathfrak{p}\text{ finite}} \lambda_{\mathfrak{p}}(\mathrm{Tr}_{\mathfrak{p}}(\alpha)) = \sum_{\mathfrak{p}\text{ finite}} \{\mathrm{Tr}_{\mathfrak{p}}(\alpha) \bmod \mathbb{Z}\} = [F:\mathbb{Q}]\,\mathrm{Trace}_{F/\mathbb{Q}}(\alpha),$$

where we note that in the last sum almost all summands are 0 modulo $\mathbb{Z}$. Hence,

$$\tau(\alpha) = \exp\{2\pi i[F:\mathbb{Q}]\,\mathrm{Trace}_{F/\mathbb{Q}}(\alpha) - 2\pi i[F:\mathbb{Q}]\,\mathrm{Trace}_{F/\mathbb{Q}}(\alpha)\} = 1. \qquad \blacksquare$$

LEMMA

1. Let f be a bounded integrable function on $\mathbb{A}_0$, supported on $M(\prod_{\mathfrak{p}} \mathfrak{o}_{\mathfrak{p}})$, for some integer M. Then the integral expressing the Fourier transform is absolutely convergent, uniformly so on compacta, and is (locally) constant on cosets of $\mathfrak{f}^{-1}M^{-1}(\prod_{\mathfrak{p}} \mathfrak{o}_{\mathfrak{p}})$, where $\mathfrak{f} \subset \mathfrak{o}$ is the different of F over $\mathbb{Q}$.

2. Let f be integrable and (locally) constant on cosets of $M(\prod_{\mathfrak{p}} \mathfrak{o}_{\mathfrak{p}})$ in $\mathbb{A}_0$. Then the Fourier transform of f is supported on $\mathfrak{f}^{-1}M^{-1}(\prod_{\mathfrak{p}} \mathfrak{o}_{\mathfrak{p}})$.
(This is elementary and is the finite-adele analogue of the relation between differentiability and growth which archimedean Fourier transforms enjoy.) ■

LEMMA (RIEMANN-LEBESGUE) Adelic or $\mathfrak{p}$-adic Fourier transforms of integrable functions go to zero. The Fourier coefficients of integrable functions on $\mathbb{A}/F$ go to zero. (Corollaries of Bessel's inequality.) ■

LEMMA For $\mathfrak{p}$ a finite prime of F, the $\mathfrak{p}$-adic Fourier transform of the characteristic function of the integers $\mathfrak{o}_{\mathfrak{p}}$ of $F_{\mathfrak{p}}$ is $[\mathfrak{o}_{\mathfrak{p}} : \mathfrak{f}\mathfrak{o}_{\mathfrak{p}}]^{-1/2}$ times the characteristic function of $\mathfrak{f}^{-1}\mathfrak{o}_{\mathfrak{p}}$, where $\mathfrak{f}$ is the different of $F/\mathbb{Q}$. (Direct calculation.) ■

Definition For a finite prime $\mathfrak{p}$, the *Schwartz functions* on $F_{\mathfrak{p}}$ are defined to be the continuous compactly supported uniformly locally constant complex valued functions. The Schwartz functions on an archimedcan vectorspace are those functions which are infinitely differentiable, and all of whose derivatives are rapidly decreasing. The Schwartz functions on an adelic vectorspace are *finite* linear combinations of functions of the form

$$f(x) = \prod_{\mathfrak{p}} f_{\mathfrak{p}}(x_{\mathfrak{p}}),$$

where $f_{\mathfrak{p}}$ is a Schwartz function on $F_{\mathfrak{p}}$.

Definition The Schwartz functions on $\mathbb{A}/F$ are those functions which, considered as functions on $\mathbb{A}$, are infinitely differentiable in the infinite factors and are uniformly locally constant in the finite factors.

THEOREM Let f be a Schwartz function on $\mathbb{A}/F$. The Fourier series

$$\sum_{\xi \in F} \tau(x\xi) \int_{\mathbb{A}/F} f(u)\bar{\tau}(u\xi)\, du$$

converges absolutely and uniformly to $f(x)$.

Proof First, we note that it suffices to prove this result for $F = \mathbb{Q}$, since

$$\mathbb{A} = F \otimes_{\mathbb{Q}} (\text{adeles of } \mathbb{Q}).$$

One should check that the normalization of measures above is the appropriate one so that this reduction is legitimate. For notational simplicity, therefore, we take $F = \mathbb{Q}$. By immediate formal properties of the Fourier transform, analogous to the lemma just above, one finds that this series is, in any case, nicely convergent. As in the classical case, it suffices to prove that

$$f(0) = \sum_{\xi \in \mathbb{Q}} \int_{\mathbb{A}/\mathbb{Q}} f(u)\bar{\tau}(u\xi)\, du = \sum_{\xi \in \mathbb{Q}} \hat{f}(\xi).$$

Further, as constant functions are clearly represented by their Fourier series, we may suppose that $f(0) = 0$. For an integer m, let

$$X(m, M) = \{r/M : r \in \mathbb{Z}, |r| \leq mM\}.$$

We will use the Riemann-Lebesgue lemma to see that

$$\sum_{\xi \in X(m, M)} \hat{f}(\xi) \to 0 \qquad (\text{as } m \to \infty, \text{ for large-enough fixed } M).$$

Summing finite geometric series,

$$\sum_{\xi \in X(m, M)} \hat{f}(\xi) = \sum_{r} \int_{\mathbb{A}/\mathbb{Q}} f(u)\bar{\tau}(ur/M)\, du =$$

(with $|r| \leq mM$)

$$= \int_{\mathbb{A}/\mathbb{Q}} \Phi_m(u)\{\bar{\tau}(u(mM + 1)/M) - \bar{\tau}(-umM/M)\}\, du,$$

where

$$\Phi_m(u) = f(u)/(1 - \bar{\tau}(M^{-1}x)).$$

By the Riemann-Lebesgue lemma applied to Φ_m (for fixed large-enough M),

$$\lim_{m \to \infty} \int_{\mathbb{A}/\mathbb{Q}} \Phi_m(u)\bar{\tau}(u(mM + 1)/M)\, du = 0,$$

and likewise for the second term. This proves the theorem. ■

THEOREM (FOURIER INVERSION) For a Schwartz function f on $\mathbb{A}$ or on $F_\mathfrak{p}$

$$\hat{f}^{\vee} = f = \check{f}^{\wedge}.$$

Proof We prove just the first equality. Without loss of generality, we may suppose that f is a product of Schwartz functions $f_\mathfrak{p}$ on the local vector spaces $F_\mathfrak{p}$. It is easy to see from the definitions that

$$\left(\prod_{\mathfrak{p}} f_\mathfrak{p}\right)^{\wedge} = \prod_{\mathfrak{p}} \hat{f}_\mathfrak{p}.$$

Thus, we need only prove the local version. We already know this in the archimedean case. (Note the normalization of measure and of exponential functions in the complex case.)

From the formal relations satisfied by Fourier transforms, translating and dilating the given test function commutes with $f \to \hat{f}^{\vee}$ in the following manner:

$$\{f(x_0 + \lambda x)\}^{\wedge\vee}(x) = \|\lambda\|_p^{-1}\{\tau(\xi x_0/\lambda)\hat{f}(\xi/\lambda)\}^{\vee}(x) = \hat{f}^{\vee}(x_0 + \lambda x)).$$

Then the inversion formula may be checked immediately by direct calculation: $\hat{f}$ is a constant c_1 times the characteristic function φ of the local inverse different, and $(c_1\varphi)^{\vee}(x)$ is again $c_1 c_2 f(x)$, for some constant c_2. The normalization of the measures was designed exactly so as to make $c_1 c_2 = 1$, so we have Fourier Inversion. ∎

THEOREM (ADELIC POISSON SUMMATION FORMULA) For a Schwartz function f on $\mathbb{A}$,

$$\sum_{v \in F} f(v) = \sum_{v \in F^n} \hat{f}(v).$$

Proof Consider the function

$$\varphi(x) = \sum_{v \in F} f(x + v).$$

This is a function on $\mathbb{A}/F$ meeting all our hypotheses, so is represented by its Fourier series:

$$\varphi(x) = \sum_{v \in F} \tau(xv) \int_{\mathbb{A}/F} \bar{\tau}(u\xi)\varphi(u)\,du.$$

For $x = 0$, unwinding the integrals into integrals against f over $\mathbb{A}$, this is

$$\sum_{v \in F} f(v) = \sum_{v \in F} \int_{\mathbb{A}} \bar{\tau}(u\xi) f(u)\,du = \sum_{v \in F} \hat{f}(v).$$

This gives the result. ∎

Appendix A.3
Strong Approximation for $SL(n)$

Let F be a number field, with adeles $\mathbb{A}$, and with ring of integers $\mathfrak{o}$. For a place $\mathfrak{p}$ of F, let $F_\mathfrak{p}$ be the $\mathfrak{p}$-adic completion of F. Let $F_\infty = \prod_\mathfrak{p} F_\mathfrak{p}$ where $\mathfrak{p}$ runs over the infinite primes of F. We will prove a special case of the Strong Approximation Theorem for $SL(n)$.

THEOREM $SL(n, F)SL(n, F_\infty)$ is dense in $SL(n, \mathbb{A})$.

Remarks This is not quite the Strong Approximation Theorem for $SL(n)$, but is sufficient for many applications and allows some simplification of the proof. (The Strong Approximation Theorem for $SL(n)$ replaces the set of infinite primes by an arbitrary non-empty set of primes.)

Proof For a commutative ring A, and for $\alpha \in SL(n, A)$, let

$$P_\alpha(x) = x^n - c_{n-1}x^{n-1} + c_{n-2}x^{n-2} - \cdots \pm c_1 x + (-1)^n$$

be the characteristic polynomial of α. For a field Ω say that $\alpha \in SL(n, \Omega)$ is *regular* if the roots of P_α (in an algebraic closure of Ω) are all of multiplicity one. Define

$$f\colon SL(n, A) \to A^{n-1}$$

by sending $\alpha \in SL(n, A)$ to the $(n-1)$-tuple $(c_{n-1}, \ldots, c_1)$.

By the nature of the topology on $SL(n, \mathbb{A})$, it suffices to show that $\alpha \in SL(n, \mathbb{A})$ is in the closure of $SL(n, F)SL(n, F_\infty)$ when all components of α are 1 except for one component, say at a finite prime $\mathfrak{p}_0$. Further, we may suppose that α is regular, since the regular elements are dense. Let U be a neighborhood of α, which we may take (without loss of generality) to be a product $\prod_\mathfrak{p} U_\mathfrak{p}$ of opens $U_\mathfrak{p} \subset SL(n, F_\mathfrak{p})$. Of course, almost all $U_\mathfrak{p}$ are $SL(n, \mathfrak{o}_\mathfrak{p})$. We may take $U_{\mathfrak{p}_0}$ sufficiently small so that $\beta \in U_{\mathfrak{p}_0} \subset SL(n, F_\mathfrak{p})$ implies that β is regular.

First, we claim that (in $\mathbb{A}$)

$$F \cap (f(U) + F_\infty) \neq \varnothing.$$

By the classical approximation theorem, we are claiming just that for every finite prime the subset $f(U_\mathfrak{p})$ of $F_\mathfrak{p}^{n-1}$ has non-trivial interior, and that $\mathfrak{o}_\mathfrak{p}^{n-1} \subset f(U_\mathfrak{p})$ for almost all primes $\mathfrak{p}$. This follows from two lemmas:

LEMMA Let α be regular in $SL(n, F_\mathfrak{p})$, let $Z = F_\mathfrak{p}[\alpha] \cap SL(n, F_\mathfrak{p})$, and let U be a neighborhood of α in Z. Then $f(U)$ is a neighborhood of $f(\alpha)$ in $F_\mathfrak{p}^{n-1}$.

Proof Take $y \in F_\mathfrak{p}[\alpha]$. Write $P_y(x) = P(x; y)$. We have a Taylor expansion

$$P_y(x) = P_{\alpha+(y-\alpha)}(x) = P(x; \alpha) + Q_1(x; y - \alpha) + Q_2(x; y - \alpha) + \cdots.$$

where Q_1 is $F_\mathfrak{p}$-linear in $y - \alpha$ and of degree $\leq n - 1$ in x, Q_2 is $F_\mathfrak{p}$-quadratic in $y - \alpha$ and of degree $\leq n - 2$ in x, etc. Using Taylor's theorem,

$$P_y(x + x_0) = P(x + x_0; \alpha) + Q_1(x + x_0; y \quad \alpha) + \cdots -$$

$$= \{P(x_0; \alpha) + P'(\alpha; \alpha)x + \tfrac{1}{2}P''(x_0; \alpha)x^2 + \cdots\} +$$

$$+ \{Q_1(x_0; y - \alpha) + Q_1'(x_0; y - \alpha)x + \tfrac{1}{2}Q_1''(x_0; y - \alpha)x^2 + \cdots\} + \cdots$$

where $P'(x)$ is the derivative of a polynomial P in x, etc. From the Cayley-Hamilton theorem, this gives

$$0 = P_y(y) = P(y; \alpha) + Q_1(y; y - \alpha) + \cdots =$$

$$= \{P(\alpha; \alpha) + P'(\alpha; \alpha)(y - \alpha)\cdots\} + \{Q_1(\alpha; y - \alpha) + Q_1'(\alpha; y - \alpha)(y - \alpha) + \cdots\} + \cdots =$$

$$= \{0 + P'(\alpha; \alpha)(y - \alpha)\cdots\} + \{Q_1(\alpha; y - \alpha) + Q_1'(\alpha; y - \alpha)(y - \alpha) + \cdots\} + \cdots =$$

$$= P'(\alpha; \alpha)(y - \alpha) + Q_1(\alpha; y - \alpha) + (\text{higher-order terms in } y - \alpha).$$

We must have equality of the $F_\mathfrak{p}$-linear terms in $y - \alpha$:

$$0 = P'(\alpha; \alpha)(y - \alpha) + Q_1(\alpha; y - \alpha).$$

Therefore, $Q_1(\alpha; y - \alpha) = 0$ only if $y = \alpha$ or if $P'(\alpha; \alpha) = 0$; the latter cannot occur for α regular. That is, the $F_\mathfrak{p}$-linear map $y \to Q_1(x, y - \alpha)$ yields the 0 polynomial (in x) only for $y = \alpha$. From this, it is clear that $y \to P_y$ covers a sufficiently small neighborhood of $\mathrm{Tr}(\alpha)$ in $F_\mathfrak{p}^{n-1}$. ■

LEMMA For all commutative $\mathfrak{o}$-algebras A, $f(SL(n, A)) = A^{n-1}$.

Proof We may take an element of $SL(n, A)$ in rational canonical form. ■

Certainly

$f(SL(n, F)) = F^{n-1}$

$F_\infty + f(U) \subset f(SL(n, F_\infty)U)$.

Therefore, the previous proposition yields

$$f(SL(n,F)) \cap f(SL(n,F_\infty)U) \neq \varnothing.$$

Next, we claim that

$$\left(\bigcup_{h \in GL(n,\mathbb{A})} hSL(n,F)h^{-1}\right) \cap SL(n,F_\infty)U \neq \varnothing.$$

From above, we have found $x \in SL(n,F)$ and $y \in SL(n,F_\infty)U$ with $f(x) = f(y)$. That is, for all primes $\mathfrak{p}$ we have $f(x)_\mathfrak{p} = f(y_\mathfrak{p})$. For almost all primes $\mathfrak{p}$, x is in $U_\mathfrak{p}$, so we only need to conjugate x at finitely many primes. At these remaining (finitely many) primes $\mathfrak{p}$, from

LEMMA Let k be any field. If $\alpha, \beta \in SL(n,k)$ are regular and $f(\alpha) = f(\beta)$, then there is $x \in GL(n,k)$ so that $\alpha = x\beta x^{-1}$. (Elementary linear algebra.) ■

There is $h_\mathfrak{p} \in GL(n,F_\mathfrak{p})$ so that $h_\mathfrak{p} x_\mathfrak{p} h_\mathfrak{p}^{-1} = y_\mathfrak{p}$; take $h_\mathfrak{p} = 1$ otherwise. Taking $h = (\ldots, h_\mathfrak{p}, \ldots)$ gives the desired outcome: $(hxh^{-1})_\mathfrak{p} \in SL(n,\mathfrak{o}_\mathfrak{p}) = U_\mathfrak{p}$ at almost all primes, and $(hxh^{-1})_\mathfrak{p} = y_\mathfrak{p} \in U_\mathfrak{p}$ at the finitely many remaining primes. This proves the claim.

LEMMA $GL(n,\mathbb{A}) = GL(n,F)EGL(n,F_\infty)$ with some compact subset E of $GL(n,\mathbb{A})$.

Proof Let Δ be a (finite) set of finite ideles representing the (absolute) ideal classes of $\mathfrak{o}$. Let $\mathbb{A}_0$ be the finite adeles of F, and put

$$K_0 = \prod_{\mathfrak{p}\,\text{finite}} GL(n,\mathfrak{o}_\mathfrak{p}).$$

First, $GL(n,\mathbb{A}_0)/K_0$ has a collection of representatives g whose finite-prime components are upper triangular; componentwise, let $GL(n,\mathfrak{o}_\mathfrak{p})$ act on the bottom row of an element of $GL(n,F_\mathfrak{p})$ to obtain zeros at all but the lower right entry, where we obtain a greatest common divisor of the entries of this row. Note that $\mathfrak{o}_\mathfrak{p}$ is a principal ideal domain; an induction on n gives the result. By left multiplication by diagonal matrices in $GL(n,F)$ and right multiplication by diagonal elements of K_0 we may adjust the diagonal entries of g to be in Δ, since

$$(\text{finite ideles of } F)\Big/\left(F^x \prod_{\mathfrak{p}\,\text{finite}} \mathfrak{o}_\mathfrak{p}^x\right) \approx \text{ideal class group of } \mathfrak{o}.$$

Finally, by suitable left multiplication by upper triangular unipotent matrices in $GL(n,F)$ and right multiplication by upper triangular unipotent matrices in K_0 we may adjust the upper right entry of g to 0, using the classical (additive) additive approximation theorem. Thus, we have shown that

$$GL(n,F)GL(n,F_\infty)\backslash GL(n,\mathbb{A})/K_0 \approx GL(n,F)\backslash GL(n,\mathbb{A}_0)/K_0$$

has a collection of diagonal representatives whose diagonal entries are in the finite set Δ. That is, this double coset space is finite. Since K is compact, this gives the result. ∎

Therefore,

$$\text{closure}(SL(n, F)SL(n, F_\infty)) \cap \left(\bigcup_{h \in E} hUh^{-1} \right) \neq \varnothing.$$

As U varies over (arbitrarily small) *compact* neighborhoods of α, we have

$$\bigcap_U \left(\text{closure}(SL(n, F)SL(n, F_\infty)) \cap \left(\bigcup_{h \in E} hUh^{-1} \right) \right) \neq \varnothing,$$

since each $\{hUh^{-1} : h \in E\}$ is compact. From this it follows that

$$\text{closure}(SL(n, F)SL(n, F_\infty)) \cap \{h\alpha h^{-1} : h \in E\} \neq \varnothing.$$

Now take some $\beta = h\alpha h^{-1}$ in this intersection; note that the $\mathfrak{p}$-component of β is 1 for $\mathfrak{p} \neq \mathfrak{p}_0$, since α has this property; in particular, $\beta_\mathfrak{p} = \alpha_\mathfrak{p}$ for $\mathfrak{p} \neq \mathfrak{p}_0$. Take $k \in GL(n, F)$ very close to h at $\mathfrak{p}_0$. Then still $k^{-1}\beta k$ lies in the closure of $SL(n, F)SL(n, F_\infty)$. We can make a sequence of such points $k^{-1}\beta k$ whose $\mathfrak{p}_0$-component approaches that of α, and whose $\mathfrak{p} \neq \mathfrak{p}_0$ components are all equal to that of α (being 1). Therefore, α is in the closure of $SL(n, F)SL(n, F_\infty)$. ∎

Appendix A.4
Invariant Differential Operators

Here we compute, in coordinates, the Casimir operator for $SL(2, \mathbb{R})$ and the Maass raising and lowering operators. The Casimir operator is a left and right $SL(2, \mathbb{R})$-invariant second-order differential operator; the Maass operators are left $SL(2, \mathbb{R})$-invariant first order operators. Among other things, the Casimir operator gives rise to the $SL(2, \mathbb{R})$-invariant Laplacian on $\mathfrak{H}$.

The computation here has a straightforward interpretation in terms of Lie theory; indeed, it is exactly a Lie-theoretic computation, though no results from that theory are needed. In effect, by defining the differential operators in coordinates, we have given the outcome of the computation; in the course of the proof we introduce the naturally defined objects, which we then determine to be identical to the operators defined in coordinates.

We use ("Iwasawa") coordinates on $SL(2, \mathbb{R})$ as follows: Let

$$u(x) = \begin{pmatrix} 1 & x \\ 0 & 1 \end{pmatrix} \qquad \delta(y) = \begin{pmatrix} y^{1/2} & 0 \\ 0 & y^{-1/2} \end{pmatrix} \qquad k(\theta) = \begin{pmatrix} \cos\theta & \sin\theta \\ -\sin\theta & \cos\theta \end{pmatrix},$$

for $x \in \mathbb{R}$, $y > 0$, $\theta \in \mathbb{R}/2\pi\mathbb{R}$; then every element of $SL(2, \mathbb{R})$ can be written in the form $u(x)\delta(y)k(\theta)$; we refer to x, y, θ as the Iwasawa coordinates on $SL(2, \mathbb{R})$. For a differentiable function f on $SL(2, \mathbb{R})$, write

$$f(u(x)\delta(y)k(\theta)) = F(x, y, \theta).$$

Then define the *Maass* (raising and lowering) *operators* by

$$Rf(u(x)\delta(y)k(\theta)) = e^{2i\theta}(y\partial/\partial x - iy\partial/\partial y - \tfrac{1}{2}\partial/\partial\theta)F(x, y, \theta),$$

$$Lf(u(x)\delta(y)k(\theta)) = e^{-2i\theta}(y\partial/\partial x + iy\partial/\partial y - \tfrac{1}{2}\partial/\partial\theta)F(x, y, \theta).$$

Define the *Casimir operator* by

$$Z = y^2((\partial/\partial x)^2 + (\partial/\partial y)^2) - y\partial^2/\partial x\,\partial\theta.$$

Say that f is of *weight* k ($k \in \mathbb{Z}$) if

$$f(gk(\theta)) = f(g)e^{ik\theta}.$$

THEOREM The differential operator Z is left and right $SL(2, \mathbb{R})$-invariant, and commutes with the raising and lowering operators R and L. Further, we have

$$2Z = R \circ L + L \circ R - \tfrac{1}{2}(\partial/\partial\theta)^2.$$

The operators R and L are left $SL(2, \mathbb{R})$-invariant, i.e., for $g, g_0 \in SL(2, \mathbb{R})$, and for a differentiable function f on $SL(2, \mathbb{R})$, we have

$$(Rf)(gg_0) = R(f \circ g)(g_0)$$

$$(Lf)(gg_0) = L(f \circ g)(g_0),$$

where $(f \circ g)(g_0) = f(gg_0)$. Further, for $\theta \in \mathbb{R}$, if f is of weight k, we have

$$Rf(gk(\theta)) = e^{i(k+2)\theta}Rf(g)$$

$$Lf(gk(\theta)) = e^{i(k-2)\theta}Lf(g).$$

That is, R raises the weight by 2, and L lowers the weight by 2.

Remarks In the course of the proof, we will give the natural, *intrinsic* definitions of Z, R, and L, and show that they are the same as the above operators given in coordinates. The intrinsic definitions, by their nature, make the invariance and other properties completely formal.

Proof First, we will treat the operators R and L. Let

$$H = \begin{pmatrix} -1 & 0 \\ 0 & 1 \end{pmatrix}$$

$$SU(1,1) = \{g \in SL(2, \mathbb{C})\colon g^*Hg = H\},$$

$$\mathfrak{G}(\mathbb{R}) = \{2 \times 2 \text{ complex matrices } \gamma\colon \text{Trace}(\gamma) = 0,\ H^{-1}\gamma^*H = -\gamma\} =$$

$$(= \text{the Lie algebra of } SU(1,1))$$

$$\mathfrak{G}(\mathbb{C}) = \{2 \times 2 \text{ complex matrices } \gamma\colon \text{Trace}(\gamma) = 0\}$$

$$(= \text{the Lie algebra of } SL(2, \mathbb{C}) =$$

$$= \text{the complexification of the Lie algebra of } SU(1,1))$$

$$C = 2^{-1/2}\begin{pmatrix} 1 & i \\ i & 1 \end{pmatrix} = \text{the Cayley element}$$

$$K = \left\{\begin{pmatrix} \mu & 0 \\ 0 & \mu^{-1} \end{pmatrix} : \mu \in \mathbb{C} \text{ so that } |\mu| = 1\right\}.$$

Then $CSU(1,1)C^{-1} = SL(2,\mathbb{R})$, and $CKC^{-1} = SO(2)$. Let

$$\eta = \begin{pmatrix} \zeta_8 & 0 \\ 0 & \zeta_8^{-1} \end{pmatrix} \in K$$

where $\zeta_8 = e^{2\pi i/8}$. Note that K is exactly the centralizer in $SU(1,1)$ of η. The operator

$$\mathrm{Ad}(\eta): \gamma \to \eta\gamma\eta^{-1}$$

maps $\mathfrak{G}(\mathbb{R})$ and $\mathfrak{G}(\mathbb{C})$ to themselves; since $\eta^4 = -1_2$ is in the center of $SU(1,1) \subset SL(2,\mathbb{C})$, $\mathrm{Ad}(\eta)^4$ acts by the identity on $\mathfrak{G}(\mathbb{R})$ and $\mathfrak{G}(\mathbb{C})$, and we conclude that $\mathrm{Ad}(\eta)$ has eigenvalues ± 1, $\pm i$ on the complexification $\mathfrak{G}(\mathbb{C}) = \mathfrak{G}(\mathbb{R}) \otimes \mathbb{C}$ of $\mathfrak{G}(\mathbb{R})$. One sees directly that

$$i\text{-eigenspace of } \mathrm{Ad}(\eta) = \left\{\begin{pmatrix} 0 & \zeta \\ 0 & 0 \end{pmatrix} : \zeta \in \mathbb{C}\right\}$$

$$(-i)\text{-eigenspace of } \mathrm{Ad}(\eta) = \left\{\begin{pmatrix} 0 & 0 \\ \zeta & 0 \end{pmatrix} : \zeta \in \mathbb{C}\right\}$$

and that there are no non-zero (-1)-eigenvectors. Let

$$\alpha = \begin{pmatrix} 0 & 1 \\ 0 & 0 \end{pmatrix} \qquad \beta = \begin{pmatrix} 0 & 0 \\ 1 & 0 \end{pmatrix}.$$

Then, in fact, for

$$h(\theta) = \begin{pmatrix} e^{i\theta} & 0 \\ 0 & e^{-i\theta} \end{pmatrix} = C^{-1}k(\theta)C \in K$$

we have

$$h(\theta)\alpha h(\theta)^{-1} = e^{2i\theta}\alpha \qquad h(\theta)\beta h(\theta)^{-1} = e^{-2i\theta}\beta.$$

For a differentiable function f on $SL(2,\mathbb{R})$, define

$$R'f(g) = (\partial/\partial t)|_{t=0} f(g\exp(tC\alpha C^{-1}))$$

$$L'f(g) = (\partial/\partial t)|_{t=0} f(g\exp(tC\beta C^{-1})).$$

Remarks Since $\exp C\alpha C^{-1}$ and $\exp C\beta C^{-1}$ are not in $SL(2, \mathbb{R})$, but only in $SL(2, \mathbb{C})$, these definitions might appear to be unjustifiable. However, $C\alpha C^{-1}$ and $C\beta C^{-1}$ are certainly in the *complexified* tangent space to $SL(2, \mathbb{R})$. Therefore, for example, given a *real analytic* function f on a neighborhood of a point g in $SL(2, \mathbb{R})$, we can extend f to a *holomorphic* function on a sufficiently small neighborhood of g in $SL(2, \mathbb{C})$, thereby allowing differentiation in complex tangent directions as well. However, reflection will make clear that not even this artifice is really necessary.

It is immediate from the definition that R' and L' are left-invariant. We will show that R' and L' have the desired raising and lowering properties, and that $R = R'$, $L = L'$. If f is of weight k, for $\theta \in \mathbb{R}$ we have

$$R'f(gCh(\theta)C^{-1}) =$$

$$= (\partial/\partial t)|_{t=0} f(gCh(\theta)C^{-1}\exp(tC\alpha C^{-1})) =$$

$$= (\partial/\partial t)|_{t=0} f(g\exp(e^{2i\theta}tC\alpha C^{-1})Ch(\theta)C^{-1}) =$$

$$= e^{ik\theta}(\partial/\partial t)|_{t=0} f(g\exp(e^{2i\theta}tC\alpha C^{-1}) = e^{ik\theta}e^{2i\theta}(\partial/\partial t)|_{t=0} f(g\exp(tC\alpha C^{-1}) =$$

$$= e^{i(k+2)\theta}R'f(g);$$

similarly, we have

$$L'f(gCh(\theta)C^{-1}) = e^{i(k-2)\theta}L'f(g).$$

That is, *R' raises the weight by 2, and L' lowers the weight by 2*. We compute R' and L' in the "Iwasawa coordinates" on $SL(2, \mathbb{R})$ to see that they coincide with R and L. To express R' in the x, y, θ coordinates, we may suppose that $t^2 = 0$, and for given x, y, θ we determine x_1, y_1, θ_1 so that

$$u(x)\delta(y)k(\theta)\exp(tC\alpha C^{-1}) = u(x_1)\delta(y_1)k(\theta_1),$$

To this end, we first express $\exp(tC\alpha C^{-1})$ in the Iwasawa coordinates, keeping in mind that $t^2 = 0$. We have

$$\exp(tC\alpha C^{-1}) = \exp\left(\tfrac{1}{2}t\begin{pmatrix} -i & 1 \\ 1 & i \end{pmatrix}\right) =$$

$$= \begin{pmatrix} 1 - \frac{1}{2}it & \frac{1}{2}t \\ \frac{1}{2}t & 1 + \frac{1}{2}it \end{pmatrix} = \begin{pmatrix} 1 - \frac{1}{2}it & t \\ 0 & 1 + \frac{1}{2}it \end{pmatrix}\begin{pmatrix} 1 & -\frac{1}{2}t \\ \frac{1}{2}t & 1 \end{pmatrix} =$$

$$= \begin{pmatrix} 1 & t \\ 0 & 1 \end{pmatrix}\begin{pmatrix} 1 - \frac{1}{2}it & 0 \\ 0 & 1 + \frac{1}{2}it \end{pmatrix}\begin{pmatrix} 1 & -\frac{1}{2}t \\ \frac{1}{2}t & 1 \end{pmatrix} =$$

$$= u(t)\delta(1 - it)k(-\tfrac{1}{2}t).$$

Therefore,

$$u(x)\delta(y)k(\theta)\exp(tC\alpha C^{-1}) = u(x)\delta(y)\exp(e^{2i\theta}tC\alpha C^{-1})k(\theta) =$$

$$= u(x)\delta(y)u(e^{2i\theta}t)\delta(1 - ie^{2i\theta}t)k(-\tfrac{1}{2}e^{2i\theta}t)k(\theta) =$$

$$= u(x)(\delta(y)u(e^{2i\theta}t)\delta(y)^{-1})\delta(y)\delta(1 - ie^{2i\theta}t)k(\theta - \tfrac{1}{2}e^{2i\theta}t) =$$

$$= u(x)u(ye^{2i\theta}t)\delta(y(1 - ie^{2i\theta}t))k(\theta - \tfrac{1}{2}e^{2i\theta}t) =$$

$$= u(x + e^{2i\theta}ty)\delta(y(1 - ie^{2i\theta}t))k(\theta - \tfrac{1}{2}e^{2i\theta}t).$$

Therefore, writing

$$f(u(x)\delta(y)k(\theta)) = F(x, y, \theta),$$

we have

$$R'f(u(x)\delta(y)k(\theta)) = (\partial/\partial t)|_{t=0}F(x + e^{2i\theta}ty,\ y(1 - ie^{2i\theta}t),\ \theta - \tfrac{1}{2}e^{2i\theta}t) =$$

$$= e^{2i\theta}(y\partial/\partial x - iy\partial/\partial y - \tfrac{1}{2}\partial/\partial\theta)F =$$

$$= Rf(u(x)\delta(y)k(\theta)).$$

Similarly,

$$\exp(tC\beta C^{-1}) = \exp\left(\tfrac{1}{2}t\begin{pmatrix} i & 1 \\ 1 & -i \end{pmatrix}\right) =$$

$$= \begin{pmatrix} 1 + \tfrac{1}{2}it & \tfrac{1}{2}t \\ \tfrac{1}{2}t & 1 - \tfrac{1}{2}it \end{pmatrix} = \begin{pmatrix} 1 + \tfrac{1}{2}it & t \\ 0 & 1 - \tfrac{1}{2}it \end{pmatrix}\begin{pmatrix} 1 & -\tfrac{1}{2}t \\ \tfrac{1}{2}t & 1 \end{pmatrix} =$$

$$= \begin{pmatrix} 1 & t \\ 0 & 1 \end{pmatrix}\begin{pmatrix} 1 + \tfrac{1}{2}it & 0 \\ 0 & 1 - \tfrac{1}{2}it \end{pmatrix}\begin{pmatrix} 1 & -\tfrac{1}{2}t \\ \tfrac{1}{2}t & 1 \end{pmatrix} =$$

$$= u(t)\delta(1 + it)k(-\tfrac{1}{2}t),$$

and

$$u(x)\delta(y)k(\theta)\exp(tC\beta C^{-1}) = u(x + e^{-2i\theta}ty)\delta(y(1 + ie^{-2i\theta}t))k(\theta - \tfrac{1}{2}e^{-2i\theta}t).$$

Therefore, we have

$$L'f(u(x)\delta(y)k(\theta)) = (\partial/\partial t)|_{t=0}F(x + e^{-2i\theta}ty, y(1 + ie^{-2i\theta}t),\ \theta - \tfrac{1}{2}e^{-2i\theta}t) =$$

$$= e^{-2i\theta}(y\partial/\partial x + iy\partial/\partial y - \tfrac{1}{2}\partial/\partial\theta)F =$$

$$= Lf(u(x)\delta(y)k(\theta)).$$

Remarks Since the $\pm i$ eigenspaces of η are one-dimensional, up to normalizing constants there really are no other possible choices for R' and L'.

Now we verify the properties of the Casimir operator. For brevity, write $G = SL(2, \mathbb{R})$, and now let

$$\mathfrak{G} = \{2 \times 2 \text{ real matrices with trace } 0\} \ (= \text{the Lie algebra of } SL(2, \mathbb{R})).$$

The group $SL(2, \mathbb{R})$ acts on $\mathfrak{G}$ by (the "Adjoint representation")

$$\mathrm{Ad}(g)(\gamma) = g\gamma g^{-1}$$

(for $g \in SL(2, \mathbb{R})$, $\gamma \in \mathfrak{G}$). For $\gamma \in \mathfrak{G}$, we have an associated left-invariant first-order differential operator X_γ on $SL(2, \mathbb{R})$ defined by

$$(X_\gamma f)(g) = (\partial/\partial t)|_{t=0} f(g \exp(t\gamma)).$$

Let T be the tensor algebra over the real vectorspace $\mathfrak{G}$, and let $\mathscr{A}$ be the algebra of left-invariant differential operators on $SL(2, \mathbb{R})$; then we have a homomorphism of (associative) algebras $\iota: T \to \mathscr{A}$ defined on the algebra generators γ of T by

$$\iota: \gamma \to X_\gamma.$$

Let $\mathfrak{U}$ (the "universal enveloping algebra" of $SL(2, \mathbb{R})$) be the image of T in $\mathscr{A}$ via ι. Let Ad also denote the natural extension of Ad: $G \to \mathrm{Aut}(\mathfrak{G})$.

PROPOSITION Let $Y_0 \in T$ be $SL(2, \mathbb{R})$-invariant under Ad. Then $Y = \iota(Y_0)$ is a left and right $SL(2, \mathbb{R})$-invariant differential operator, and is in the center of $\mathfrak{U}$.

Proof of Proposition Every element of $\iota(T)$ is already (by definition) a left-invariant differential operator. Also, immediately from the definition of ι, it is clear that Ad-invariance of Y_0 implies the invariance of Y under $g \to hgh^{-1}$; since Y is already left-invariant, this yields the two-sided invariance of Y.

If Y commutes with $SL(2, \mathbb{R})$, then it certainly commutes with elements of $SL(2, \mathbb{R})$ of the form $\exp(t\gamma)$ for $\gamma \in \mathfrak{G}$. That is, Y commutes with $\iota(\mathfrak{G})$; the latter image generates $\mathfrak{U}$, so Y is in the center of $\mathfrak{U}$. ■

We have a bilinear form B (up to a constant, the "Killing form" for $SL(2, \mathbb{R})$) on $\mathfrak{G}$ defined by

$$B(\alpha, \beta) = \mathrm{Trace}(\alpha\beta).$$

Let $\tilde{B}$ be the element in the dual $(\mathfrak{G} \otimes \mathfrak{G})^*$ of $\mathfrak{G} \otimes \mathfrak{G}$ induced by the bilinear form. It is immediate that for $g \in SL(2, \mathbb{R})$, $\alpha, \beta \in \mathfrak{G}$,

$$B(\mathrm{Ad}(g)(\alpha), \mathrm{Ad}(g)(\beta)) = B(\alpha, \beta).$$

By the invariance property of B, we see that $\tilde{B}$ is invariant under the natural action of $SL(2, \mathbb{R})$. Now the non-degenerate bilinear map B on $\mathfrak{G}$ induces an isomorphism φ of $\mathfrak{G} \otimes \mathfrak{G}$ to $(\mathfrak{G} \otimes \mathfrak{G})^*$ by

$$\varphi(u \otimes v)(u' \otimes v') = B(u, u')B(v, v')$$

(for u, v, u', v' in $\mathfrak{G}$). Let $Q = \varphi^{-1}(\tilde{B})$; this is an $SL(2, \mathbb{R})$-invariant element (under the extended Ad) of $\mathfrak{G} \otimes \mathfrak{G} \subset T$, so gives rise to a right- and left-invariant differential operator (the Casimir operator) $Z = \frac{1}{2}\imath(Q)$ in the center of $\mathfrak{U}$. We will compute Z in coordinates.

We take

$$e_1 = C\alpha C^{-1} \in \mathfrak{G} \otimes \mathbb{C} \qquad e_2 = C\beta C^{-1} \in \mathfrak{G} \otimes \mathbb{C}$$

$$e_0 = \begin{pmatrix} 0 & 1 \\ -1 & 0 \end{pmatrix} \in \mathfrak{G}$$

where

$$\alpha = \begin{pmatrix} 0 & 1 \\ 0 & 0 \end{pmatrix} \qquad \beta = \begin{pmatrix} 0 & 0 \\ 1 & 0 \end{pmatrix}$$

are as before and C is the Cayley element as above; note that it does no harm to take these elements in $\mathfrak{G} \otimes \mathbb{C}$. Then

$$0 = B(e_1, e_1) = B(e_2, e_2) = B(e_1, e_0) = B(e_2, e_0)$$

$$B(e_1, e_2) = 1$$

$$B(e_0, e_0) = -2.$$

Let $\{e_i^*\}$ be the basis for $\mathfrak{G}^*$ so that $e_i^*(e_j) = B(e_i, e_j)$; then $\{e_i \otimes e_j\}$ is a basis for $\mathfrak{G} \otimes \mathfrak{G}$ and $\{e_i^* \otimes e_j^*\}$ is naturally identifiable with a basis for $(\mathfrak{G} \otimes \mathfrak{G})^*$. We have

$$\tilde{B} = e_1^* \otimes e_2^* + e_2^* \otimes e_1^* - \tfrac{1}{2} e_0^* \otimes e_0^*.$$

It is easy to see that

$$\varphi^{-1}(\tilde{B}) = e_1 \otimes e_2 + e_2 \otimes e_1 - \tfrac{1}{2} e_0 \otimes e_0.$$

Therefore,

$$2Z = \imath(e_1 \otimes e_2 + e_2 \otimes e_1 - \tfrac{1}{2} e_0 \otimes e_0) = R \circ L + L \circ R - \tfrac{1}{2}(\partial/\partial\theta)^2,$$

since

$$\imath(e_1) = R \qquad \imath(e_2) = L \qquad \imath(e_0) = \partial/\partial\theta.$$

We can also make use of the fact that $\bar{R} = L$, so that $R \circ L$ is the complex conjugate of $L \circ R$, etc.

We compute $R \circ L$. This is

$$R \circ L = (e^{2i\theta}(y\partial/\partial x - iy\partial/\partial y - \tfrac{1}{2}\partial/\partial\theta))(e^{-2i\theta}(y\partial/\partial x + iy\partial/\partial y - \tfrac{1}{2}\partial/\partial\theta)) =$$

$$= e^{2i\theta}(y(\partial/\partial x)(e^{-2i\theta}(y\partial/\partial x + iy\partial/\partial y - \tfrac{1}{2}\partial/\partial\theta)) -$$

$$- iy(\partial/\partial y)(e^{-2i\theta}(y\partial/\partial x + iy\partial/\partial y - \tfrac{1}{2}\partial/\partial\theta)) -$$

$$- \tfrac{1}{2}(\partial/\partial\theta)(e^{-2i\theta}(y\partial/\partial x + iy\partial/\partial y - \tfrac{1}{2}\partial/\partial\theta))) =$$

$$= e^{2i\theta}\{e^{-2i\theta}(y^2(\partial/\partial x)^2 + iy^2\partial^2/\partial x\partial y - \tfrac{1}{2}y\partial^2/\partial x\partial\theta) -$$

$$- ie^{-2i\theta}y(y\partial^2/\partial y\partial x + \partial/\partial x + i\partial/\partial y + iy\partial^2/\partial y\partial y - \tfrac{1}{2}\partial^2/\partial y\partial\theta)) -$$

$$-\tfrac{1}{2}e^{-2i\theta}(y\partial^2/\partial\theta\partial x - 2iy\partial/\partial x + iy\partial^2/\partial\theta\partial y + 2y\partial/\partial y - \tfrac{1}{2}\partial^2/\partial\theta\partial\theta + i\partial/\partial\theta)\} =$$

$$= (y^2(\partial/\partial x)^2 + iy^2\partial^2/\partial x\partial y - \tfrac{1}{2}y\partial^2/\partial x\partial\theta) -$$

$$- iy(y\partial^2/\partial y\partial x + \partial/\partial x + i\partial/\partial y + iy\partial^2/\partial y\partial y - \tfrac{1}{2}\partial^2/\partial y\partial\theta)) -$$

$$- \tfrac{1}{2}(y\partial^2/\partial\theta\partial x - 2iy\partial/\partial x + iy\partial^2/\partial\theta\partial y + 2y\partial/\partial y - \tfrac{1}{2}\partial^2/\partial\theta\partial\theta + i\partial/\partial\theta).$$

Therefore, in $R \circ L + L \circ R$ all terms with purely imaginary coefficients will disappear:

$$R \circ L + L \circ R = 2(y^2(\partial/\partial x)^2 - \tfrac{1}{2}y\partial^2/\partial x\partial\theta +$$

$$+ y\partial/\partial y + y^2\partial^2/\partial y\partial y - \tfrac{1}{2}(y\partial^2/\partial\theta\partial x + 2y\partial/\partial y - \tfrac{1}{2}\partial^2/\partial\theta\partial\theta)) =$$

$$= 2y^2\partial^2/\partial x\partial x + 2y^2\partial^2/\partial y\partial y - 2y\partial^2/\partial\theta\partial x + \tfrac{1}{2}\partial^2/\partial\theta\partial\theta.$$

Then

$$2Z = R \circ L + L \circ R - \tfrac{1}{2}\partial^2/\partial\theta\partial\theta$$

$$= 2y^2(\partial^2/\partial x\partial x + \partial^2/\partial y\partial y) - 2y\partial^2/\partial\theta\partial x.$$

■

Appendix A.5
Dirichlet L-Functions over $\mathbb{Q}$

For the sake of completeness, here we quickly review the "well-known" properties of special values of Dirichlet L-functions at positive integers of the same parity as the character. First, using one type of integral representation of an L-function, we can evaluate the L-function at all negative integers. Then, via another integral representation, we obtain the functional equation. Together, these give a description of the values at positive integers (and behavior under Galois automorphisms), which suffices for our purposes.

Let $1 < N \in \mathbb{Z}$. Let ω be a group homomorphism $(\mathbb{Z}/N\mathbb{Z})^\times \to \mathbb{C}^\times$. Suppose that ω is *primitive* modulo N, i.e., that there is no proper divisor M of N so that ω induces a well-defined function on $(\mathbb{Z}/M\mathbb{Z})^\times$. Extend ω to a function on $\mathbb{Z}/N\mathbb{Z}$ by defining it to be 0 off $(\mathbb{Z}/N\mathbb{Z})^\times$. By composition with $\mathbb{Z} \to \mathbb{Z}/N\mathbb{Z}$, consider ω as a function on $\mathbb{Z}$. Such ω is a *Dirichlet character* of *conductor* N. For $s \in \mathbb{C}$ with $\mathrm{Re}(s) > 1$, define the *Dirichlet L-function* attached to ω by

$$L(s, \omega) = \sum_{1 \leq n \in \mathbb{Z}} \omega(n)/n^s.$$

By unique factorization in $\mathbb{Z}$, and by the multiplicative property $\omega(xy) = \omega(x)\omega(y)$ of ω,

$$L(s, \omega) = \prod_{p \text{ prime}} (1 - \omega(p)p^{-s})^{-1} = \prod_{p \text{ prime not dividing } N} (1 - \omega(p)p^{-s})^{-1}.$$

Let $\mathbb{Q}(\omega)$ be the field extension of $\mathbb{Q}$ obtained by adjoining the values of ω to $\mathbb{Q}$. For $\sigma \in \mathrm{Gal}(\bar{\mathbb{Q}}/\mathbb{Q})$, define another Dirichlet character ω^σ by $\omega^\sigma(\alpha) = \omega(\alpha)^\sigma$. We say that ω is *even* if $\omega(-1) = 1$, and *odd* if $\omega(-1) = -1$. Define the Gauss sum attached to ω by

$$\gamma(\omega) = \sum_{x \in \mathbb{Z}/N\mathbb{Z}} \omega(x) \exp(2\pi i x/N).$$

THEOREM If ω is odd, let $\delta = 1$ and $\varepsilon = -i$, and if ω is even, put $\delta = 0$ and $\varepsilon = 1$. Then, for $1 \leq k \in \mathbb{Z}$,

$$L(2k+\delta,\omega)/\pi^{2k+\delta}\varepsilon\gamma(\omega) \in \mathbb{Q}(\omega).$$

Let $\sigma \in \mathrm{Gal}(\bar{\mathbb{Q}}/\mathbb{Q})$. We have

$$\{L(2k+\delta,\omega)/\pi^{2k+\delta}\varepsilon\gamma(\omega)\}^{\sigma} = L(2k+\delta,\omega^{\sigma})/\pi^{2k+\delta}\varepsilon\gamma(\omega^{\sigma}).$$

THEOREM 1 Suppose that ω has conductor N. $L(s,\omega)$ has an analytic continuation to a meromorphic function of $s \in \mathbb{C}$. For $0 \le k \in \mathbb{Z}$, we have

$$L(-k,\omega) \in \mathbb{Q}(\omega)$$

and for $\sigma \in \mathrm{Gal}(\bar{\mathbb{Q}}/\mathbb{Q})$,

$$L(-k,\omega)^{\sigma} = L(-k,\omega^{\sigma}).$$

THEOREM 2 Let $\varepsilon = \varepsilon(\omega) = 1$ if ω is even, and $\varepsilon = -i$ if ω is odd. Then

$$\tilde{L}(s,\omega) = (N/\pi)^{(s+\delta)/2}\Gamma(\tfrac{1}{2}(s+\delta))L(s,\omega)$$

has an analytic continuation to an entire function of $s \in \mathbb{C}$, and has functional equation

$$\tilde{L}(s,\omega) = (\varepsilon N^{-1/2}\gamma(\omega))\tilde{L}(1-s,\bar{\omega}).$$

PROPOSITION Take $z \in \mathbb{Z}$. We have

$$\sum_{x \in (\mathbb{Z}/N\mathbb{Z})^{\times}} \omega(x)\exp(2\pi i x z/N) = \bar{\omega}(z)\gamma(\omega).$$

Proof If z is prime to N, then we simply replace x by x/z modulo N in the sum, and the equality is obtained, as $\bar{\omega}(z) = \omega(z)^{-1}$. Now suppose that $d > 1$ divides both z and N. Since ω is primitive modulo N, there are $\alpha, \beta \in (\mathbb{Z}/N\mathbb{Z})^{\times}$ so that $\alpha \equiv \beta$ modulo N/d but $\omega(\alpha) \ne \omega(\beta)$. Therefore, $\zeta = \alpha\beta^{-1}$ (modulo N) is congruent to 1 modulo N/d, but $\omega(\zeta) \ne 1$. Upon replacing x by $x\zeta$ in the sum,

$$\sum_{x \in (\mathbb{Z}/N\mathbb{Z})^{\times}} \omega(x)\exp(2\pi i x z/N) = \sum_{x \in (\mathbb{Z}/N\mathbb{Z})^{\times}} \omega(x\zeta)\exp(2\pi i x\zeta z/N) =$$

$$= \omega(\zeta) \sum_{x \in (\mathbb{Z}/N\mathbb{Z})^{\times}} \omega(x)\exp(2\pi i x\zeta z/N).$$

Now

$$x\zeta z - xz = xz(\zeta - 1) = x(z/d)d(\zeta - 1) \equiv 0 \text{ modulo } N,$$

since $\zeta - 1 \equiv 0$ modulo N/d. Therefore, $\exp(2\pi i x\zeta z/N) = \exp(2\pi i x z/N)$ for all x, and

$$\sum_{x \in (\mathbb{Z}/N\mathbb{Z})^\times} \omega(x) \exp(2\pi i x z/N) = \omega(\zeta) \sum_{x \in (\mathbb{Z}/N\mathbb{Z})^\times} \omega(x) \exp(2\pi i x z/N).$$

This implies that the sum is 0. ∎

PROPOSITION We have $|\gamma(\omega)| = N^{1/2}$.

Proof We have

$$|\gamma(\omega)|^2 = \sum_{x,y \in (\mathbb{Z}/N\mathbb{Z})^\times} \omega(x)\bar{\omega}(y) \exp(2\pi i x/N) \exp(-2\pi i y/N).$$

Replacing x by x/y modulo N:

$$|\gamma(\omega)|^2 = \sum_{x,y \in (\mathbb{Z}/N\mathbb{Z})^\times} \omega(x) \exp(2\pi i y(x-1)/N).$$

By the previous proposition, for fixed y, if y is not prime to N we have

$$\sum_{x \in (\mathbb{Z}/N\mathbb{Z})^\times} \omega(x) \exp(2\pi i x y/N) = 0.$$

Therefore, we can write

$$\sum_{x,y \in (\mathbb{Z}/N\mathbb{Z})^\times} \omega(x) \exp(2\pi i y(x-1)/N) =$$

$$= \sum_{x \in (\mathbb{Z}/N\mathbb{Z})^\times} \sum_{y \in \mathbb{Z}/N\mathbb{Z}} \omega(x) \exp(2\pi i y(x-1)/N).$$

For fixed x, the sum over y is 0 unless $x - 1 \equiv 0$ modulo N, in which case it is just N. ∎

PROPOSITION We have $\gamma(\omega)\gamma(\bar{\omega}) = \omega(-1)N$, and $\gamma(\bar{\omega}) = \omega(-1)\gamma(\omega)^-$.

Proof Upon replacing x by $-x$ in the sum defining $\gamma(\bar{\omega})$, we have

$$\gamma(\bar{\omega}) = \sum_{x \in (\mathbb{Z}/N\mathbb{Z})^\times} \bar{\omega}(-x) \exp(-2\pi i x/N) =$$

$$= \bar{\omega}(-1) \left\{ \sum_{x \in (\mathbb{Z}/N\mathbb{Z})^\times} \omega(x) \exp(2\pi i x/N) \right\}^- =$$

$$= \omega(-1)\gamma(\omega)^-.$$

Then the first assertion follows from this by the previous proposition. ∎

Again, let ω have conductor $N > 1$. Fix $\varepsilon > 0$. Let C be the path in $\mathbb{C}$ which goes from $+\infty$ to ε along $\mathbb{R}$, then around 0 in the clockwise direction to ε, and then goes along $\mathbb{R}$ from ε to $+\infty$. For $t \in \mathbb{C}$, define a function of t by

$$B(t;\omega) = (1 - e^{-Nt})^{-1} \sum_{b=1,\ldots,N} \omega(b)e^{-bt}.$$

PROPOSITION For $s \notin \mathbb{Z}$, $s \in \mathbb{C}$,

$$\Gamma(s)L(s,\omega) = (1 - e^{2\pi is})^{-1} \int_C t^s B(t;\omega)\,dt/t,$$

where $t^s = e^{s\log t}$ as t goes from ε to ∞ on the last part of C, and $t^s = e^{s(\log t + 2\pi i)}$ on the first part of C. Therefore, as the integral on the right is convergent for all $s \in \mathbb{C}$, $L(s,\omega)$ has an analytic continuation to a meromorphic function of $s \in \mathbb{C}$.

Proof Take $\mathrm{Re}(s) > 1$, $s \notin \mathbb{Z}$. We have

$$\int_0^\infty e^{-tn}t^s\,dt/t = \Gamma(s)/n^s.$$

By elementary complex variables, one finds that

$$\int_C e^{-tn}t^s\,dt/t = (1 - e^{2\pi is})\int_0^\infty e^{-tn}t^s\,dt/t.$$

Therefore,

$$\Gamma(s)L(s,\omega) = \sum_{b=1,\ldots,N} \omega(b) \sum_{n\geq 0} \int_0^\infty e^{-t(Nn+b)}t^s\,dt/t =$$

$$= \sum_{b=1,\ldots,N} \omega(b) \sum_{n\geq 0} (1 - e^{2\pi is})^{-1} \int_C e^{-t(Nn+b)}t^s\,dt/t =$$

$$= \sum_{b=1,\ldots,N} \omega(b)(1 - e^{2\pi is})^{-1} \int_C e^{-tb}(1 - e^{-Nt})^{-1}t^s\,dt/t =$$

$$= (1 - e^{2\pi is})^{-1} \int_C B(t;\omega)t^s\,dt/t.$$

■

PROPOSITION As a function of t, $B(t;\omega)$ is holomorphic near $t = 0$. The power series expansion $B(t;\omega) = \sum_{n\geq 0} b_n(\omega)t^n$ near $t = 0$ has coefficients $b_n(\omega) \in \mathbb{Q}(\omega)$. For $\sigma \in \mathrm{Gal}(\bar{\mathbb{Q}}/\mathbb{Q})$, $b_n(\omega)^\sigma = b_n(\omega^\sigma)$.

Proof By definition,

$$B(t;\omega) = \sum_{b=1,\ldots,N} \omega(b)e^{-bt}(1 - e^{-Nt})^{-1}.$$

Now $(1 - e^{-Nt})^{-1}$ is meromorphic near 0, with just a simple pole of residue N^{-1}, and has rational Laurent coefficients at 0. Therefore, the same is true for $f(t; b) = (1 - e^{-Nt})^{-1}e^{-tb}$ for $b = 1, \ldots, N$. Then $B(t; \omega) = \sum_b f(t; b)\omega(b)$ is meromorphic at 0, with at worst a simple pole. The residue of this pole is

$$\sum_{b=1,\ldots,N} \omega(b) \times (\text{Residue of } f(t; b) \text{ at } t = 0) =$$

$$= \sum_{b=1,\ldots,N} \omega(b)N^{-1} = N^{-1} \sum_{b=1,\ldots,N} \omega(b) = N^{-1} \times 0 = 0,$$

since ω is non-trivial. Thus, $B(t; \omega)$ is holomorphic at $t = 0$.

Take $\sigma \in \mathrm{Gal}(\bar{\mathbb{Q}}/\mathbb{Q})$. Let $c_n(b)$ be the n-th Laurent coefficient of $f(t; b)$ at 0; it is rational. Then

$$b_n(\omega) = \sum_{b=1,\ldots,N} \omega(b)c_n(b) \in \mathbb{Q}(\omega),$$

and

$$b_n(\omega)^\sigma = \sum_{b=1,\ldots,N} \omega(b)^\sigma c_n(b)^\sigma = \sum_{b=1,\ldots,N} \omega(b)^\sigma c_n(b) =$$

$$= \sum_{b=1,\ldots,N} \omega^\sigma(b)c_n(b) = b_n(\omega^\sigma).$$ ■

Proof of Theorem 1 We can evaluate $L(s, \omega)$ at non-positive integers via the integral representation above. Take $0 \le k \in \mathbb{Z}$. We have

$$\{\Gamma(s)(1 - e^{2\pi i s})\}^{-1}|_{s=-k} = (-1)^k(-2\pi i)^{-1}k!$$

On the other hand, by residues,

$$\int_C t^{-k}B(t; \omega)\,dt/t = (-2\pi i) \times (\text{residue of } t^{-k-1}B(t; \omega) \text{ at } t = 0) =$$

$$= (-2\pi i) \times (k\text{-th power series coefficient of } B(t; \omega) \text{ at } t = 0) =$$

$$= (-2\pi i) \times b_n(\omega),$$

in the notation above. Therefore,

$$L(-k, \omega) = b_n(\omega)(-1)^k k! \in b_n(\omega)\mathbb{Q} \subset \mathbb{Q}(\omega),$$

and the Galois action assertion follows as well. This finishes the proof of Theorem 1. ■

For $t > 0$, define theta series

$$\theta(t;\omega) = \sum_{n\in\mathbb{Z}} \omega(n)\exp(-\pi t n^2) \qquad \text{(for } \omega \text{ even)}$$

$$\theta(t;\omega) = \sum_{n\in\mathbb{Z}} \omega(n) n t^{1/2}\exp(-\pi t n^2) \qquad \text{(for } \omega \text{ odd).}$$

THEOREM (FUNCTIONAL EQUATIONS OF THETA SERIES) For $t > 0$, we have

$$\theta(t;\omega) = t^{-1/2}N^{-1}\gamma(\omega)\theta(t^{-1}N^{-2};\bar{\omega}) \qquad \text{(for } \omega \text{ even)}$$

$$= -it^{-1/2}N^{-1}\gamma(\omega)\theta(t^{-1}N^{-2};\bar{\omega}) \qquad \text{(for } \omega \text{ odd).}$$

Proof For $x \in \mathbb{R}$, let $f(x) = \exp(-\pi x^2)$ if ω is even, and $f(x) = x\exp(-\pi x^2)$ if ω is odd. Then (e.g., using Cauchy's theorem), we have

$$\hat{f}(x) = f(x) \qquad \text{(for } \omega \text{ even)}$$

$$= -if(x) \qquad \text{(for } \omega \text{ odd).}$$

Let $\varepsilon = 1$ if ω is even, and $\varepsilon = -i$ if ω is odd. Also, letting $F(x) = f(ax + b)$ with $a \neq 0$, we have

$$\hat{F}(x) = \exp(2\pi i x b/N)a^{-1}\hat{f}(x/a).$$

Applying the Poisson Summation Formula, we compute:

$$\theta(t;\omega) = \sum_{b=1,\ldots,N} \omega(b) \sum_{n\in\mathbb{Z}} f(t^{1/2}(nN + b)) =$$

$$= \sum_{b=1,\ldots,N} \omega(b)t^{-1/2}N^{-1} \sum_{n\in\mathbb{Z}} \exp(2\pi i n b/N)\hat{f}(n/Nt^{1/2}) =$$

$$= \varepsilon t^{-1/2}N^{-1} \sum_{n\in\mathbb{Z}} f(n/Nt^{1/2})\left\{\sum_{b=1,\ldots,N} \omega(b)\exp(2\pi i n b/N)\right\}.$$

From our results on Gauss sums, the inner sum is $\bar{\omega}(n)\gamma(\omega)$. Therefore,

$$\theta(t;\omega) = \varepsilon t^{-1/2}N^{-1}\gamma(\omega) \sum_{n\in\mathbb{Z}} f(n/Nt^{1/2})\bar{\omega}(n) =$$

$$= \varepsilon t^{-1/2}N^{-1}\gamma(\omega)\theta(t^{-1}N^{-2};\bar{\omega}).$$

■

Proof of Theorem 2 Let δ and ε be as above. From the identity

$$\int_0^\infty e^{-\pi n t}t^{s/2}\,dt/t = \pi^{-s/2}\Gamma(\tfrac{1}{2}s)n^{-s},$$

for $\mathrm{Re}(s) > 1$ we obtain (integrating term by term)

$$\tfrac{1}{2}\int_0^\infty \theta(t/N;\omega)t^{s/2}\,dt/t = (N/\pi)^{(s+\delta)/2}\Gamma(\tfrac{1}{2}(s+\delta))L(s,\omega).$$

Let $\tilde{L}(s,\omega)$ denote the righthand side. We have

$$\tilde{L}(s,\omega) = \tfrac{1}{2}\int_0^\infty \theta(t/N;\omega)t^{s/2}\,dt/t =$$

$$= \tfrac{1}{2}\int_1^\infty \theta(t/N;\omega)t^{s/2}\,dt/t + \tfrac{1}{2}\int_0^1 \theta(t/N;\omega)t^{s/2}\,dt/t.$$

It is easy to see that the first integral is an entire function of $s \in \mathbb{C}$, since the integrand is rapidly decreasing at ∞. In the second integral, use the functional equation of the theta series and replace t by t^{-1} to obtain

$$\tfrac{1}{2}\int_0^1 \theta(t/N;\omega)t^{s/2}\,dt/t = \varepsilon\tfrac{1}{2}\int_0^1 (t/N)^{-1/2}N^{-1}\gamma(\omega)\theta((t/N)^{-1}N^{-2};\bar{\omega})t^{s/2}\,dt/t =$$

$$= \varepsilon N^{-1/2}\gamma(\omega)\tfrac{1}{2}\int_0^1 t^{-1/2}\theta(t^{-1}N^{-1};\bar{\omega})t^{s/2}\,dt/t =$$

$$= \varepsilon N^{-1/2}\gamma(\omega)\tfrac{1}{2}\int_1^\infty t^{1/2}\theta(t^{-1}N^{-2};\bar{\omega})t^{-s/2}\,dt/t.$$

The latter integral is also an entire function of $s \in \mathbb{C}$.

Using $\gamma(\omega)\gamma(\bar{\omega}) = \omega(-1)N$, we have $N^{-1/2}\gamma(\omega) = \omega(-1)N^{1/2}/\gamma(\bar{\omega})$. Also, it is immediate that $\varepsilon^{-1} = \omega(-1)\varepsilon$. Therefore,

$$\tilde{L}(s,\omega) =$$

$$= \tfrac{1}{2}\int_1^\infty \theta(t/N;\omega)t^{s/2}\,dt/t + \varepsilon N^{-1/2}\gamma(\omega)\tfrac{1}{2}\int_1^\infty t^{1/2}\theta(t^{-1}N^{-2};\bar{\omega})t^{-s/2}\,dt/t =$$

$$= \tfrac{1}{2}\varepsilon N^{-1/2}\gamma(\omega)\left\{\varepsilon\gamma(\bar{\omega})N^{-1/2}\int_1^\infty \theta(t/N;\omega)t^{s/2}\,dt/t + \int_1^\infty t^{1/2}\theta(t^{-1}N^{-2};\bar{\omega})t^{-s/2}\,dt/t\right\} =$$

$$= \varepsilon N^{-1/2}\gamma(\omega)\tilde{L}(1-s,\bar{\omega}).$$

This is the asserted functional equation. ■

Proof of Theorem For an integer m, let ζ_m be a primitive m-th root of unity. Choose n so that $\mathbb{Q}(\omega) = \mathbb{Q}(\zeta_n)$. Let M be the least common multiple of n and N. Let k be a positive integer, and $\sigma = \sigma_\alpha$ with $\alpha \in (\mathbb{Z}/M\mathbb{Z})^\times$. From the functional equation, with

δ and ε as before, we obtain

$$L(2k+\delta,\omega)/\varepsilon\pi^{2k+\delta} = \{\Gamma(k)^{-1}N^{-2k-\delta}\}\{\Gamma(\tfrac{1}{2}-k)\pi^{-1/2}\}\gamma(\omega)L(1-2k-\delta,\bar{\omega}) =$$

$$= \rho_k\gamma(\omega)L(1-2k-\delta,\bar{\omega})$$

where ρ_k is a rational number depending only on k. That is

$$L(2k+\delta,\omega)/\varepsilon\pi^{2k+\delta}\gamma(\omega) = \rho_k L(1-2k-\delta,\bar{\omega}).$$

The righthand side lies in $\mathbb{Q}(\omega)$. Therefore, using the Galois behavior of $L(1-2k-\delta)$,

$$\{L(2k+\delta,\omega)/\varepsilon\pi^{2k+\delta}\gamma(\omega)\}^\sigma = \rho_k L(1-2k-\delta,\bar{\omega}^\sigma) =$$

$$= L(2k+\delta,\omega^\sigma)/\varepsilon\pi^{2k+\delta}\gamma(\omega^\sigma).$$

■

Bibliography

[A] A.O.L. Atkin and J. Lehner, "Hecke operators on $\Gamma_0(m)$," *Math. Ann. 185* (1970), pp. 134–160.

[B1] O. Blumenthal, "Uber Modulfunktionen von mehreren Veranderlichen," *Math. Ann. 56* (1903), pp. 509–548, *Math. Ann. 58* (1904), pp. 497–527.

[B2] O. Blumenthal, "Uber Thetafunktionen und Modulfunktionen von mehreren Veranderlichen," *Jahresb. Deutsh. Math.* (1904), pp. 120–132.

[Br1] H. Braun, "Zur Theorie der Modulformen n-ten Grades," *Math. Ann. 115* (1938), pp. 507–517.

[Br2] H. Braun, "Convergenz verallgemeinerter Eisensteinscher Reihen," *Math. Z. 44* (1939), pp. 387–397.

[C] W. Casselman, "On some results of Atkin and Lehner," *Math Ann. 201* (1973), pp. 301–314.

[CF] *Algebraic Number Theory*, ed. J.W.S. Cassels and A. Frohlich, *Thompson Inc., Washington, 1967.*

[D] R.M. Damerell, "L-functions of elliptic curves with complex multiplication I, II," *Acta Arith. 17* (1970), pp. 287–301, *19* (1971), pp. 311–317.

[De] P. Deligne, "Valeurs de fonctions L et periodes d'integrales," in *Proc. Symp. Pure Math. 33*, vol. 2 (1977), A.M.S., pp. 313–342.

[E1] M. Eichler, "Allgemeine Kongruenzklasseneinteilungen der Ideale einfacher Algebren uber algebraischen Zahlkorpern und ihre L-Reihen," *J. Reine Angew. Math. 179* (1938), pp. 227–251.

[E2] M. Eichler, "Die Ahnlichkeitsklassen indefiniter Gitter," *Math. Zeit.* 55 (1951–52), pp. 216–252.

[G1] P.B. Garrett, "Pullbacks of Eisenstein series; applications," in *Automorphic Forms in Several Complex Variables*, Birkhauser, 1984.

[G2] P.B. Garrett, "Decomposition of Eisenstein series: Rankin triple products," *Ann. of Math. 125* (1987), pp. 209–235.

[G3] P.B. Garrett, "Integral representations of Eisenstein series and L-functions," in *Number Theory, Trace Formulas, and Discrete Groups*, Academic Press, 1987.

[Gu] R. Gunning, *Lectures on Modular Forms*, Princeton Univ. Press, 1962.

[Ge] S. Gelbart, *Automorphic Forms on Adele Groups*, Stud. Math. no. 83, Princeton Univ. Press, 1975.

[GPR] S. Gelbart, I. Piatetski-Shapiro, S. Rallis, *Explicit Constructions of Automorphic L-functions*, Springer Lecture Notes in Math. no. 1254, 1987.

[GGP] I.M. Gelfand, M.I. Graev, I.I. Piatetski-Shapiro, *Representation Theory and Automorphic Functions* (English translation), W.B. Saunders Co., 1969.

[Go] R. Godement, "The decomposition of $L^2(\Gamma\backslash G)$ for $\Gamma = SL(2, \mathbb{Z})$," in *Algebraic Groups and Discontinuous Subgroups*, vol. 9, *Proc. Symp. Pure Math*, A.M.S., 1966, pp. 211–224.

[H1] E. Hecke, "Hohere Modulfunktionen und ihre Anwendung auf die Zahlentheorie," *Math. Ann. 71* (1912), pp. 1–37.

[H2] E. Hecke, "Theorie der Eisensteinschen Reihen hoherer Stufe und ihre Anwendung auf Funktiontheorie und Arithmetik," *Abh. Math. Sem. Hamb. 5* (1927), pp. 199–224.

[H3] E. Hecke, "Uber Modulfunktionen und die Dirichletshen Reihen mit Eulersche Produktentwicklung, II," *Math Ann. 114* (1937), pp. 316–351.

[JL] H. Jacquet, R. Langlands, *Automorphic Forms on GL*(2), Springer Lecture Notes in Math. no. 114.

[K1] H. Klingen, "Uber die Werte der Dedekindschen Zetafunktion," *Math. Ann. 145* (1962), pp. 265 272.

[K2] H. Klingen, "Uber den arithmetischen Charakter der Fourierkoeffizienten von Modulformen," *Math. Ann. 147* (1962), pp. 176–188.

[KL1] H.D. Kloosterman, "Theorie der Eisensteinschen Reihen von mehreren Veranderlichen," *Abh. Math. Sem. Hamb. 6* (1928), pp. 163–188.

[KL2] H.D. Kloosterman, "The behavior of general theta functions under the modular group and the characters of binary congruence groups I, II," *Ann. of Math. 47* (1946), pp. 317–447.

[Kn] M. Kneser, "Starke Approximation in algebraischen Gruppen I," *J. Reine Angew. Math. 218* (1965), pp. 190–203.

[L] E. Landau, "Uber die Anzahl der Gitterpunkte in gewissen Bereichen, II," *Nachr. Ges. Wiss. Gottingen* (1915), pp. 209–243.

[La] S. Lang, *Algebraic Number Theory*, Addison-Wesley, 1970.

[M] H. Maass, "Zur Theorie der automorphen Funktionen von n Veranderlichen," *Math. Ann. 117* (1940), pp. 538–578.

[Mi] T. Miyake, "On automorphic forms on GL_2 and Hecke operators," *Ann of Math. 94* (1971), pp. 174–189.

[P] H. Petersson, "Uber die Berechnung der Skalarprodukte ganzer Modulformen," *Comm. Math. Helv. 22* (1949), pp. 168–199.

[PR] I. Piatetski-Shapiro and S. Rallis, "Rankin triple *L*-functions," *Comp. Math. 64* (1987), pp. 31–115.

[R1] R. Rankin, "Contributions to the theory of Ramanujans's function $\tau(n)$ and similar arithmetic functions, I, II," *Proc. Cambridge Phil. Soc. 35* (1939), pp. 351–372.

[R2] R. Rankin, "The scalar product of modular forms," *Proc. Lond. Math. Soc. 3, 2* (1952), pp. 198–217.

[Sc] B. Schoeneberg, "Das Verhalten von Mehrfachen Thetareihen bei Modulsubstitutionen," *Math. Ann. 116* (1939).

[Se] A. Selberg, "Bemerkungen uber eine Dirichletshe Reihe die mit der Theorie der Modulformen nahe verbunden ist," *Arch. Math. Naturvid. 43* (1940), pp. 47–50.

[S1] F. Shahidi, "Functional equation satisfied by certain *L*-functions," *Comp. Math. 37* (1978), pp. 171–201.

[S2] F. Shahidi, "On certain *L*-functions," *Amer. J. Math. 103* (1981), pp. 297–355.

[Sh1] G. Shimura, "On canonical models of arithmetic quotients of bounded symmetric domains I, II," *Ann. of Math. 91* (1970), pp. 144–222, *92* (1970), pp. 528–549.

[Sh2] G. Shimura, *Introduction to the Arithmetic Theory of Automorphic Functions*, Iwanami-Shoten and Princeton University Press, 1971.

[Sh3] G. Shimura, "On some arithmetic properties of modular forms in one and several variables," *Ann. of Math. 102* (1975), p. 491–515.

[Sh4] G. Shimura, "The special values of the zeta functions associated with cuspforms," *Comm. Pure and Applied Math. 29* (1976), pp. 783–804.

[Sh5] G. Shimura, "On the periods of modular forms," *Math. Ann. 229* (1977), pp. 211–221.

[Sh6] G. Shimura, "The special values of the zeta functions associated with Hilbert modular forms," *Duke Math. J. 45* (1978), pp. 637–679.

[Sh7] G. Shimura, "On Eisenstein series," *Duke J. Math. 50* (1983), pp. 417–476.

[Si1] C.L. Siegel, *Advanced Analytic Number Theory*, Lecture Notes, Tata Institute, 1960.

[Si2] C.L. Siegel, "Einfuhrung in die Theorie der Modulfunktionen *n*-ten Grades," *Math. Ann. 116* (1939), pp. 617–657.

[Si2] C.L. Siegel, "Berechnung von Zetafunktionen an ganzzahligen Stellen," *Gott. Nachr. Akad. Wiss. 1969*, pp. 87–102.

[Si2] C.L. Siegel, "Uber die Fourierschen Koeffizienten von Modulformen," *Gott. Nachr. Akad. Wiss. 1970*, pp. 15–56.

[W] A. Weil, *Adeles and Algebraic Groups* (1962 notes at I.A.S.), Birkhauser, Boston, 1982.

Index